AF785332

Advances in Brassica Crops Genomics and Breeding

Advances in Brassica Crops Genomics and Breeding

Editors

Xiaowu Wang
Jian Wu
Xu Cai

MDPI

Basel • Beijing • Wuhan • Barcelona • Belgrade • Novi Sad • Cluj • Manchester

Editors

Xiaowu Wang
Chinese Academy of Agricultural Sciences
Beijing
China

Jian Wu
Chinese Academy of Agricultural Sciences
Beijing
China

Xu Cai
Chinese Academy of Agricultural Sciences
Beijing
China

Editorial Office
MDPI
St. Alban-Anlage 66
4052 Basel, Switzerland

This is a reprint of articles from the Special Issue published online in the open access journal *Horticulturae* (ISSN 2311-7524) (available at: https://www.mdpi.com/journal/horticulturae/special_issues/Brassica_Genomics).

For citation purposes, cite each article independently as indicated on the article page online and as indicated below:

Lastname, A.A.; Lastname, B.B. Article Title. *Journal Name* **Year**, *Volume Number*, Page Range.

ISBN 978-3-7258-0079-7 (Hbk)
ISBN 978-3-7258-0080-3 (PDF)
doi.org/10.3390/books978-3-7258-0080-3

© 2024 by the authors. Articles in this book are Open Access and distributed under the Creative Commons Attribution (CC BY) license. The book as a whole is distributed by MDPI under the terms and conditions of the Creative Commons Attribution-NonCommercial-NoDerivs (CC BY-NC-ND) license.

Contents

horticulturae

Editorial

Harvesting Knowledge: Illuminating Advances in *Brassica* Crops Genomics and Breeding

Xu Cai, Jian Wu * and Xiaowu Wang *

Institute of Vegetables and Flowers, Chinese Academy of Agricultural Sciences, No. 12, Haidian District, Beijing 100081, China; caixu@caas.cn
* Correspondence: wujian@caas.cn (J.W.); wangxiaowu@caas.cn (X.W.)

Brassica crops encompass a diverse array, including vegetables, oil crops, ornamentals, and condiments. In particular, some *Brassica* crops such as turnip, Chinese cabbage, cabbage, and rapeseed have solidified their status as global agricultural cash crops. Beyond their economic significance, *Brassica* species stand out due to the domestication of extreme morphological types, offering a distinctive spectrum of leafy heading, root/stem enlarging, and florescence heading. Moreover, the occurrence of polyploidization events within *Brassica* species positions them as ideal models for investigating the complexities of polyploidization. With these distinctive features, research on *Brassica* crops has remained a hot research area. In the ever-evolving landscape of agricultural research, the Special Issue (SI) on "Advances in *Brassica* Crops Genomics and Breeding" emerges as a crucial guide for understanding the key research trends in the genomics and breeding of *Brassica* crops.

Genomic Exploration: The rapid progress in sequencing technologies has propelled *Brassica* crops into the genomic spotlight. Numerous genomes of *Brassica* crop species have been meticulously sequenced, yielding high-quality chromosome-scale assemblies. This groundbreaking achievement has been complemented by the availability of large-scale resequencing data of germplasm resources in *B. rapa*, *B. oleracea*, *B. juncea*, and *B. napus*. This wealth of data empowers researchers with the tools necessary for genome-wide association studies (GWAS) and domestication analyses, unlocking the genetic mechanisms embedded for the agricultural traits of these vital crops. One feature paper [1] conducted a reanalysis of previously published resequencing data encompassing various *B. rapa* morphotypes. It revealed a significant biased distribution of haplotypes for variations on *BrFLC1* across diverse population accessions, including turnip, Chinese cabbage, Pak choi, Caixin, Wutacai, and Taicai. This observation implies a close association between the evolutionary patterns of *BrFLC1* haplotypes and the dissemination of different *B. rapa* morphotypes, highlighting the vital role of natural variation in the diversification of *Brassica* crops.

Functional Genomics: One of the primary objectives of the *Brassica* community is to showcase the functional aspects of important genes within *Brassica* crops. The investigation into complex *Brassica* genomes extends beyond mere sequencing, seeking to reveal the functions of key genes that contribute to the adaptability and unique characteristics of these crops. The intersection of genomics and functionality provides a holistic understanding that is pivotal for informed breeding strategies. Nine research papers, exploring various aspects of *Brassica* crops, contribute valuable insights into their genetic regulation, stress responses, and biochemical composition of *Brassica* crops. The featured analyses cover a diverse range of topics, including exploring the regulatory role of miR398a and its target gene *BraCSD1-1* in Chinese cabbage's response to heat stress [2], uncovering the significance of *BnA.JAZ5* in *B. napus* and highlighting its key role as a genetic regulator in drought stress [3], and investigating the involvement of *PARP1* in *B. rapa*'s response to drought stress, with a specific emphasis on its impact on root growth and stress-related gene expression [4]. Additionally, the issue delves into the effects of TuMV on *B. rapa*, conducting detailed analyses of the volatile metabolome and transcriptome in resistant and

Citation: Cai, X.; Wu, J.; Wang, X. Harvesting Knowledge: Illuminating Advances in *Brassica* Crops Genomics and Breeding. *Horticulturae* **2023**, *9*, 1332. https://doi.org/10.3390/horticulturae9121332

Received: 23 November 2023
Revised: 29 November 2023
Accepted: 30 November 2023
Published: 12 December 2023

Copyright: © 2023 by the authors. Licensee MDPI, Basel, Switzerland. This article is an open access article distributed under the terms and conditions of the Creative Commons Attribution (CC BY) license (https://creativecommons.org/licenses/by/4.0/).

susceptible lines [5]. The physiological and biochemical responses in *B. juncea* cultivars under drought stress influenced by brassinolide are explored as well [6]. Furthermore, there is an analysis of the poorly understood group of enzymes, *BrPAPs*, in *B. rapa*, with a focus on their roles and potential implications [7]. Other studies identify a key mutation (Pe1 + 58) in the Br*FLC1* gene that impacts flowering time variation in *B. rapa* [1], explore the glucosinolate (GSL) profile of *B. nigra*, and emphasize the identification of sinigrin as the predominant GSL [8]. Lastly, the issue investigates pollen abortion and abnormal development in Chinese cabbage under Ogura CMS, providing comprehensive insights [9]. Collectively, these studies contribute to advancing our understanding of *Brassica* crops, encompassing genetic mechanisms, stress responses, and biochemical characteristics.

Molecular Marker-Assisted Breeding: Advancements in genomics are not confined to the laboratory; they have a tangible impact on the field through molecular marker-assisted breeding. It is very important to explore how the fusion of genomics and breeding is accelerating the development of *Brassica* cultivars with enhanced traits. This marks a transformative era in agriculture, where precision breeding aligns with the specific needs of farmers and global food security. The SI features three papers, each focusing on essential aspects of genetic mapping and improvement in *B. rapa*. The first paper places a primary emphasis on constructing a comprehensive genetic map with 105 intragenic SSR markers distributed across 10 linkage groups [10]. This map serves as a valuable resource, providing insights into the intricate genetic architecture governing essential agronomic traits in Chinese cabbage. The research identified 48 QTLs associated with various traits, offering significant contributions to molecular breeding and marker-assisted genetic enhancement initiatives for Chinese cabbage accessions. The second paper highlights the successful application of molecular marker-assisted gene pyramiding and backcrossing techniques to improve Chinese cabbage accessions. This study stands out for its focus on integrating self-compatibility, multilocular ovaries, and resistance to clubroot through precise molecular marker identification [11]. Together, these papers contribute to advancing genetic knowledge and practical applications for the genetic improvement of Chinese cabbage. The third paper maps 102 QTLs associated with biochemical traits, primarily located on the fifth, sixth, seventh, and ninth linkage groups [12]. These findings offer insights for genetic and breeding work, aiding in the development of genotypes with desirable biochemical compositions adapted to specific photoperiodic conditions through marker-assisted selection in *B. rapa*.

Additionally, a review is focused on discussing the emerging use of rapeseed (*B. napus*) as an ornamental crop, particularly in China where tourism centered around fields of blooming yellow flowers has become an economic opportunity [13]. Additionally, there is a paper that addresses the crucial issue of seed quality in vegetable production. Employing instrumental automated methods, the study specifically focuses on the digital X-ray analysis of seeds from *B. oleracea*, *Raphanus sativus*, and *Lepidium sativum* [14]. These works are of particular interest to breeders of *Brassica* crops.

Conclusion: "Advances in *Brassica* Crops Genomics and Breeding" shows the collective efforts of researchers unraveling the genetic intricacies of *Brassica* crops. From the economic significance of these diverse crops to the exploration of their complex genomes, these research findings will contribute to advancements in breeding research on *Brassica* crops. As we glean insights from the latest breakthroughs in genomics and breeding, the future of *Brassica* crops appears brighter, promising enhanced productivity, adaptability, and nutritional value.

Author Contributions: Writing—original draft preparation, X.C.; Writing—review and editing, J.W. X.W. All authors have read and agreed to the published version of the manuscript.

Funding: This research received no external funding.

Informed Consent Statement: Not applicable.

Data Availability Statement: Not applicable.

Conflicts of Interest: The authors declare no conflict of interest.

References

1. Liu, J.; Cai, X.; Li, Y.; Chen, Y.; Gao, B.; Lin, R.; Liang, J.; Wang, X.; Wu, J. Selection on *BrFLC1* Is Related to Intraspecific Diversity of *Brassica rapa* Vegetables. *Horticulturae* **2021**, *7*, 247. [CrossRef]
2. Cao, B.; Jiang, J.; Bai, J.; Wang, X.; Li, Y.; Shao, W.; Hu, S.; He, Y.; Yu, X. miR398 Attenuates Heat-Induced Leaf Cell Death via Its Target CSD1 in Chinese Cabbage. *Horticulturae* **2022**, *8*, 299. [CrossRef]
3. Cao, B.; Bai, J.; Wang, X.; Zhang, Y.; Yu, X.; Hu, S.; He, Y. *BnA.JAZ5* Attenuates Drought Tolerance in Rapeseed through Mediation of ABA–JA Crosstalk. *Horticulturae* **2022**, *8*, 131. [CrossRef]
4. Cao, G.; Jiang, W.; Shi, G.; Tian, Z.; Shang, J.; Xie, Z.; Chen, W.; Tian, B.; Wei, X.; Wei, F.; et al. BrPARP1, a Poly (ADP-Ribose) Polymerase Gene, Is Involved in Root Development in *Brassica rapa* under Drought Stress. *Horticulturae* **2022**, *8*, 78. [CrossRef]
5. Lu, X.; Zhang, L.; Huang, W.; Zhang, S.; Zhang, S.; Li, F.; Zhang, H.; Sun, R.; Zhao, J.; Li, G. Integrated Volatile Metabolomics and Transcriptomics Analyses Reveal the Influence of Infection TuMV to Volatile Organic Compounds in *Brassica rapa*. *Horticulturae* **2022**, *8*, 57. [CrossRef]
6. Naveen, N.; Kumari, N.; Avtar, R.; Jattan, M.; Ahlawat, S.; Rani, B.; Malik, K.; Sharma, A.; Singh, M. Evaluation of Effect of Brassinolide in *Brassica juncea* Leaves under Drought Stress in Field Conditions. *Horticulturae* **2021**, *7*, 514. [CrossRef]
7. Cai, Y.; Qi, J.; Li, C.; Miao, K.; Jiang, B.; Yang, X.; Han, W.; Wang, Y.; Gao, J.; Dong, X. Genome-Wide Analysis of Purple Acid Phosphatase Genes in *Brassica rapa* and Their Association with Pollen Development and Phosphorus Deprivation Stress. *Horticulturae* **2021**, *7*, 363. [CrossRef]
8. Li, Y.; Yu, Y.; Xu, L.; Guo, E.; Zang, Y.; He, Y.; Zhu, Z. Transcriptome Profiling Reveals Candidate Key Genes Involved in Sinigrin Biosynthesis in *Brassica nigra*. *Horticulturae* **2021**, *7*, 173. [CrossRef]
9. Hu, L.; Zhang, X.; Yuan, Y.; Wang, Z.; Yang, S.; Li, R.; Kumar Nath, U.; Zhao, Y.; Tian, B.; Shi, G.; et al. Comparative Transcriptome Identifies Gene Expression Networks Regulating Developmental Pollen Abortion in Ogura Cytoplasmic Male Sterility in Chinese Cabbage (*Brassica rapa* ssp. *pekinensis*). *Horticulturae* **2021**, *7*, 157. [CrossRef]
10. Gao, H.; Yang, X.; Wang, H.; Qiu, N.; Chen, Y.; Wang, F.; Zhang, Y.; Li, H.; Li, J.; Gao, J. Construction of an Intragenic SSR-Based Linkage Map and QTL Mapping for Agronomic Traits in Chinese Cabbage (*Brassica rapa* L. ssp. *pekinensis*). *Horticulturae* **2022**, *8*, 165. [CrossRef]
11. Zheng, J.; Zhao, H.; Ma, Y.; Jiang, M.; Zhan, Z.; Li, X.; Piao, Z. Marker-Assisted Pyramiding of Genes for Multilocular Ovaries, Self-Compatibility, and Clubroot Resistance in Chinese Cabbage (*Brassica rapa* L. ssp. *pekinensis*). *Horticulturae* **2022**, *8*, 139. [CrossRef]
12. Egorova, K.V.; Sinyavina, N.G.; Artemyeva, A.M.; Kocherina, N.V.; Chesnokov, Y.V. QTL Analysis of the Content of Some Bioactive Compounds in Brassica rapa L. Grown under Light Culture Conditions. *Horticulturae* **2021**, *7*, 583. [CrossRef]
13. Xiao, M.; Wang, H.; Li, X.; Mason, A.S.; Fu, D. Rapeseed as an Ornamental. *Horticulturae* **2021**, *8*, 27. [CrossRef]
14. Musaev, F.; Priyatkin, N.; Potrakhov, N.; Beletskiy, S.; Chesnokov, Y. Assessment of Brassicaceae Seeds Quality by X-ray Analysis. *Horticulturae* **2021**, *8*, 29. [CrossRef]

Disclaimer/Publisher's Note: The statements, opinions and data contained in all publications are solely those of the individual author(s) and contributor(s) and not of MDPI and/or the editor(s). MDPI and/or the editor(s) disclaim responsibility for any injury to people or property resulting from any ideas, methods, instructions or products referred to in the content.

horticulturae

MDPI

Article

Selection on *BrFLC1* Is Related to Intraspecific Diversity of *Brassica rapa* Vegetables

Jiahe Liu †, Xu Cai †, Yufang Li †, Yue Chen, Baozhen Gao, Runmao Lin, Jianli Liang, Xiaowu Wang and Jian Wu *

Institute of Vegetables and Flowers, Chinese Academy of Agricultural Sciences, No.12, Haidian District, Beijing 100081, China; 13121270799@163.com (J.L.); caixu0518@163.com (X.C.); liyufang326815@163.com (Y.L.); caascy@163.com (Y.C.); rongruonuanxinfei@163.com (B.G.); linrunmao@caas.cn (R.L.); liangjianli@caas.cn (J.L.); wangxiaowu@caas.cn (X.W.)

* Correspondence: wujian@caas.cn

† These authors contributed equally to this work.

Citation: Liu, J.; Cai, X.; Li, Y.; Chen, Y.; Gao, B.; Lin, R.; Liang, J.; Wang, X.; Wu, J. Selection on *BrFLC1* Is Related to Intraspecific Diversity of *Brassica rapa* Vegetables. *Horticulturae* **2021**, *7*, 247. https://doi.org/10.3390/horticulturae7080247

Academic Editor: Larisa Garkava-Gustavsson

Received: 9 July 2021
Accepted: 12 August 2021
Published: 14 August 2021

Publisher's Note: MDPI stays neutral with regard to jurisdictional claims in published maps and institutional affiliations.

Copyright: © 2021 by the authors. Licensee MDPI, Basel, Switzerland. This article is an open access article distributed under the terms and conditions of the Creative Commons Attribution (CC BY) license (https://creativecommons.org/licenses/by/4.0/).

Abstract: Flowering time is important for *Brassica rapa* vegetables because premature bolting before harvest can lower yield and quality. *FLOWERING LOCUS C* (*FLC*) acts as a key repressor of flowering. In this study, we identified a nonsynonymous mutation at the 58th nucleotide of exon1 in *BrFLC1* (named as Pe1+58 (A/C)) by screening resequencing data of 199 *B. rapa* accessions and verified this mutation as being related to flowering time variation. Strong linkage inheritance was detected between this locus and a previously reported splicing site mutation at intron 6 of *BrFLC1* (Pi6+1 (G/A)), showing as co-occurrence of BrFLC1Pe1+58(A) and BrFLC1Pi6+1(G), named as haplotype H1: AG, or co-occurrence of BrFLC1Pe1+58(C) and BrFLC1Pi6+1(A), named as haplotype H2: CA. The frequency distribution of *BrFLC1* haplotypes skewed to the haplotype H1 in turnip, broccoletto, mizuna, komatsuna, and taicai, while it was skewed to the haplotype H2 in caixin, pak choi, zicaitai, and wutacai. The frequencies of the two haplotypes were comparable in Chinese cabbage. This indicated that *BrFLC1* haplotypes were related to *B. rapa* intraspecific diversification. Further analysis of a Chinese cabbage collection revealed that accessions from the spring ecotype preferred to keep H1: AG and almost all accessions from the summer ecotype were H2: CA. The early flowering haplotype of *BrFLC1* was purified in summer Chinese cabbage, indicating that BrFLC1 had been strongly selected during genetic improvement of summer Chinese cabbages. A significant difference in flowering time of F_2 individuals with the homologous BrFLC1Pi6+1(G) allele but different BrFLC1Pe1+58 (A/C) alleles, indicated that this locus had independent genetic effects on flowering time. The newly identified allelic diversity of *BrFLC1* can be used for breeding of resistance to premature bolting in *B. rapa* vegetables.

Keywords: *Brassica rapa*; *FLOWERING LOCUS C*; haplotype; flowering time; intraspecific diversification

1. Introduction

Brassica rapa comprises vegetables, fodders, and oilseed crops. *B. rapa* responds to vernalization as early as at the stage of germinated seed. This characteristic is not favorable for producing leafy *B. rapa* vegetables because premature bolting often happens due to low temperatures during early growth and leads to loss of commercial value. Breeding of *B. rapa* varieties with strict requirements on vernalization conditions, such as longer period and lower temperature, could be an efficient way to avoid premature bolting in the cultivation of *B. rapa* vegetables.

Plants have evolved a complex genetic network to ensure flowering and seed set during favorable environmental conditions [1]. *FLOWERING LOCUS C* (*FLC*) is one of the key genes involved in controlling the conditions for vernalization. *FLC* encodes a MADS-box transcription factor that inhibits flowering by directly binding to floral promoting genes such as *FLOWERING LOCUS T* (*FT*), *SUPPRESSPR OF COSTANS 1* (*SOC1*), and

SQUAMOSA PROMOTER-BINDNG PROTEIN-LIKE 15 (*SPL15*) to block their transcription [2] In *Arabidopsis thaliana* most of the variation in flowering time is controlled by *FLC* and variation of the *FRI* allele that activates the expression of *FLC* [3,4].

The ancestor of *Brassica* species underwent a genome triplication event after divergence from *A. thaliana* [5]. Based on syntenic analysis between *B. rapa* and *A. thaliana*, three copies of *FLC* (*BrFLC1*, *BrFLC2*, and *BrFLC3*) were identified in *B. rapa* collinear regions [5,6]. The three syntenic copies were confirmed being functional in the regulation of flowering time by transferring the genes into *A. thaliana* [7]. Moreover, genetic analysis using different germplasm materials and segregating populations further confirmed that *BrFLC1*, *BrFLC2*, and the copy located in the non-collinear region, *BrFLC5*, are flowering repressors [8–13]. These studies revealed a few naturally occurred sequence variations of *FLC* orthologous genes in *Brassica* that affected their functions in flowering repression. SNPs located at splicing sites and resulting in alternative splicing patterns were reported for two *B. rapa FLC* homologues, *BrFLC1 and BrFLC5* [10,13]. Okazaki et al. identified a single-base deletion in exon 4 of *BoFLC2* that produced a non-functional allele in *B. oleracea* [14]. Wu et al. detected a 57-bp deletion across exon 4 to intron 4 that resulted in loss-of-function of *BrFLC2* [9]. Large InDels have often been identified in introns of *FLC* genes. In *B. napus* a 2.833 kb fragment insertion in the first intron of *BnFLC.A2* generates a loss-of-function allele that can promote flowering when this allele is introgressed into chromosome C2 [15] In contrast, Kitamoto et al. reported a large insertion of 5037 bp near the 5′ end of the first intron of *BrFLC2* related to the late flowering in *B. rapa* [12]; however, there was no apparent correlation between a 5678 bp insertion in the first intron of *BrFLC3* and bolting time. These function-related sequence variations could potentially be used in marker-assisted selection (MAS) for breeding of premature bolting resistant *B. rapa* vegetables.

Flowering time is an important environmental adaption and thus has been under selection during crop domestication. The domestication of *B. rapa* vegetables has been analyzed regarding leafy head formation in Chinese cabbage and tuber formation in turnips [16]. It is interesting that genes related to tuber formation in turnips were found to be involved in the flowering pathway [17] Flowering time was analyzed in different Chinese cabbage ecotypes to dissect the effect of modern breeding selection in ecotype improvement [18] In Chinese cabbage, incorporation of the elite alleles *BrFLC1* and *BrVIN3.1* is the determining genetic factor of the spring ecotype [18] However, it is still unclear whether *BrFLC* genes were selected during intraspecific diversification of *B. rapa*.

Here, we report a nonsynonymous mutation in exon 1 of *BrFLC1*, named as BrFLC1Pe1+58 (A/C), related to flowering time variation observed in three *B. rapa* germplasm collections containing 199, 236, and 829 accessions. Linkage inheritance was found between BrFLC1Pe1+58 (A/C) and a previously identified splicing site mutation, BrFLC1Pi6+1 (G/A). Further analysis revealed that BrFLC1Pe1+58 (A/C) had an independent effect on flowering time in an F_2 population segregating only at the BrFLC1Pe1+58 (A/C) locus. The frequency distribution of haplotypes comprising BrFLC1Pe1+58 (A/C) and BrFLC1Pi6+1 (G/A) loci in different subspecies varied according to their flowering pattern; late-flowering had a high frequency of the H1 (AG), while early-flowering had a high frequency of H2 (CA). We demonstrate that there was strong selection on *BrFLC1* haplotype H2:CA during summer Chinese cabbage improvement.

2. Materials and Methods

2.1. Sequence Variation in Genome Region of BrFLC1

To identify naturally occurring mutations in *BrFLC1*, we extracted the genomic sequence of the *BrFLC1* genic region (including 2 kb upstream and downstream of the gene body) from the Chiifu v3.0 Assembly (http://brassicadb.cn (accessed on 1 August 2021)) and then mapped all resequencing data to the reference sequence. Reseuqencing data was collected from a total of 391 *B. rapa* accessions. Among them, 199 accessions were from our previous work [16]; 192 were collected from a previously reported study [18]. First, raw reads were filtered using fastp (version 0.12.3) [19] with parameters '-z 4-q 20-u 30-n 5'.

Then, all of the clean reads were mapped to the extracted sequence using BWA-MEM (version 0.7.5a-r405) [20] with the default parameters. Finally, variants were called using SAMtools (version 0.1.19-44428cd) [21].

2.2. Plant Materials and Growth Conditions

A total of 236 *B. rapa* accessions from 11 cultivar groups (Supplementary Table S4) were used to investigate flowering time. The 236 *B. rapa* accessions were sown in pots in a greenhouse on 19 February 2016. The temperature varied from 15 to 28 °C. After 21 days of growth, these seedlings were transplanted into an open field in the Shunyi District of Beijing, China, in a randomized complete block design. Each of these accessions was grown in five replicates.

A large-scale *B. rapa* germplasm collection with 785 accessions, including 212 from the previous 236 accessions, was used to analyze the *BrFLC1* haplotype distribution frequency among *B. rapa* cultivar groups. The 785 accessions were sown in plug trays in a greenhouse on 16 August 2017. The temperature was 24–31 °C. Twenty-two days later, leaf samples were harvested for DNA isolation.

A segregating population was constructed from an accession, that had a heterozygous allele of BrFLC1Pe1+58 (AC) and a homologous BrFLC1Pi6+1 (G) allele. In total, 166 F_2 lines were used for analyzing flowering phenotype differences between the individuals with different alleles of BrFLC1Pe1+58 (A/C). Germinated seeds were sown into plug trays in a greenhouse in the Haidian District of Beijing, China on 22 January 2021. After 18 days, the seedlings were transferred into pots and grown in the greenhouse without climate control until they flowered. The photoperiod was set to a 14 h:10 h day: night rhythm with supplementary lighting by high pressure sodium lamp.

2.3. Evaluation of Bolting Time and Flowering Time

Bolting time and flowering time were evaluated for the germplasm collection using 236 accessions and the F_2 population segregating at the BrFLC1Pe1+58 (A/C) allele. Bolting time was recorded as the number of days from sowing to the first flower bud appeared (days to bolting, DTB). Flowering time was recorded as the number of days from sowing to the first flower opening. The survey ended at 150 days after sowing for the germplasm collection. Bolting time and flowering time were recorded as 150 DTB and 160 DTF, respectively, for the accessions that did not bolt or flower until this time point. The survey ended at 120 days after sowing for the F_2 population. Bolting time and flowering time were recorded as 120 DTB and 140 DTF, respectively, for the individuals that did not bolt or flower before this time point.

2.4. KASP Assay for BrFLC1 Genotyping

Genomic DNA was extracted from fresh young leaves using a modified CTAB protocol [22]. The DNA concentration was measured by NanoDrop (ND-1000, Thermor Fisher Scientific) and normalized to a concentration of 15 ng/μL. The allele-specific primers were designed carrying the standard FAM (5′-GAAGGTGACCAAGTTCATGCT-3′) and VIC (5′-GAAGGTCGGAGTCAACG-GATT-3′) tails and with the targeted SNP at the 3′ end. The forward primers BrFLC1Pe1+58_F1 and BrFLC1Pe1+58_F2 for the A and C alleles, and the common reverse primer BrFLC1Pe1+58_R were designed to produce a PCR product with a length of 48 bp (for the A allele) or 50 bp (for the C allele) for the KASP assay. The primers used to discriminate the alleles at the locus BrFLC1Pi6+1 were the same as those reported by Xi et al. [13]. The KASP assay procedure followed Xi et al. [13].

2.5. Statistical Analysis

Analysis of variance (ANOVA) by independent samples *t*-tests was performed in the SPSS version 22.0 statistical package (SPSS Inc., Chicago, IL, USA). One-way ANOVA was performed with flowering time as the test variable and genotype as the grouping variable

3. Results

3.1. A Nonsynonymous Mutation Pe1+58(A/C) of BrFLC1 Differently Appeared among B. rapa Subspecies

A nonsynonymous mutation at the nucleotide 58 in the first exon of *BrFLC1*, named as BrFLC1Pe1+58 (A/C), was identified from analyzing resequencing data of a natural population comprising 199 diverse *B. rapa* accessions. The population included accessions from groups of turnip, Chinese cabbage, pak choi, wutacai, caixin, and zicaitai, and it had been used to represent the variation landscape of *B. rapa* in our previous study [16]. A total of 49 SNPs and six InDels in the genic regions of *BrFLC1* (including 2 kb upstream and downstream of the gene body) were identified (Supplementary Table S1). Among these mutations, only two SNPs occurred in the CDS regions, one of which resulted in a non-synonymous mutation (ACC → cCC) and the other that led to a synonymous mutation. This non-synonymous mutation locus was named as BrFLC1Pe1+58 (A/C). Furthermore, the genotype frequency of BrFLC1Pe1+58 (A/C) in the population with 199 diverse accessions was analyzed (Supplementary Table S1). There was a clear trend of genotype frequency in the subspecies with different flowering patterns. Among the 54 turnip accessions, 92.59% were of the A genotype. In contrast, none of the 30 caixin accessions and 13 zicaitai accessions were of the A genotype (Figure 1). This result suggested that the nonsynonymous mutation at BrFLC1Pe1+58 (A/C) was related to flowering time variation, since the turnip accessions were later flowering, while accessions from caixin and zicaitai were early flowering.

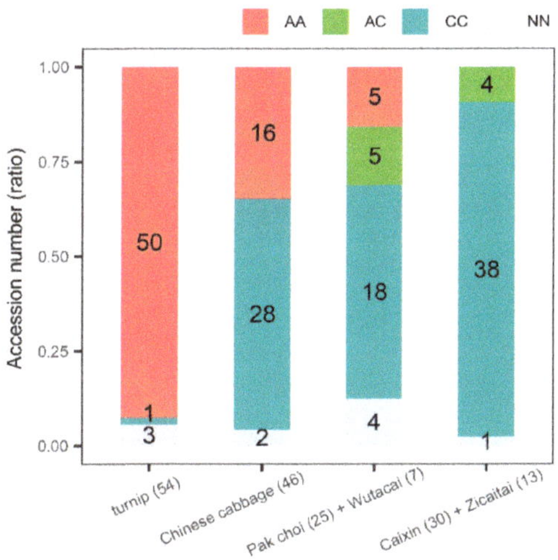

Figure 1. Frequency distribution of Pe1+58(A/C) genotypes in different *B. rapa* cultivar groups.

3.2. Pe1+58(A/C) of BrFLC1 Is Related to Flowering Time Variations among a Germplasm Collection of 236 B. Rapa

To determine whether BrFLC1Pe1+58 (A/C) was related to variation in flowering time, a germplasm collection with 236 *B. rapa* accessions from 11 subspecies or varieties was screened for allelic variations using a KASP assay. Among the 236 accessions, 123 carried the A allele; 100 carried the C allele; and 13 were heterozygous at this locus (Supplementary Table S2). Flowering time was assessed for the 236 accessions in an open field in the spring of 2016. The bolting time varied between 32 DTB to not bolting until the end of investigation (recorded as 150 DTB), while the flowering time varied between 45 DTF to not flowering at the end of investigation (recorded as 160 DTF). The DTB of plants homozygous for the C allele was mainly distributed between 60 and 100 days, while this was clustered at 80–150 days for plants homozygous for the A allele (Figure 2a). The accessions with a mutated C allele bolted significantly earlier than those with the wild type A allele ($p < 0.001$,

Figure 2b, Supplementary Table S2). This result indicated that the Pe1+58(A/C) allelic variation of *BrFLC1* was related to the flowering time variation in *B. rapa*.

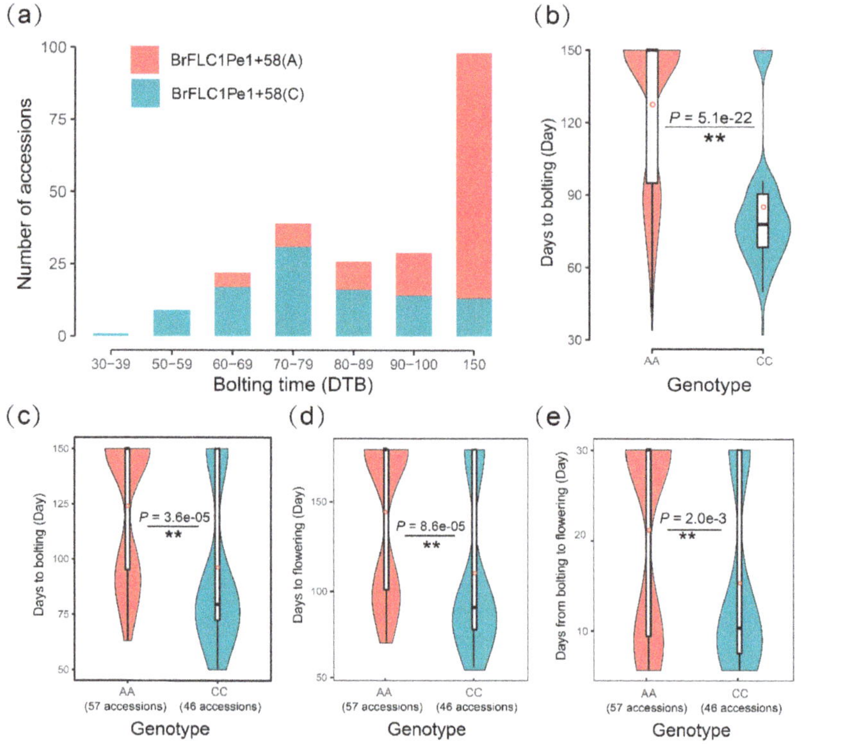

Figure 2. Frequency distribution of *BrFLC1* Pe1+58 alleles and bolting time distribution in a *B. rapa* germplasm collection ($n = 236$). (**a**) Distribution of BrFLC1Pe1+58 alleles with their associated bolting time. The heterozygous genotype was not included in the analysis because the number of accessions was too few; (**b**) distribution of bolting time for the accessions with different *BrFLC1* Pe1+58 alleles; bolting time (**c**), flowering time (**d**), and the time from bolting to flowering (**e**) according to their *BrFLC1* Pe1+58 alleles in the group of Chinese cabbage. Comparisons are done by Student-Newman-Keuls test with $\alpha = 0.01$, and the marker (**) indicates the p-value <= 0.01.

Among the 236 accessions, 112 were Chinese cabbages (ssp. *pekinensis*). In contrast to the distribution bias for the BrFLC1Pe1+58 (A/C) allele in the groups of turnip, caixin, or pak choi, the Chinese cabbage had comparable frequencies of the two alleles. The frequencies in the Chinese cabbage group were 50%, 42%, and 8% for AA, CC, and AC plants, respectively. The DTB and DTF of Chinese cabbages with AA allele were significantly longer than those of accessions carried CC allele (Figure 2c,d). The accessions with AA alleles also took significantly longer from bolting to flowering (Figure 2e). This result further confirmed that this natural mutation is related to variation in flowering time.

3.3. BrFLC1 Haplotype Consisting of BrFLC1Pe1+58 (A/C) and BrFLC1Pi6+1(G/A) Was Associated with the Intraspecific Diversification and Ecotype Diversification in B. rapa

A linkage inheritance between BrFLC1Pe1+58 and a previously reported splicing site mutation in intron 6 [10] BrFLC1Pi6+1 (G/A) was detected (Figure 3a) from analysis of *BrFLC1* sequence variations. The linkage inheritance manifested as the co-occurrence of genotypes BrFLC1Pe1+58(A) and BrFLC1Pi6+1(G) or co-occurrence of genotypes BrFLC1Pe1+58(C) and BrFLC1Pi6+1(A), and the linkage inheritance was strongest in the turnip and caixin (Figure 3a). Based on the strong linkage inheritance of the two loci, two haplotypes of the combinations of BrFLC1Pe1+58 and BrFLC1Pi6+1 were proposed as H1: AG and H2: CA (Figure 3c). All of the turnip accessions showed haplotype 1 (H1: AG), and all of caixin and zicaitai accessions showed haplotype 2 (H2: CA) (Figure 3a,d). However, the Chinese cabbage group showed a mixture of two haplotypes.

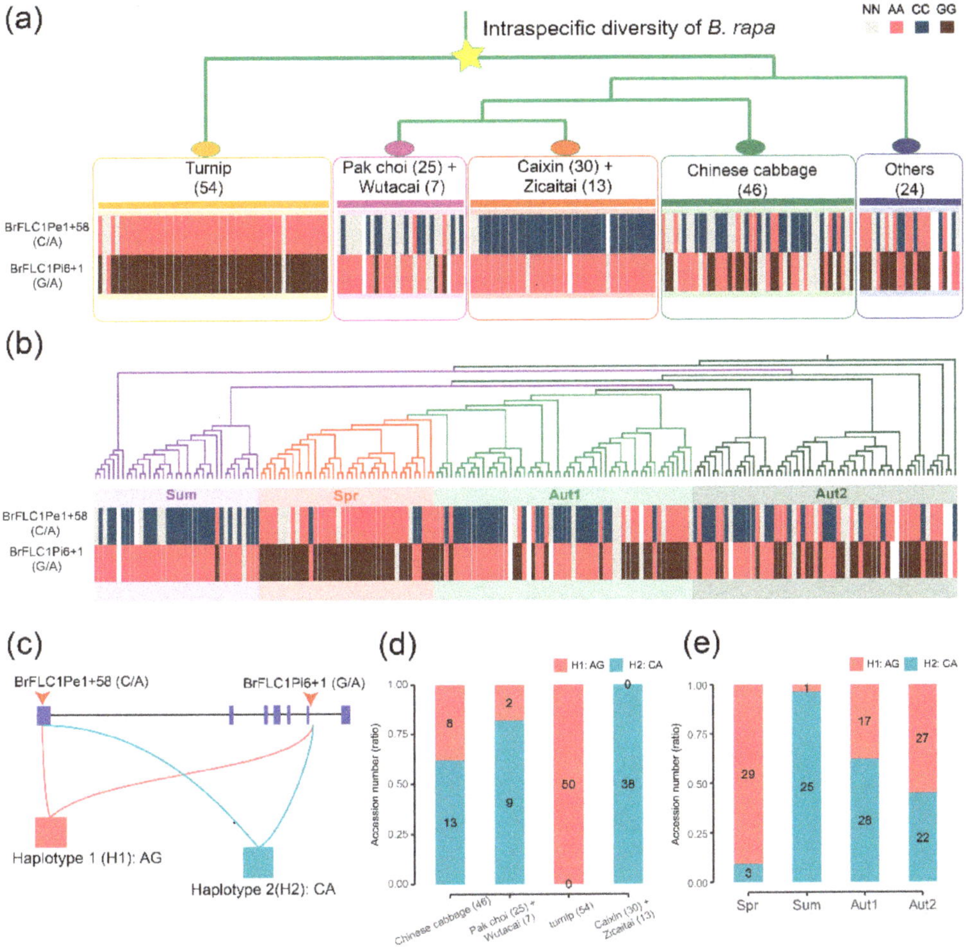

Figure 3. Distribution of haplotypes comprising BrFLC1Pe1+58 (A/C) and BrFLC1Pi6+1 (G/A) in different *B. rapa* subspecies. (**a**) Genotype distribution of BrFLC1Pe1+58 (A/C) and BrFLC1Pi6+1 (G/A) in a germplasm collection of 199 accessions [16]; (**b**) Genotype distribution of BrFLC1Pe1+58 (A/C) and BrFLC1Pi6+1 (G/A) in a collection of 192 Chinese cabbage accessions [18]; (**c**) Schematic model of haplotypes BrFLC1Pe1+58 (A/C) and BrFLC1Pi6+1 (G/A); (**d**) haplotype distribution in 120 of 199 *B. rapa* germplasm accessions; (**e**) haplotype distribution in 152 of 192 Chinese cabbage accessions. In (**d**,**e**), the accessions carrying heterozygous alleles in either BrFLC1Pe1+58 (A/C) or BrFLC1Pi6+1 (G/A) were not included.

To analyze the *BrFLC1* haplotype-phenotype association in Chinese cabbages, different ecotypes of Chinese cabbage were analyzed for the two haplotypes. The 192 Chinese cabbage accessions were divided into four ecotypes according to their growing seasons, including spring, summer, autumn 1, and autumn 2 [18]. In the present study, a phylogenetic tree was constructed based on the genome-wide SNPs (Figure 3b). The tree was in line with the ecotype division by growing season. Strong linkage inheritance and distribution bias were also clear in Chinese cabbage. Around 90.63% of the accessions from spring Chinese cabbage showed haplotype 1 (H1: AG), and 96.15% of accessions from summer Chinese cabbage showed haplotype 2 (H2: CA) (Figure 3b,e). This result indicated that the strong linkage inheritance of the two loci was tightly associated with the diversification of different Chinese cabbage ecotypes.

To further explore the relationship between *BrFLC1* haplotypes and intraspecific diversity of *B. rapa*, an expanded germplasm collection with 829 diverse accessions was screened for genotypes of these two loci. These 829 accessions belonged to 11 subspecies or varieties, covering *B. rapa* crops as much as possible (Supplementary Table S3). Accessions of Chinese cabbage (252), pak choi (162), turnip (152), and caixin (62) occupied about 76% of this collection, because these varieties are rich in *B. rapa* germplasm. The genotyping results showed that among the 829 accessions, 390 were haplotype 1 (H1: AG) and 395 were haplotype 2 (H2: CA); 42 were heterozygous for both loci (Supplementary Table S3). The distribution of haplotype frequency differed among *B. rapa* cultivar groups (Figure 4). Two types could be observed. Type 1 was represented by turnip along with mizuna, komatsuna, taicai, and broccoletto, with a high proportion of haplotype 1. Type 2 was represented by pak choi together with caixin, wutacai, and zicaitai, with a high ratio of haplotype 2. The results did suggest that the *BrFLC1* haplotypes consisting of these two loci were associated with intraspecific diversification and have been under strong selection during *B. rapa* domestication. In terms of Chinese cabbage, as shown by the 192 accessions, *BrFLC1* haplotypes were strongly selected in the recent breeding procedure.

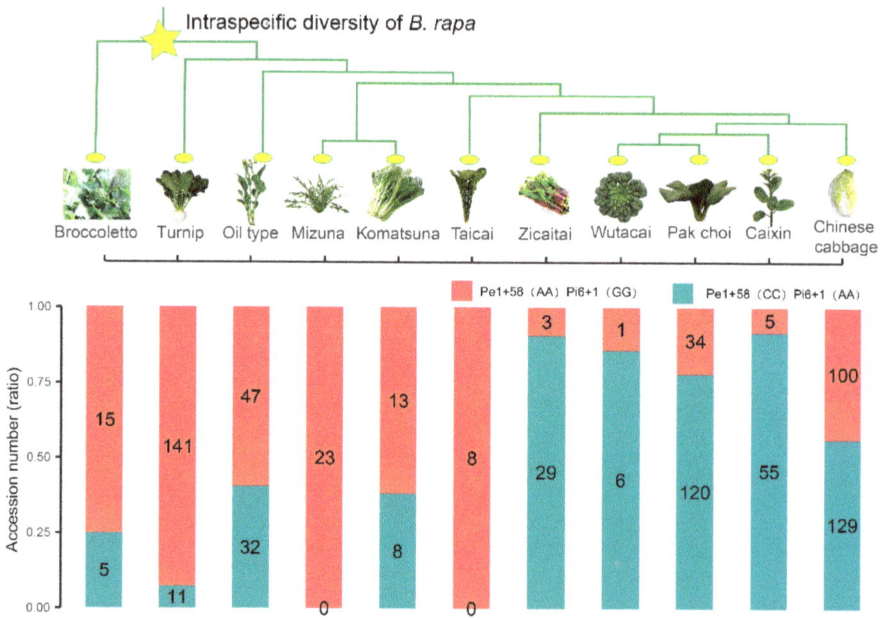

Figure 4. The frequencies of *BrFLC1* haplotypes in a germplasm collection with 829 *B. rapa* accessions. The phylogenetic tree was constructed based on the neighbor joining tree constructed with resequencing data of 524 accessions [23].

3.4. The BrFLC1Pe1+58 (A/C) Locus Poses Independent Genetic Effects on Flowering Time in B. rapa

The locus BrFLC1Pe1+58 (A/C) was revealed to have an independent genetic effect on flowering by analyzing an F_2 population segregating for this locus. From screening of 785 germplasm accessions, an oil-type accession HNAJHCZ was identified carrying heterozygous alleles BrFLC1P1e+58 (A/C) and homozygous on BrFLC1Pi6+1 (GG). The plant was self-pollinated to generate a population segregating at the BrFLC1e+58 (A/C) locus to evaluate the genetic effect on variation in flowering time. In total, 166 F_2 individuals were investigated for bolting time and flowering time and were genotyped for BrFLC1P1e+58 (A/C) using the KASP assay. Among the 166 individuals, 39 carried the A allele; 37 carried the C allele, and 90 were heterozygous (Supplementary Table S4). The genotype ratio of A/H/C was 1:2.3:1, which was in line with the segregation ratio of the F_2 population. HNAJHCZ plants bolted at 63 DTB and flowered at 76 DTB. Among the F_2 individuals, the average bolting and flowering times of individuals with the A allele (62.0 DTB and 78.9 DTF) were later than those of CC individuals (55.4 DTB, 69.0 DTF) ($p < 0.05$ for DTB, $p < 0.05$ for DTF). However, there was no significant difference between individuals with AA and AC (58.1 DTB and 72.9 DTF), or between individuals with CC and AC genotypes. It is interesting that when considering flowering time as the period from the first bud appearing to the first flower blooming (DTF-DTB), the difference between AA (16.9 days) and CC (13.2 days) or between AA and AC (14.8 days) were both significant (Figure 5c). These results indicated that the nonsynonymous mutation BrFLC1Pe1+58 (A/C) affected *BrFLC1* function in repressing flowering in *B. rapa*. Although there was strong genetic linkage between the BrFLC1Pe1+58 (A/C) and BrFLC1Pi6+1 (G/A) loci, BrFLC1Pe1+58 (A/C) had an independent genetic effect and could contribute to the breeding of late bolting *B. rapa* vegetables.

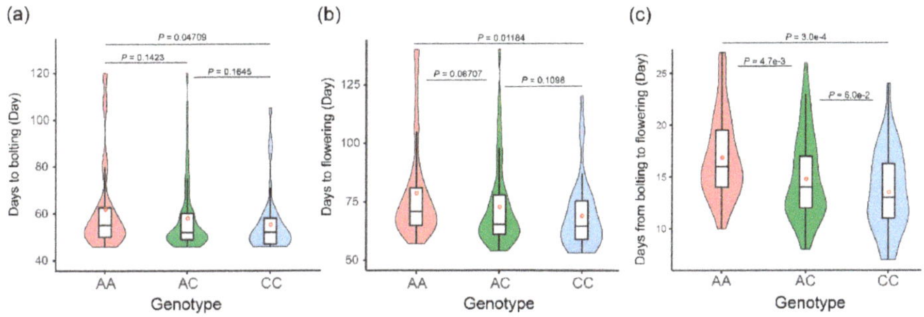

Figure 5. Distribution of bolting time (**a**), flowering time (**b**), and time from bolting to flowering (**c**) in an F_2 population. AA, homozygous A allele of BrFLC1Pe1+58; CC, homozygous C allele of BrFLC1Pe1+58; AC, heterozygous for BrFLC1Pe1+58. Comparisons are done by Student-Newman-Keuls test with $\alpha = 0.01$.

4. Discussion

In this study, a naturally occurring single nucleotide mutation, BrFLC1Pe1+58 (A/C), was detected in *BrFLC1*, and this mutation led to an amino acid shift from Thr to Pro. This mutation was significantly associated with flowering time variation in *B. rapa*.

The newly detected mutation BrFLC1Pe1+58 (A/C) was found to be in inherited linkage with a previously reported splicing site mutation, BrFLC1Pi6+1 (G/A) [10]. Mutated BrFLC1Pi6+1 (A) allele led to alternative splicing patterns and early flowering. To test whether the BrFLC1Pe1+58 (A/C) locus had an independent effect on flowering, a population segregating at this locus, but homologous to BrFLC1Pi6+1(G) was developed. By analyzing differences between flowering time of F_2 individuals with different BrFLC1Pe1+58 genotypes, we found that this locus affected flowering time independent of the BrFLC1Pi6+1 (G/A) locus. This result gives a hint for breeding of *B. rapa* crops that both loci need to be considered for enhancing premature bolting resistance.

Flowering time is one of the major traits under strong selection during crop domestication [24,25]. Previous research indicated that selection for flowering-related genes, especially *FLC* was a powerful force during the domestication of *Brassica* crops. *FLC*, encodes a core regulator in *B. napus* ecotypes improvement [26]. By pan-genome and large-scale re-sequencing of *B. rapa*, phylogenetic relationships have been established [23]. Therefore, we could analyze the relationship between the distribution of the *BrFLC1* haplotype in different *B. rapa* subspecies and intraspecific diversification. The haplotypes from combining these two loci were revealed to be associated with *B. rapa* diversity by analyzing a total of 829 germplasm accessions. Based on the previously constructed phylogenetic tree [23], different *B. rapa* subspecies were divided into two groups according to the position of taicai (Figure 4). The more ancient subspecies were skewed to carrying the haplotype of late flowering; however, the subspecies or varieties differentiated in China after the turnip was introduced from the Mediterranean–Middle Eastern area were biased to carrying the early flowering haplotype. This result was in line with the proposition that China is the secondary origin and differentiation center of *B. rapa* [27] Pak choi, wutacai, as well as caixin and zicaitai are prevalently cultivated in southern regions of the Yangzi River basin. Caixin flowers extremely early because its commercial organ is the flower stem. Our results indicate that selection, including natural and artificial, has played a significant role in *B. rapa* diversification.

In the group of Chinese cabbage, the case was more complicated. Su et al. divided 194 Chinese cabbage accessions into four ecotypes, spring, summer, autumn 1 and autumn 2. Spring Chinese cabbage has stronger winterness compared with the autumn Chinese cabbage cultivars, and the elite allelic assembly of *BrVIN3.1* and *BrFLC1* was found to be a major genetic source of variation during selection [18]. In contrast to spring Chinese cabbage showing a very uniform genetic background, summer Chinese cabbage was more complex in its genetic composition; therefore, no clear genetic resource was identified [18]. However, in this study, resequencing data from Su et al. [18] was analyzed for the *BrFLC1* haplotype, in which summer Chinse cabbage showed a pure haplotype (H2: CA). The general characteristics of summer Chinese cabbage are more heat resistance and faster growth. In addition, summer Chinese cabbages are cultivated in Southern China where the temperature varies within a range of warm to high all year round; therefore, weak winterness ensures that they flower and produce seeds in this environment. Consistent with this growing habitat, the summer ecotype showed an opposite haplotype distribution from that of the spring ecotype. These results also indicated that the natural and artificial selection had posed very strong purification effects on the *BrFLC1* haplotype during the breeding of summer Chinese cabbage, and this might also be the case for caixin.

5. Conclusions

In summary, in this work, we demonstrated that *BrFLC1* haplotypes were related to *B. rapa* intraspecific diversification. Moreover, we showed that naturally occurred variations in *BrFLC1*, BrFLC1Pe1+58 (A/C), and BrFLC1Pi6+1 (G/A), both relate to flowering time, but each has an independent genetic effect. Our work will help to improve pre-mature bolting resistance in *B. rapa* vegetables.

Supplementary Materials: The following are available online at https://www.mdpi.com/article/10.3390/horticulturae7080247/s1, Table S1. SNPs and InDels in the BrFLC1 region detected in 199 *B. rapa* germplasm accessions. Table S2. DTB and DTF of the germplasm collection and the corresponding *BrFLC1*Pe1+58 genotype. Table S3. The genotype of *BrFLC1*Pe1+58 locus and *BrFLC1*Pi6+1 locus of 829 *B. rapa* germplasm accessions. Table S4. DTB and DTF of the F_2 population and the corresponding *BrFCL1* Pe1+58 allele genotype.

Author Contributions: Project designing, J.W. and X.W.; Data curation, J.L. (Jiahe Liu); Formal analysis, X.C. and J.W.; Investigation, J.L. (Jiahe Liu), Y.L., Y.C. and B.G.; Visualization, X.C.; Writing—original draft, J.L. (Jiahe Liu); Writing—review and editing, R.L., J.L. (Jianli Liang), X.W. and J.W. All authors have read and agreed to the published version of the manuscript.

Funding: This work was supported by the Agricultural Science and Technology Innovation Program (ASTIP), the Key Laboratory of Biology and Genetic Improvement of Horticultural Crops, Ministry of Agriculture, P.R. China.

Institutional Review Board Statement: Not applicable.

Informed Consent Statement: Not applicable.

Acknowledgments: This research work was supported the Science and Technology Innovation Program of the Chinese Academy of Agricultural Sciences, and the Key Laboratory of Biology and Genetic Improvement of Horticultural Crops, Ministry of Agriculture, P.R. China.

Conflicts of Interest: The authors declare no conflict of interest.

References

1. Fornara, F.; de Montaigu, A.; Coupland, G. SnapShot: Control of flowering in Arabidopsis. *Cell* **2010**, *141*, 550. [CrossRef]
2. Deng, W.W.; Ying, H.; Helliwell, C.A.; Taylor, J.M.; Peacock, W.J.; Dennis, E.S. *FLOWERING LOCUS C* (*FLC*) regulates development pathways throughout the life cycle of Arabidopsis. *Proc. Natl. Am. Sci. USA* **2011**, *108*, 6680–6685. [CrossRef]
3. Michaels, S.D.; Amasino, R.M. *FLOWERING LOCUS C* encodes a novel MADS domain protein that acts as a repressor of flowering. *Plant Cell* **1999**, *11*, 949–956. [CrossRef]
4. Johanson, U.; West, J.; Lister, C.; Michaels, S.; Amasino, R.; Dean, C. Molecular analysis of FRIGIDA, a major determinant of natural variation in Arabidopsis flowering time. *Science* **2000**, *290*, 344–347. [CrossRef] [PubMed]
5. Wang, X.; Wang, H.; Wang, J.; Sun, R.; Wu, J.; Liu, S.; Bai, Y.; Mun, J.H.; Bancroft, I.; Cheng, F.; et al. The genome of the mesopolyploid crop species *Brassica rapa*. *Nat. Genet.* **2011**, *43*, 1035–1039. [CrossRef] [PubMed]
6. Yang, T.J.; Kim, J.S.; Kwon, S.J.; Lim, K.B.; Choi, B.S.; Kim, J.A.; Jin, M.; Park, J.Y.; Lim, M.H.; Kim, H.I.; et al. Sequence-level analysis of the diploidization process in the triplicated *FLOWERING LOCUS C* region of *Brassica rapa*. *Plant Cell* **2006**, *18*, 1339–1347. [CrossRef]
7. Kim, S.Y.; Park, B.S.; Kwon, S.J.; Kim, J.; Lim, M.H.; Park, Y.D.; Kim, D.Y.; Suh, S.C.; Jin, Y.M.; Ahn, J.H.; et al. Delayed flowering time in Arabidopsis and Brassica rapa by the overexpression of FLOWERING LOCUS C (FLC) homologs isolated from Chinese cabbage (*Brassica rapa* ssp. *pekinensis*). *Plant Cell Rep.* **2007**, *26*, 327–336. [CrossRef] [PubMed]
8. Lou, P.; Zhao, J.; Kim, J.S.; Shen, S.; Carpio, D.P.D.; Song, X.; Jin, M.; Vreugdenhil, D.; Wang, X.; Koornneef, M.; et al. Quantitative trait loci for flowering time and morphological traits in multiple populations of *Brassica rapa*. *J. Exp. Bot.* **2007**, *58*, 4005–4016. [CrossRef]
9. Wu, J.; Wei, K.; Cheng, F.; Li, S.; Wang, Q.; Zhao, J.; Bonnema, G.; Wang, X. A naturally occurring InDel variation in *BraA.FLC.b* (*BrFLC2*) associated with flowering time variation in *Brassica rapa*. *BMC Plant Biol.* **2012**, *12*, 151. [CrossRef]
10. Yuan, Y.X.; Wu, J.; Sun, R.F.; Zhang, X.W.; Xu, D.H.; Bonnema, G.; Wang, X.W. A naturally occurring splicing site mutation in the *Brassica rapa FLC1* gene is associated with variation in flowering time. *J. Exp. Bot.* **2009**, *60*, 1299–1308. [CrossRef]
11. Zhao, J.J.; Kulkarni, V.; Liu, N.; Carpio, D.P.D.; Bucher, J.; Bonnema, G. *BrFLC2* (*FLOWERING LOCUS C*) as a candidate gene for a vernalization response QTL in *Brassica rapa*. *J. Exp. Bot.* **2010**, *61*, 1817–1825. [CrossRef]
12. Kitamoto, N.; Yui, S.; Nishikawa, K.; Takahata, Y.; Yokoi, S. A naturally occurring long insertion in the first intron in the *Brassica rapa* FLC2 gene causes delayed bolting. *Euphytica* **2014**, *196*, 213–223. [CrossRef]
13. Xi, X.; Wei, K.; Gao, B.; Liu, J.; Liang, J.; Feng, C.; Wang, X.; Wu, J. BrFLC5: A weak regulator of flowering time in *Brassica rapa*. *Theor. Appl. Genet.* **2018**, *131*, 2107–2116. [CrossRef] [PubMed]
14. Okazaki, K.; Sakamoto, K.; Kikuchi, R.; Saito, A.; Togashi, E.; Kuginuki, Y.; Matsumoto, S.; Hirai, M. Mapping and characterization of *FLC* homologs and QTL analysis of flowering time in *Brassica oleracea*. *Theor. Appl. Genet.* **2007**, *114*, 595–608. [CrossRef]
15. Chen, L.; Dong, F.; Cai, J.; Xin, Q.; Fang, C.; Liu, L.; Wan, L.; Yang, G.; Hong, D. A 2.833-kb insertion in *BnFLC.A2* and its homeologous exchange with *BnFLC.C2* during breeding selection generated early-flowering rapeseed. *Mol. Plant* **2017**, *11*, 222–225. [CrossRef] [PubMed]
16. Cheng, F.; Sun, R.; Hou, X.; Zheng, H.; Zhang, F.; Zhang, Y.; Liu, B.; Liang, J.; Zhuang, M.; Liu, Y.; et al. Subgenome parallel selection is associated with morphotype diversification and convergent crop domestication in *Brassica rapa* and *Brassica oleracea*. *Nat. Genet.* **2016**, *48*, 1218–1224. [CrossRef] [PubMed]
17. Liu, M.; Bassetti, N.; Petrasch, S.; Zhang, N.; Bucher, J.; Shen, S.; Zhao, J.; Bonnema, G. What makes turnip: Anatomy, physiology and transcriptome during early stages of its hypocotyl-tuber development. *Horti Res.* **2019**, *6*, 38. [CrossRef]
18. Su, T.; Wang, W.; Li, P.; Zhang, B.; Li, P.; Xin, X.; Sun, H.; Yu, Y.; Zhang, D.; Zhao, X.; et al. A genomic variation map provides insights into the genetic basis of spring Chinese Cabbage (*Brassica rapa* ssp. *pekinensis*) selection. *Mol. Plant* **2018**, *11*, 1360–1376. [CrossRef] [PubMed]
19. Chen, S.F.; Zhou, Y.Q.; Chen, Y.R.; Gu, J. fastp: An ultra-fast all-in-one FASTQ preprocessor. *Bioinformatics* **2018**, *34*, 884–890. [CrossRef]
20. Li, H.; Durbin, R. Fast and accurate short read alignment with Burrows-Wheeler transform. *Bioinformatics* **2009**, *25*, 1754–1760. [CrossRef] [PubMed]
21. Li, H.; Handsaker, B.; Wysoker, A.; Fennell, T.; Ruan, J.; Homer, N.; Marth, G.; Abecasis, G.; Durbin, R. The Sequence Alignment/Map format and SAMtools. *Bioinformatics* **2009**, *25*, 2078–2079. [CrossRef] [PubMed]
22. Wang, X.; Lou, P.; Bonnema, G.; Yang, B.; He, H.; Zhang, Y.; Fang, Z. Linkage mapping of a dominant male sterility gene *Ms-cd1* in *Brassica oleracea*. *Genome* **2005**, *48*, 848–854. [CrossRef] [PubMed]
23. Cai, X.; Chang, L.; Zhang, T.; Chen, H.; Zhang, L.; Lin, R.; Liang, J.; Wu, J.; Freeling, M.; Wang, X. Impacts of allopolyploidization and structural variation on intraspecific diversification in *Brassica rapa*. *Genome Biol.* **2021**, *22*, 166. [CrossRef] [PubMed]
24. Fuller, D.Q. Contrasting patterns in crop domestication and domestication rates: Recent archaeobotanical insights from the Old World. *Ann. Bot.* **2007**, *100*, 903–924. [CrossRef]

25. Izawa, T. Adaptation of flowering time by natural and artificial selection in Arabidopsis and rice. *J. Exp. Bot.* **2007**, *58*, 3091–3097. [CrossRef]
26. Lu, K.; Wei, L.; Li, X.; Wang, Y.; Wu, J.; Liu, M.; Zhang, C.; Chen, Z.; Xiao, Z.; Jian, H.; et al. Whole-genome resequencing reveals *Brassica napus* origin and genetic loci involved in its improvement. *Nat. Commun.* **2019**, *10*, 1154. [CrossRef]
27. Warwick, S.I. Brassicaceae in Agriculture. In *Genetics and Genomics of the Brassicaceae*; Schmidt, R., Bancroft, I., Eds.; Plant Genetics and Genomics: Crops and Models; Springer: New York, NY, USA, 2011; Volume 9, pp. 34–38.

horticulturae

MDPI

Article

Integrated Volatile Metabolomics and Transcriptomics Analyses Reveal the Influence of Infection TuMV to Volatile Organic Compounds in *Brassica rapa*

Xinxin Lu [1,†], Lei Zhang [1,†], Wenyue Huang [1], Shujiang Zhang [1], Shifan Zhang [1], Fei Li [1], Hui Zhang [1], Rifei Sun [1], Jianjun Zhao [2,*] and Guoliang Li [1,*]

1 State Key Laboratory of North China Crop Improvement and Regulation, Institute of Vegetables and Flowers, Chinese Academy of Agricultural Sciences, Beijing 100081, China; 82101195074@caas.cn (X.L.); zhanglei950301@163.com (L.Z.); huangWY688@163.com (W.H.); zhangshujiang@caas.cn (S.Z.); zhangshifan@caas.cn (S.Z.); lifei@caas.cn (F.L.); zhanghui05@caas.cn (H.Z.); sunrifei@caas.cn (R.S.)

2 State Key Laboratory of North China Crop Improvement and Regulation, Department of Horticulture, Hebei Agricultural University, Baoding 071000, China

* Correspondence: yyzjj@hebau.edu.cn (J.Z.); liguoliang@caas.cn (G.L.)

† These authors contributed equally to this work.

Abstract: Turnip mosaic virus (TuMV), which is distributed almost all over the world and has a wide range of hosts, mainly brassica crops, was first described in *Brassica rapa* in the USA. Plant volatile compounds play an important role in the host searching behavior of natural enemies of herbivorous insects. In this study, TuMV-inoculated resistant and susceptible *B. rapa* lines were tested using volatile metabolome and transcriptome analyses. In volatile metabolome analysis, the volatile organic compounds (VOCs) were different after inoculation with TuMV in resistant B80124 and susceptible B80461, and the degree of downregulation of differentially expressed metabolites was more obvious than the degree of upregulation. Through transcriptome analysis, 70% of differentially expressed genes were in biological process, especially focusing on defense response, flavonoid biosynthetic process, and toxin metabolic process, which indicates that TuMV stress maybe accelerate the increase of VOCs. Integrating the metabolome and transcriptome analyses, after inoculating with TuMV, auxin regulation was upregulated, and ARF, IAA and GH3 were also upregulated, which accelerated cell enlargement and plant growth in tryptophan metabolism. The different genes in zeatin biosynthesis pathways were downregulated, which reduced cell division and shoot initiation. However, the metabolite pathways showed upregulation in brassinosteroid biosynthesis and α-linolenic acid metabolism, which could cause cell enlargement and a stress response. This study determined the difference in volatiles between normal plants and infected plants and may lay a foundation for anti-TuMV research in *B. rapa*.

Keywords: TuMV; *Brassica rapa*; GC-MS; VOCs; HIPVs

Citation: Lu, X.; Zhang, L.; Huang, W.; Zhang, S.; Zhang, S.; Li, F.; Zhang, H.; Sun, R.; Zhao, J.; Li, G. Integrated Volatile Metabolomics and Transcriptomics Analyses Reveal the Influence of Infection TuMV to Volatile Organic Compounds in *Brassica rapa*. *Horticulturae* **2022**, *8*, 57. https://doi.org/10.3390/horticulturae8010057

Academic Editor: Dilip R. Panthee

Received: 29 November 2021
Accepted: 5 January 2022
Published: 8 January 2022

Publisher's Note: MDPI stays neutral with regard to jurisdictional claims in published maps and institutional affiliations.

Copyright: © 2022 by the authors. Licensee MDPI, Basel, Switzerland. This article is an open access article distributed under the terms and conditions of the Creative Commons Attribution (CC BY) license (https://creativecommons.org/licenses/by/4.0/).

1. Introduction

Turnip mosaic virus (TuMV), which is distributed almost all over the world and has a wide range of hosts, mainly brassica crops, belongs to the *Potyviridae* family. TuMV is one of the most prevalent viruses and is threatening brassica vegetables around the world, especially in Europe, Asia, and North America [1–3]. Therefore, TuMV has become a model for potyvirus–host interactions [3]. TuMV-diseased brassica crops show various symptoms, including the development of a mosaic pattern, shrinking, slight leaf stunting, mottling, chlorosis or spotting, and in the late stages of infection, severe stunting, chlorosis, necrosis, non-heading or unwound heading, and withering of the entire plant [4]. Moreover, TuMV caused a loss of 30% in *Brassica napus* production in Canada [5], as well as seed yield losses of up to 70% in *B. napus* in the UK [6] and 50% reductions in *B. oleracea* (cabbage) head production in Kenya [7].

In addition, plants suffer from numerous pathogens and herbivore challenges in both natural and agricultural environments and often face multiple simultaneous threats [8]. Most plant viruses need mediators to transmit viruses, and insects are the most important type of mediators. Most of these vectors are hemipterans, which are a group of phloem-feeding insects, such as aphids, planthoppers, and whiteflies [9]. Aphids are the main transmission vector of TuMV among brassica plants. Special volatiles are released after plants are attacked by insects. Aphids mainly use host volatiles to identify hosts through their olfactory systems [10]. Therefore, there are more studies on plant volatiles. Plant volatile compounds can be divided into plant- and pest-induced volatiles according to the presence or absence of pest induction. The volatile odor components of plants are affected by season, plant age, physiological conditions, microorganisms, and environmental factors, such as soil and light. In addition, changes due to mechanical damage and pest feeding occur frequently.

Plant volatiles are often mixed with a variety of substances, and volatiles with different components and concentrations can be recognized by specific insects; therefore, plant volatiles are chemical signals for host recognition by herbivorous insects. They affect the host selection behavior of herbivorous insects [11,12]. Plant volatile compounds play an important role in the host selection of herbivorous insects. When brassica crops are harmed by herbivorous insects, they release plant volatiles that are different from those in healthy periods to regulate the relationship among brassica plants, herbivorous insects, and natural enemy insects. Allyl isothiocyanates released by cruciferous plants have a strong attractive effect on *Diaeretiella rape*, and the sinigrin released by these plants is a chemical clue for aphids to find hosts [13].

Plant volatile compounds play an important role in the host searching behavior of natural enemies of herbivorous insects. The composition and content of volatiles released by plants change significantly after plants are attacked by herbivorous insects. For example, terpene biosynthesis is suppressed in begomovirus–infected plants, leading to reduced plant resistance and enhanced whitefly (*Bemisia tabaci*) performance, enabling virus transmission [14]. The behavioral responses of different herbivorous insects to herbivore-induced plant volatiles (HIPVs) are different. When some plants are injured and release HIPVs, it will induce the healthy parts of the same plant and adjacent plants to release HIPVs, to reduce the damage caused by insect pests. In this study, TuMV-inoculated resistant/susceptible lines were tested by metabolome and transcriptome analyses to determine the difference in volatiles between normal plants and infected plants and may lay a foundation for anti-TuMV research in *B. rapa*.

2. Materials and Methods

2.1. Plant Materials

B80124 is a TuMV-resistant Chinese cabbage line, and B80461 is a TuMV-susceptible line. B80124-CK and B80461-CK were not inoculated with TuMV, while B80124 and B80461 were inoculated with TuMV. Three plants from each line were mixed into a sample and placed in a 15-mL cryopreservation tube. Six biological replicates were conducted in GC-MS test. Fresh plant materials were harvested, weighed, immediately frozen in liquid nitrogen, and stored at −80 °C until further use. Samples were ground to a powder in liquid nitrogen.

2.2. Detection of TuMV Resistance in Brassica rapa

The TuMV C4 isolate was used in this study, which was maintained on susceptible cultivar '113' fresh leaves at −80 °C. The second and third true leaves were mechanically inoculated with TuMV-C4 at the third true leaf stage [15]. Resistance (the absence of systemic spread) was assessed 20 d post-inoculation with an enzyme-linked immunosorbent assay (ELISA) on the non-inoculated fourth and fifth leaves [15]. ELISA reagents were purchased from Agdia Inc. (Elkhart, IN, USA), and an ELx808 microplate reader (BioTek, Winusky,

VT, USA) was used to measure absorbance at 405 nm. Segregation data were analyzed by the Chi-square test for goodness of fit.

2.3. Material Cultivation

In September 2020, the materials were planted in a 7×7 nutrient bowl in an insect-proof net in an artificial climate room. The temperature during the day was 20 °C–25 °C, and the temperature at night was 15 °C–19 °C. The humidity was about 60%. Plants were thoroughly watered, and aphids were controlled regularly.

2.3.1. TuMV Inoculation

Plants were inoculated with TuMV following our previous study [16], and the specific steps were as follows: when the third true leaf of the test material was fully expanded, a thin layer of emery was evenly sprayed on the front of the second and third leaves of the expanded plant; 1 g of TuMV-C4 infected leaves was ground in a high-temperature sterilized mortar and 4 mL phosphate buffer (0.05 mol/L, pH = 7.0) was added. Evenly ground samples were immediately used for inoculation. The TuMV homogenate was gently applied in the direction of the leaf veins; inoculation was repeated three times. Immediately after inoculation, the leaves were rinsed with clean water and shaded for 24 h. The inoculation was repeated 1 d after the first inoculation. The temperature during the day was controlled at 25 °C–28 °C, and the temperature at night was controlled at 20 °C–22 °C to cause the disease to develop. TuMV resistance identification was conducted after 3 weeks.

2.3.2. TuMV-ELISA Test

The TuMV-ELISA reagent test kit and positive control were purchased from Agdia, USA. This test was performed according to the manufacturer's instructions. Detailed steps for TuMV-ELISA testing are provided in Supplementary Materials.

2.3.3. Isolation and Concentration of Volatiles

Volatiles were detected by MetWare (http://www.metware.cn/, accessed on 21 August 2021) based on the Agilent 8890-5977B platform. Six replicates of each assay were performed. One gram or 1 mL of the sample was transferred immediately to a 20-mL head-space vial (Agilent, Palo Alto, CA, USA), containing NaCl saturated solution, to inhibit any enzyme reaction. The vials were sealed using crimp-top caps with TFE-silicone headspace septa (Agilent). At the time of SPME analysis, each vial was placed at 60 °C for 10 min; then, a 65 µm divinylbenzene/carboxen/polydimethylsiloxane fiber (Supelco, Bellefonte, PA, USA) was exposed to the headspace of the sample for 20 min at 60 °C.

GC-MS conditions. After sampling, desorption of the VOCs from the fiber coating was conducted in the injection port of the GC apparatus (Model 8890; Agilent) at 250 °C for 5 min in splitless mode. The identification and quantification of VOCs was conducted using an Agilent Model 8890 GC and a 5977B mass spectrometer (Agilent), equipped with a 30 m × 0.25 mm × 0.25 µm DB-5MS (5% phenyl-polymethylsiloxane) capillary column. Helium was used as the carrier gas at a linear velocity of 1.0 mL/min. The injector temperature was kept at 250 °C and the detector at 280 °C. The oven temperature was programmed from 40 °C (5 min), increasing at 5 °C/min to 280 °C, and held for 5 min. Mass spectra were recorded in electron impact (EI) ionization mode at 70 eV. The quadrupole mass detector, ion source, and transfer line temperatures were set at 150 °C, 230 °C, and 280 °C, respectively. Mass spectra were scanned in the range m/z 30–350 amu at 1-s intervals. Identification of volatile compounds was achieved by comparing the mass spectra with the data system library (MWGC) and linear retention index.

PCA. Unsupervised PCA was performed by the function 'prcomp' within R (www.r-project.org/, accessed on 21 August 2021). The data were unit variance scaled before unsupervised PCA.

Hierarchical Cluster Analysis and Pearson Correlation Coefficients. Hierarchical cluster analyses (HCA) of samples and metabolites were presented as heatmaps with dendrograms, while Pearson correlation coefficients (PCC) between samples were calculated by the 'cor' function in R and presented as heatmaps. Both HCA and PCC were conducted using the R package 'pheatmap.' For HCA, normalized signal intensities of metabolites (unit variance scaling) were visualized as a color spectrum.

Selection of Differential Metabolites. Significantly differentially regulated metabolites between groups were determined by VIP ≥ 1 and absolute Log2FC (fold change) ≥ 1. VIP values were extracted from OPLS-DA, which also contained score plots and permutation plots, and were generated using the R package 'MetaboAnalystR.' The data were log transformed (log2) and mean centered before OPLS-DA. To avoid overfitting, a permutation test (200 permutations) was performed.

KEGG Annotation and Enrichment Analysis Identified metabolites were annotated using the KEGG compound database (http://www.kegg.jp/kegg/compound/, accessed on 21 August 2021). Annotated metabolites were then mapped to the KEGG pathway database (http://www.kegg.jp/kegg/pathway.html, accessed on 21 August 2021). Pathways with significantly regulated metabolites were then imported into MSEA (metabolite sets enrichment analysis), and their significance was determined by hypergeometric test p-values.

2.3.4. RNA Extraction and Library Construction

This experiment used the TRIzol reagent method to extract total RNA and Agilent 2010 to detect the quality of the obtained RNA. The cDNA library was constructed on the qualified RNA samples, and the fragment size and concentration of the library were detected by an Agilent 2100 Bioanalyzer. Sequencing was performed by combinatorial Probe-Anchor Synthesis (cPAS), and a sequencing read length of 150 bp was obtained.

Data Analyses

The FASTQ Trimmer was used to filter the sequencing data, check the error rate, and check the GC content distribution to obtain clean reads. HISAT2 was used to compare the clean reads after quality control with the cabbage reference genome to obtain the position information on the reference genome or gene and the unique sequence feature information of the sequenced sample.

3. Results

3.1. *Phenotype Analysis of Resistance and Susceptibility to TuMV in B. rapa*

'B80124' and 'B80461' lines were planted in the greenhouse, and the leaves were inoculated with TuMV at the three-leaf stage. Plants with the same growth status were selected and placed in no-insect nets numbered 'B80124-CK' and 'B80461-CK'. After 20 d, the resistance index was calculated by phenotype analysis and enzyme linked immunosorbent assay (ELISA) methods. From the phenotype analysis, the resistant index was divided into six ranks, i.e., 0, 1, 3, 5, 7, and 9 (Figure 1A), and according to the ELISA instructions, the resistant index was divided into positive (susceptible) and negative (resistant). Samples with an O.D. value higher than two times the healthy average were positive, and samples with an OD value below two times the healthy average were negative. (OD value = the OD_{405} value of the sample/the OD_{405} value of the healthy average). The average OD_{405} values of the healthy plants were 0.180. Through the two methods of analysis, the B80124 line was resistant (OD value = 0.235), and the B80461 line was susceptible (OD value = 4.360) (Figure 1B). The OD_{405} values of the samples are shown in Table S1.

Figure 1. Ranks of turnip mosaic virus (TuMV) in brassica crops. From the phenotype analysis, the resistant index was divided into six ranks, i.e., 0, 1, 3, 5, 7, and 9 (**A**), B80124-CK and B80461-CK were the blank group (not inoculated with TuMV), and B80124-TuMV and B80461-TuMV were the experimental materials (inoculated with TuMV) (**B**).

3.2. Volatile Metabolome Analysis of Volatile Organic Compound Accumulation in B. rapa

3.2.1. The Volatile Organic Compound Data Quality

GC-MS analysis was used to obtain the volatile organic compounds (VOCs). Three methods, namely, total ion current (TIC), principal component analysis (PCA), and cluster analysis, were used to control the data quality. TIC analysis of different quality control (QC) samples showed that the curves of the total ion flow detected by metabolites had high overlap; that is, the retention time and peak intensity were consistent, indicating that the signal stability was good when the same sample was detected at different times by mass spectrometry (Figure S1A). In addition, PCA is a multi-dimensional data statistical analysis method for unsupervised pattern recognition. Through orthogonal transformation, a set of potentially correlated variables from B80124 and B80461 were converted into a set of linearly uncorrelated variables, indicating that the volatile metabolome data are reliable (Figure S1B). Furthermore, cluster analysis is a multivariate statistical analysis method for classification. Individuals, objects, and subjects were classified according to the four lines B80124-CK, B80124, B80461-CK, and B80461, so that individuals within the same category had a high homogeneity, and each category had a high heterogeneity, which infers that the volatile metabolome analysis is within reasonable limits (Figure S1C). The cluster analysis of different GC-MS samples is shown in Figure S2.

Before analysis, PCA was conducted on the group samples to observe the degree of variation between groups and between samples within groups. The principal components from B80124-CK completely contained those from B82104, which indicated that after inoculation with TuMV, changes to the principal components were not observed in the highly resistant line B80124 (Figure 2A). The principal components changed dramatically between B80461-CK and B80461, which was the highly susceptible line (Figure 2B). There were consistent and inconsistent principal components between B80124-CK and B80461-CK (Figure 2C); however, the principal components were different between resistant B80124 and susceptible B80461 (Figure 2D), which suggested that the VOCs were different after inoculation with TuMV in resistant B80124 and susceptible B80461.

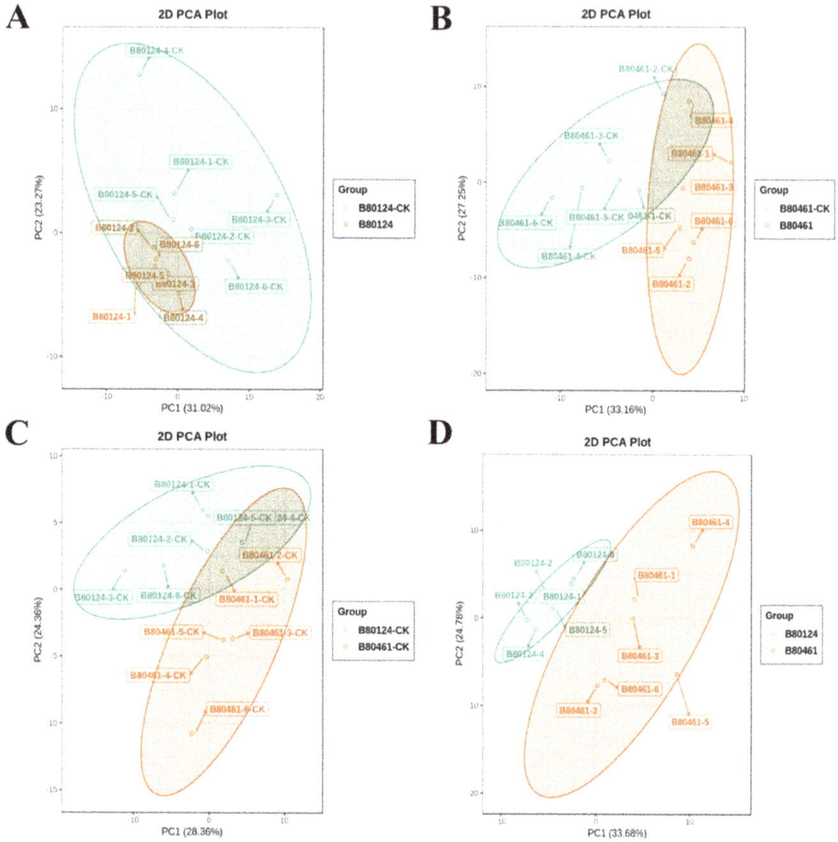

Figure 2. Principal component analysis (PCA) of samples between groups for difference comparison. The four groups were B80124-CK ("-CK" stands for the non-inoculated materials) and B80124 (**A**), B80461-CK and B80461 (**B**), B80124-CK and B80461-CK (**C**), B80124 and B80461 (**D**).

3.2.2. VOC Data Analysis

After qualitative and quantitative analysis of the detected metabolites were combined with the grouping situation of specific samples, 99 types of volatile substances were obtained from the test results (Table S2). Changes in the quantitative information of metabolites in each grouping were compared. In this study, we mainly compared the different metabolites of resistant and susceptible lines (B80124 vs. B80461) after inoculation to determine the potential processes related to TuMV resistance in *B. rapa*.

In B80124 vs. B80461, Seven metabolites, namely KMW0186 (Octanal), XMW0048 (Phenol, 2-nitro), NMW0108 (Bicyclo [2.2.1]heptane-2,5-dione, 1,7,7-trimethyl), XMW0947 ((E)-Hex-3-enyl (E)-2-methylbut-2-enoate), KMW0253 (2-octenal), KMW0158 (Benzaldehyde), and KMW0504 (Benzene, 1-ethenyl-4-methoxy) were upregulated, and the metabolites of KMW0186 (Octanal) were most elevated (Figure S3A). However, 13 metabolites were downregulated, and the decrease in KMW0110 (Allyl Isothiocyanate) was the greatest (Figure S3A), (The significantly regulated metabolites after inoculation B80124 vs. B80461 in Table S3). The results of differentially expressed metabolites ranked first in the VIP value in the OPLS-DA model showed that the decrease in metabolites was higher than the increase, and the decrease of final metabolites was still obvious (Figure S3B). By means of volcano plot, the difference in metabolite expression levels in the two samples (groups) and the statistical significance of the difference were examined. The degree of downregulation of differentially expressed metabolites was more obvious than the degree of upregulation (Figure S3C). According to the Pharmacopoeia of the People's Republic of China [17], the relative standard deviation (RSD) of the repeatability peak area of five consecutive headspace injections is less than 10%. The relative content of differential volatile metabolites in Table S3 is based on a logarithm of 10, and the RSD values of differential volatile metabolites of B80124 and B80461 are shown in Table S4. It was obvious that except KMW0499 of B80461, the RSD of all volatile substances was less than 10%. Similarities between replicates showed that repeatability was good in general. The differential metabolites were annotated using the KEGG database and classified according to pathway types in KEGG. The results showed that the differential metabolites between resistant material B80124 and susceptible material B80461 were mainly related to phenylalanine metabolism, metabolic pathways, and biosynthesis of secondary metabolites (Figure S4).

3.3. Transcriptome Analysis of Volatile Organic Compound Accumulation in *B. rapa*

RNA sequencing was conducted on 12 samples, consisting of three independent biological replicates for each of line (B80124-CK, B80124, B80461-CK, B80461). Comparing the RNA sequencing date to the *Brassica* genome and TuMV reference genome, it was found that only three samples of B80461 were mapped to the TuMV reference genome, and the remaining nine samples were mapping to the *Brassica* reference genome and not to TuMV genome (Table S5). In addition, gene expression was detected by transcriptome analysis. Based on the fragments per kilobase of transcript per million fragments mapped (FPKM), the dispersion degree of gene expression level distributed in each sample was very similar, and overall gene expression levels were similar in each sample (Figure 3A). A density map showed the trend of gene abundance in the sample with the change in gene expression and clearly reflected the range of gene expression concentrations in the sample (Figure 3B). Furthermore, Pearson's correction coefficient was analyzed, which revealed the corrections among all repetitive samples. The R^2 values were all > 0.8, implying that the corrections among the three repetitive samples were reliable, and the data could be used in the subsequent analysis (Figure 3C). In addition to R^2 values, PCA was also conducted to uncover the corrections among the repetitive samples. PCA suggested that the quality of the transcriptomic samples reach to the level for further analysis (Figure 3D).

DESeq2 software was used to count the differentially expressed genes, the total number of differential genes, the number of upregulated genes, and the number of downregulated genes in each group. The statistical results are shown in Table S6. There were 9376 differential genes between B80124 and B80461. From the MA analysis, after inoculating *B. rapa* with TuMV, there were more upregulated genes than downregulated genes (Figure 4A), which might be due to the stress response caused by the virus entering the plant. In addition, the upregulated and downregulated genes were classified through the standardized FPKM analysis. Clustering was performed with the k-means values. As shown in Figure 4B, the sub-class 3/7/8/10 obtained identical results, i.e., the FPKM values between B80124 and B80124-CK were similar, which were obviously different from the FPKM values between B80461 and B80461-CK (Figure 4B). There were changes in gene expression pathways between lines inoculated with TuMV and those that were not and between the resistant line (B80124) and the susceptible line (B80461). Notably, sub-class 1/2/4/6 showed irregular gene changes among the resistant/susceptible lines (Figure 4B). All differentially expressed genes were used to draw the cluster heat map, in which the upregulated and downregulated genes among the resistant/susceptible lines were shown (Figure 4C). The differentially expressed genes showed great changes before and after inoculation with TuMV in the resistant/susceptible lines. In addition, the Venn diagram among different combinations also showed differences in gene changes before and after inoculation of different lines with TuMV, which could indicate that biological stress can promote changes in the metabolic pathways of resistant materials (Figure 4D). There were only 95 common differentially expressed genes among B80124, B80124-CK, B80461, and B80461-CK.

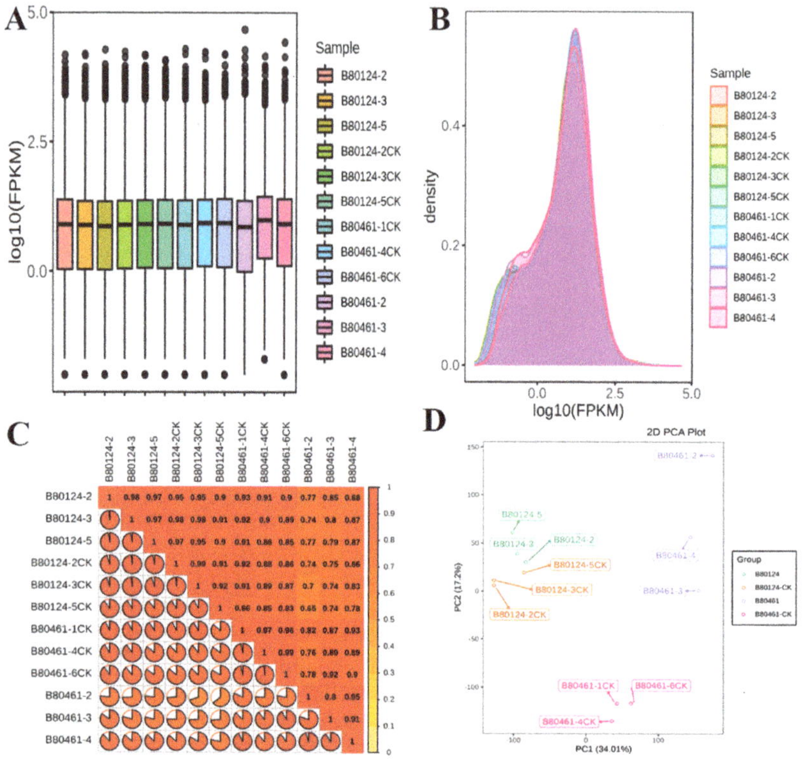

Figure 3. Transcriptome analysis of gene expression. Analyses were performed using the following four methods: fragments per kilobase of the transcript per million fragments mapped (FPKM) (**A**), density map (**B**), Pearson's correction coefficient ($R^2 > 0.8$) (**C**), and principal component analysis (**D**).

The Kyoto Encyclopedia of Genes and Genomes (KEGG) was analyzed to identify the different pathways. The differentially expressed genes were associated with five KEGG pathways (cellular processes, environment information processing, genetic information processing, metabolism, and organismal systems), and metabolic pathways in metabolism was the most frequent category (1207 different genes, accounting for 39.49%). The results were consistent with the metabolomic data (Figure 5), indicating that TuMV stress was related to metabolic pathways, especially focusing on defense response, flavonoid biosynthetic process, and toxin metabolic process (Figure 6). This indicates that TuMV stress maybe influence the release of VOCs; however, the further exploration for the mechanism of the interaction between TuMV and VOCs, are still necessary (Figure 7). In the plant–pathogen interaction after TuMV infected the plants, the hypersensitive response would upgrade, which could cause gene upregulation, such as that of NADPH oxidase, WRKY2, and MEKK (Figure 8). Notably, the *elf18* gene played an important role in the interaction between plant and pathogen (Figure 8), which demonstrated that eukaryotic translation initiation factors (eIFs), such as eIF4E and eIF(iso)4E, could help plant pathogens replicate and spread in *B. rapa*, like *retr01*/*retr02* [eIF(iso)4E], *retr03* (*eIF2Bβ*).

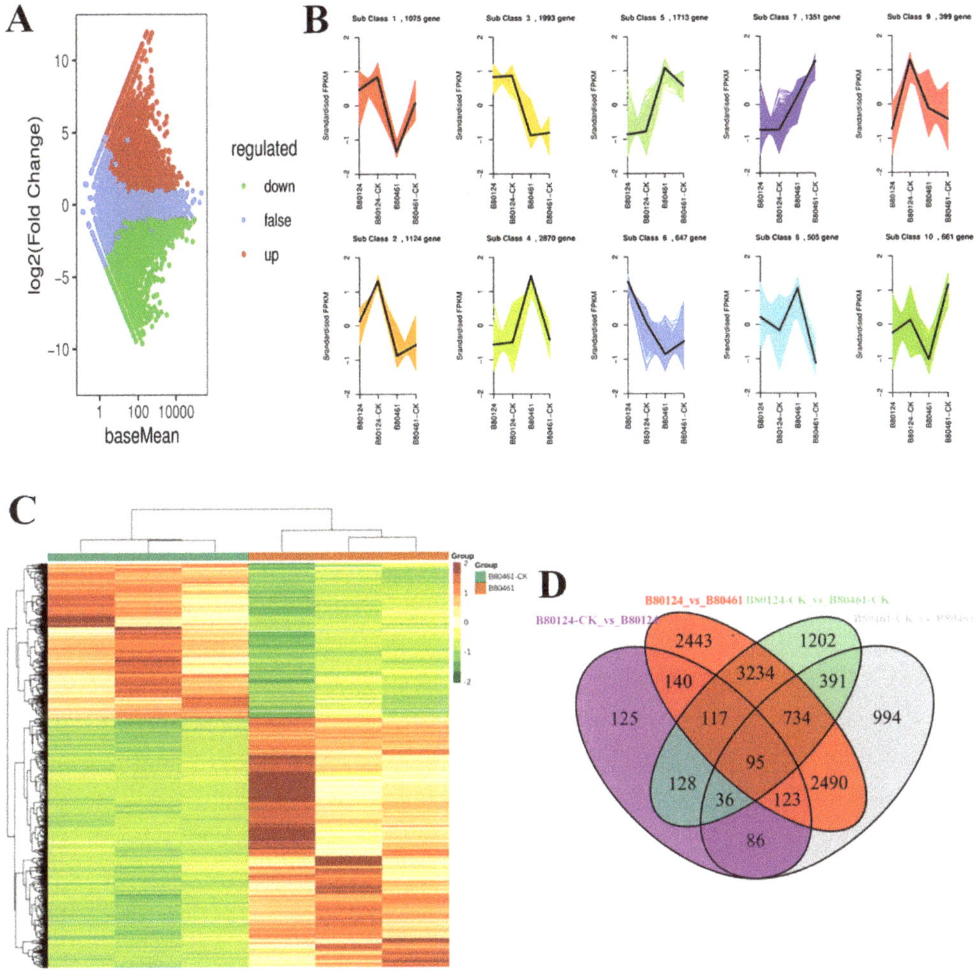

Figure 4. MA analysis showed that after inoculating *Brassica rapa* with turnip mosaic virus (TuMV), there were more upregulated genes than downregulated genes (**A**). Upregulated and downregulated genes classified through standardized FRKM analysis; the clustering was performed by the k-means values (**B**). All upregulated and downregulated genes among the resistant/susceptible lines are shown in the cluster heat map (**C**). The Venn diagram among different combinations showed the differences in gene changes before and after the inoculation of different materials with TuMV (**D**).

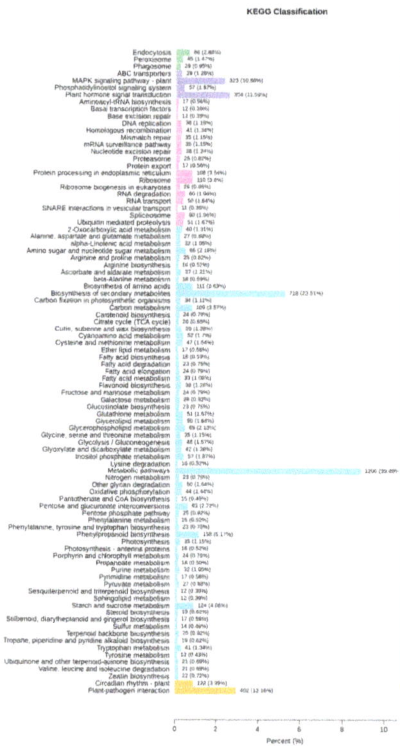

Figure 5. Differentially expressed genes divided into five different pathways based on Kyoto Encyclopedia of Genes and Genomes (KEGG).

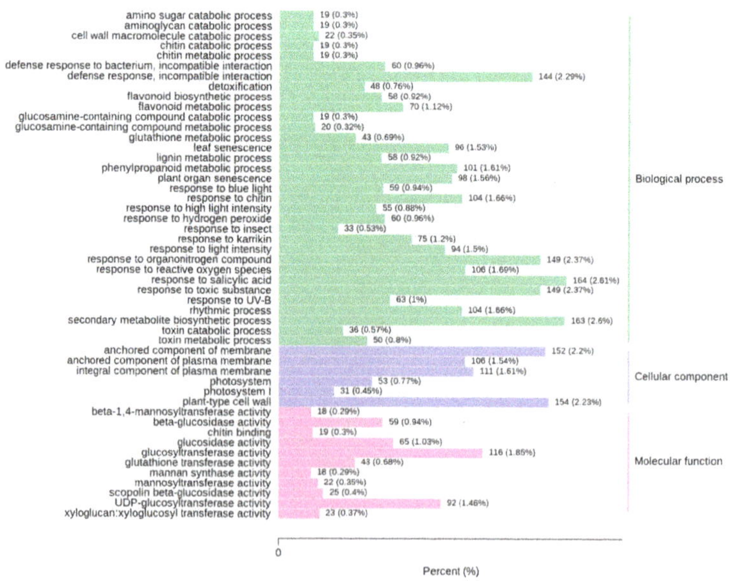

Figure 6. Differentially expressed genes divided into three different pathways based on gene ontology (GO).

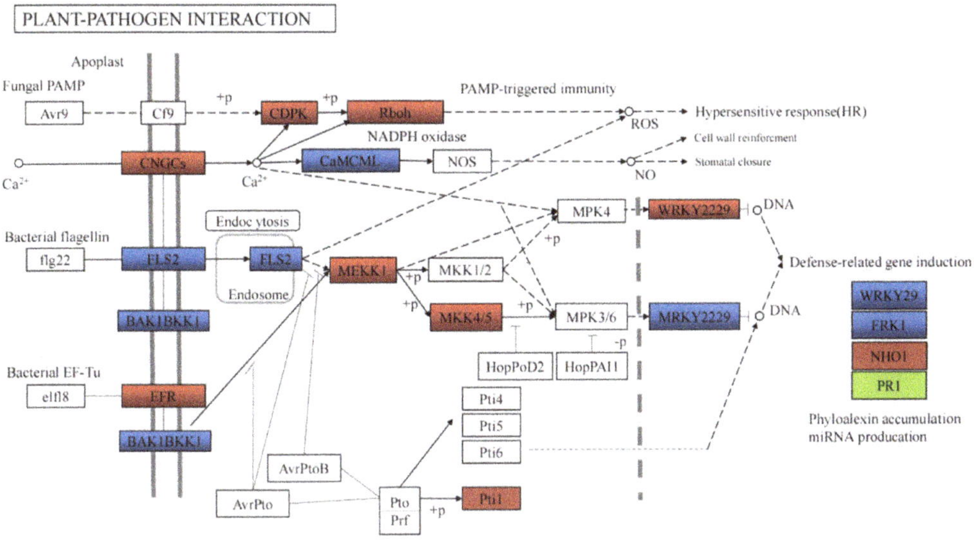

Figure 7. KEGG pathway: Plant–pathogen interaction.

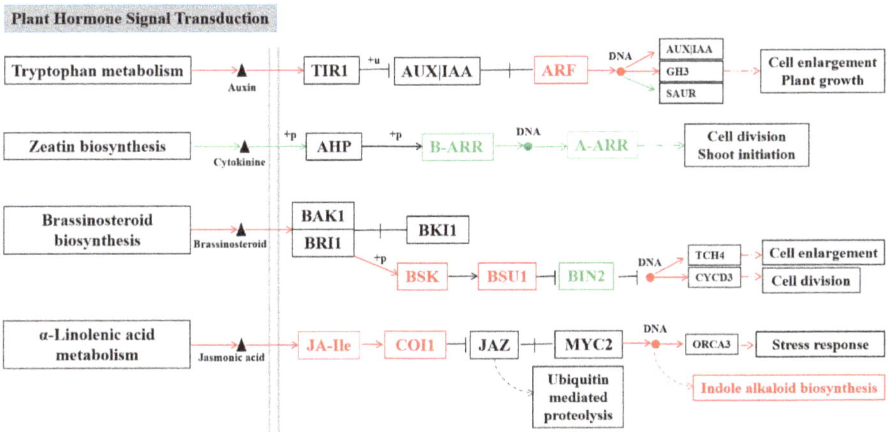

Figure 8. Differential genes and metabolites of B80124 and B80461 involved in four signaling pathways.

3.4. Integrated Volatile Metabolome and Transcriptome Analysis of VOC Accumulation in *B. rapa*

A hypergeometric test was used to determine the ratio of the number of differentially expressed metabolites, genes in the corresponding pathway to the total number of metabolites, and genes detected and annotated in the pathway:

$$P = 1 - \sum_{i=0}^{m-1} \frac{\binom{M}{i}\binom{N-M}{n-i}}{\binom{N}{n}} \quad (1)$$

'N' represents the number of volatile metabolites, genes with KEGG annotation in all volatile metabolites/genes; 'n' represents the number of differential metabolites/genes in 'N'; 'M' represents the number of volatile metabolites/genes in a KEGG pathway in

'*N*'; and '*M*' represents the number of differential volatile metabolites/genes in a KEGG pathway in '*M*'.

By integrating the transcriptome and volatile metabolome data into KEGG analysis, the accumulation of different genes was higher than the accumulation of different metabolites in the biosynthesis of secondary metabolites, cyanoamino acid metabolites, and metabolite pathways; in particular, the accumulation of different genes was 50-fold higher than that of various metabolites in metabolite pathways (Figure 9A). Different genes were highest in metabolites' pathways; however, the accumulation of different genes was highest in the cyanoamino acid metabolites (Figure 9A). Pearson's correction coefficient (PCC) analysis showed that of the genes and metabolites with inconsistent regulatory trends, metabolites were upregulated, while genes were unchanged or downregulated in the 1/2/4 quadrants (Figure 9B). However, the differential expression pattern of genes and metabolites was consistent, and the change in metabolites may be positively regulated by genes if the genes and metabolites have the same regulation trend in the 3/7 quadrants (Figure 9B). The expression abundance of metabolites was lower than that of genes, and genes and metabolites with an inconsistent regulation trend were upregulated, while metabolites were not changed or downregulated in the 6/8/9 quadrants (Figure 9B). Both genes and metabolites were not differentially expressed, and the differentially grouped genes and metabolites were not differentially expressed in the fifth quadrant (Figure 9B). For differential metabolites with a Pearson's correlation coefficient above 0.8, all correlation calculation results were selected to draw a clustering heat map, and the different genes and metabolites varied between the control and those inoculated with TuMV in the resistant/susceptible lines (Figure 9C). Five pathways showed higher corrections with TuMV inoculation, which included six genes (BraA02g001320.3C, BraA09g053440.3C, BraA01g009070.3C, BraA05g013640.3C, BraA05g029570.3C, and BraA03g045650.3C) (Figure 9D). These genes (BraA09g053440.3C, BraA01g009070.3C, and BraA05g013640.3C) could encode the aminotransferase, which could regulate VOCs in the resistant/susceptible lines after inoculating with TuMV. The VOC metabolites could attract aphids to harm plants, which could also spread the virus.

The destruction of viral infection to the normal plant developmental physiology is often related to plant hormone accumulation and changes in signaling changes. Changes in phytohormone levels have been repeatedly thought to be related to changes in virus accumulation [18]. In the process of virus–plant interactions, hormone signal-mediated plant resistance plays an important role in regulating the process of virus occurrence, such as symptom development, virus replication and virus movement [19,20]. Based on KEGG and gene ontology (GO) analysis, the different genes and metabolites between B80124 and B80461 were integrated into four signaling pathways related to plant hormone signal transduction, namely, tryptophan metabolism, zeatin biosynthesis, brassinosteroid biosynthesis, and α-linolenic acid metabolism (Figure 8). After inoculating with TuMV, auxin was upregulated, and ARF, IAA and GH3 were also upregulated, which accelerated cell enlargement and plant growth in tryptophan metabolism, AUX is mainly involved in plant growth and development, but also plays a role in plant defense [21] (Figure 8). The genes in zeatin biosynthesis pathways were downregulated, which reduced cell division and shoot initiation. However, the metabolite pathways were upregulated in brassinosteroid biosynthesis and α-linolenic acid metabolism, which could cause cell enlargement and a stress response (Figure 8). Jasmonic acid (JA) is mainly involved in plant defense against pathogens [22]. Some studies have found that jasmonic acid-mediated resistance is very important for regulating plant resistance to vector insects or plant viruses [23] (Figure 8).

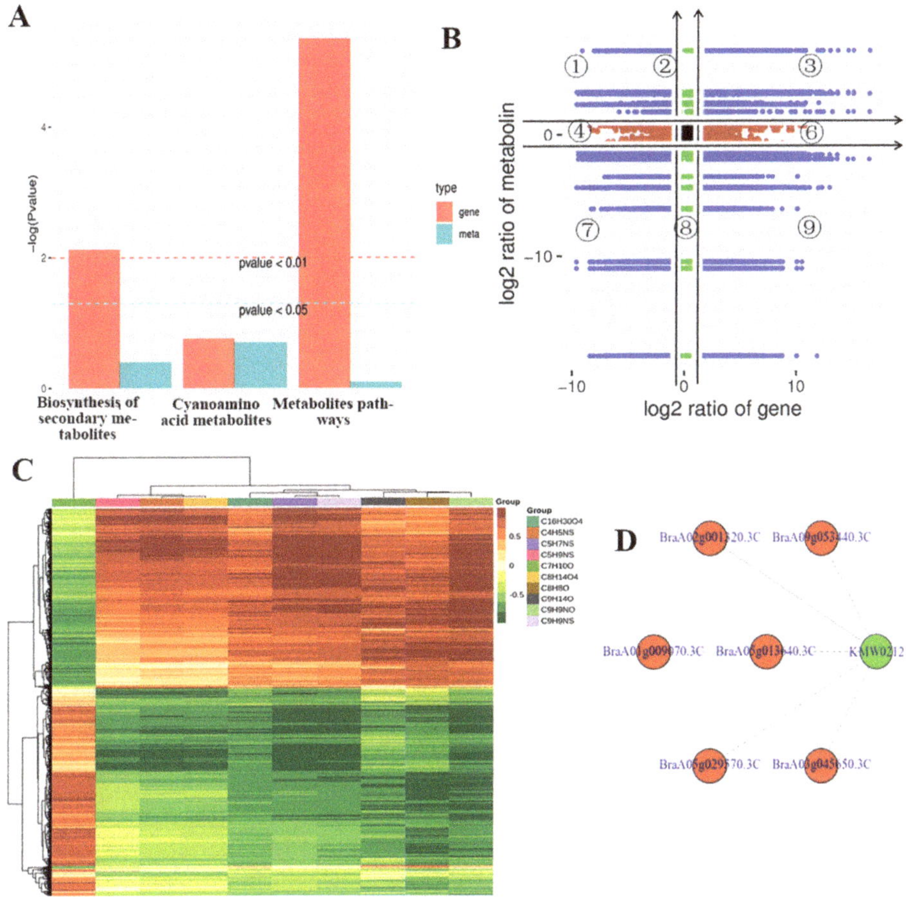

Figure 9. Integrated volatile metabolome and transcriptome analysis of volatile organic compound (VOC) accumulation in *Brassica rapa*. There were more differentially expressed genes than differentially expressed metabolites enriched in biosynthesis of secondary metabolites, cyanoamino acid metabolites, and metabolites pathways (**A**), Through Pearson's correction coefficient (PCC) analysis for the different genes and metabolites, genes and metabolites had an inconsistent regulatory trend. Red dots indicate changes only at the metabolic level; Green dots indicate changes only at the gene level; Blue dots indicate changes at the metabolic level and gene levels. Black dots indicate both genes and metabolites were not differentially expressed (**B**), Heat map with Pearson's correlation coefficient above 0.8 (**C**), Five pathways showed higher corrections after inoculation with TuMV. Green represents metabolites, red represents genes, solid lines represent positive correlations, and dotted lines represent negative correlations (**D**).

4. Discussion

TuMV disease was first described in *B. rapa* in the USA [24,25], while it was later found in *B. oleracea* in the UK [26], and in *B. napus* in China [27]. After TuMV infection of brassica crops, the veins in the plants' new leaves became increasingly obvious and gradually turned into mottled leaves, and the leaves shrank and grew slowly. This leads to dwarfing and abnormal growth, non-heading, or loose heading, and could cause serious production losses in areas with serious disease [28,29].

In this study, TuMV-resistant Chinese cabbage line B80124 and TuMV-susceptible line B80461 (inoculated with TuMV) and B80124-CK, B80461-CK (non-inoculated) were subjected to volatile metabolome and transcriptome analyses. TuMV challenge induces drastic changes in gene expression and metabolite production in *B. rapa*. This manuscript focued on comparisons between B80124 and B80461, to find out potential processes associated with resistance to TuMV in *B. rapa*.

4.1. VOCs Changed Greatly between B80124 and B80461 Inoculated with TuMV

The volatiles produced and released by plants after they are attacked by insects are called herbivore-induced plant volatiles (HIPVs). The composition of HIPVs is very complex, including alkanes, olefins, alcohols, aldehydes, ketones, ethers, esters, and hydroxy acids [30]. Brassica plant volatiles can effectively stimulate herbivorous insects to feed and lay eggs. These volatiles mainly include hydrocarbons, alcohols, aldehydes, ketones, esters, organic acids, and terpenes [31]. Herbivorous insects sense these volatile odors through their antennae for host identification and orientation. In the study of the interaction between TuMV-transmitting aphids and brassica crops, brassica crops produce some chemical volatiles to regulate the behavior of aphids. Sesquiterpenes and monoterpenes, including (E)-α-vanilene, (E)-β-butene, and camphor, affect the feeding behavior of insects. Pathogens may modulate plant volatile production to influence vector behavior. For instance, volatile terpenoids mediate direct defense against the whitefly *Bemisia tabaci* (Hemiptera: Aleyrodidae) [32,33]. Previous studies have shown that there are significant differences in the drive ability of non-toxic vector insects and virulent vector insects for healthy plants and susceptible plants, and non-toxic vector insects tend to feed on susceptible plants, while virulent vector insects tend to consume healthy plants [34,35]. In volatile metabolome analysis, the volatile organic compounds (VOCs) were different after inoculation with TuMV in resistant B80124 and susceptible B80461, and the degree of downregulation of differentially expressed metabolites was more obvious than the degree of upregulation. This was linked to the use of VOCs to attract aphids. In particular, the allyl isothiocyanate, which causes the special smell released by cruciferous plants, was not detected in B80461. Studies have shown that *Plutella xylostella* (Linnaeus) and aphids are attracted by the mustard oil emitted by cruciferous plants, namely allyl isothiocyanate [13,36]. It may also be related to the close of plant stomata, which controls the release of plant VOCs [37]. Many pathogens invade plant tissue through stomata [38], and the closure of plant stomata prevents the invasion of plant pathogens, thereby reducing the release of plant volatiles induced by herbivores, and the potential for VOC/green leaf volatiles (GLV)-mediated inter/intra-plant signaling/communication [39,40]. After the susceptible material B80461 was inoculated with TuMV, why did this substance disappear? It may be because this metabolite is an attractant for aphids. After 25 d inoculating TuMV in susceptible *B. rapa* line, the plant leaves were obviously necrotic, which caused the vegetable quality decline to deteriorate. The special smell from the deteriorate plants was no longer emitted, driving the aphids to eat other healthy plants. However, the specific interaction mechanisms among TuMV-VOCs-aphids needs to be further explored.

4.2. Integrated Volatile Metabolome and Transcriptome analysis of VOC accumulation in B. rapa

Plant volatiles are often mixed with a variety of substances, and volatiles with different components and concentrations can be recognized by specific insects, which promotes the selection of different types of insects among different types of crops [11]. The VOC compositions and concentrations were different between inoculated and non-inoculated varieties in brassica crops, indicating that there were significant differences in gene expression and metabolism. In this study, there are seven metabolites were upregulated, and 13 metabolites were downregulated. The differentially expressed genes were excavated by transcriptome analysis, revealing which genes were upregulated and which genes were downregulated. Compared with volatile metabolome analysis, the transcriptome analysis obtained the similar results, which included the upregulation of genes from the tryptophan pathway, brassinosteroid biosynthesis pathway, and α-Linolenic acid pathway, and the downregulation of genes form the zeatin biosynthesis pathway (Figure 8). Zeatin is a natural cytokinin that promotes cell division. In the TuMV-inoculated line (B80461, susceptible line), the significant decrease of zeatin may be the main cause of leaf shrinkage. In addition, jasmonates and SA are upregulated with the release of this VOCs in the brassinosteroid biosynthesis pathway, which is related to resistance to pathogen infection [22], as both are signal molecules that induce the expression of resistance genes in plants in response to external injury (mechanical injury, herbivore injury, insect injury) and pathogen infection. In the volatile metabolome, determining the final differential metabolites and the content of the metabolites would help explain the function of specific metabolites.

Supplementary Materials: The following supporting information can be downloaded at: https://www.mdpi.com/article/10.3390/horticulturae8010057/s1, Table S1. The OD_{450} values of the samples. Table S2. The 99 types of volatile substances obtained from the test results. Table S3. Significantly Regulated Metabolites After Inoculation B80124 vs B80461. Table S4. The RSD values of different volatile metabolites of B80124 and B80461. Table S5. The results of RNA sequence date mapping to the Brassica genome and TuMV reference genome. Table S6. Statistical table of different genes. Figure S1. Sample Quality control and Principal Component Analysis. Gas chromatography-mass spectrometry (GC-MS) analysis based on unsupervised total ion current (TIC) of different quality control (QC) samples, (Rt, retention time; cps, count per second) (A), principal component analysis (PCA) showed the trend of volatile metabolome separation between groups, PC1 represents the first principal component, PC2 represents the second principal component, and the percentage indicates the interpretation rate of the principal component to the data set. Each point in the figure represents a sample, and samples in the same group are represented by the same color (B), For cluster analysis, the horizontal axis is the sample name, the vertical axis is the metabolite information, and the different colors are the values obtained after the relative content standardization process (red represents high content, green represents low content) (C). Figure S2. The cluster analysis of different GC-MS samples. Figure S3. Screening of differential metabolites of B80124 and B80461. Fold change of the quantitative information of metabolites in each group based on log2FC (fold change ≥ 2 or fold change ≤ 0.5) (A), VIP value (VIP ≥ 1) (B), volcano plot (C). Figure S4. KEGG classification diagram of differential metabolites of B80124 vs. B80461. User Guide: Compound-ELISA Reagent Set.

Author Contributions: Data curation, X.L., L.Z. and W.H.; Formal analysis, X.L., L.Z. and W.H.; Investigation, S.Z. (Shifan Zhang), F.L. and H.Z.; Methodology, R.S.; Supervision, G.L.; Writing—original draft, X.L. and S.Z. (Shujiang Zhang); Writing—review and editing, G.L., J.Z. and S.Z. (Shujiang Zhang). All authors have read and agreed to the published version of the manuscript.

Funding: This work was funded by the State Key Laboratory of North China Crop Improvement and Regulation, the Beijing Natural Science Foundation (6212030), the National Natural Science Foundation of China (32102373, 31772302). This work was performed at the Key Laboratory of Biology and Genetic Improvement of Horticultural Crops, Ministry of Agriculture, Beijing, China.

Institutional Review Board Statement: Not applicable.

Informed Consent Statement: Not applicable.

Data Availability Statement: The date presented in this study are openly avaliable in NCBI with BioProject number: PRJNA764554 (https://www.ncbi.nlm.nih.gov/search/all/?term=PRJNA764554, accesed on 21 August 2021).

Acknowledgments: We thank LetPub for its linguistic assistance during the preparation of this manuscript.

Conflicts of Interest: All the authors declare that they have no conflicts of interests.

References

1. Tomlinson, J.A. Epidemiology and control of virus diseases of vegetables. *Ann. Appl. Biol.* **1987**, *110*, 661–681. [CrossRef]
2. Walkey, D.G.A.; Pink, D.A.C. Reactions of white cabbage (*Brassica oleracea* var. capitata) to four different strains of turnip mosaic virus. *Ann. Appl. Biol.* **1988**, *112*, 273–284. [CrossRef]
3. Walsh, J.A.; Jenner, C.E. Turnip mosaic virus and the quest for durable resistance. *Mol. Plant Pathol.* **2002**, *3*, 289–300. [CrossRef]

4. Walsh, J.A. Turnip mosaic virus. In *Data Sheet for Commonwealth Agriculture Bureau International Global Crop Protection Compendium*; CAB International: Wallingford, UK, 1997.
5. Shattuck, V.I.; Stobbs, L.W. Evaluation of rutabaga cultivars for turnip mosaic virus resistance and the inheritance of resistance. *Hortscience.* **1987**, *22*, 935–937.
6. Hardwick, N.V.; Davies, J.M.L.; Wright, D.M. The incidence of three virus diseases of winter oilseed rape in England and Wales in the 1991/92 and 1992/93 growing seasons. *Plant Pathol.* **1994**, *43*, 1045–1049. [CrossRef]
7. Spence, N.J.; Phiri, N.A.; Hughes, S.L.; Mwaniki, A.; Simons, S.; Oduor, G.; Chacha, D.; Kuria, A.; Ndirangu, S.; Kibata, G.N.; et al. Economic impact of Turnip mosaic virus, Cauliflower mosaic virus and Beet mosaic virus in three Kenyan vegetables. *Plant Pathol.* **2007**, *56*, 317–323. [CrossRef]
8. Casteel, C.L.; Hansen, A.K. Evaluating Insect-Microbiomes at the Plant-Insect Interface. *J. Chem. Ecol.* **2014**, *40*, 836–847. [CrossRef] [PubMed]
9. Antolinez, C.A.; Fereres, A.; Moreno, A. Risk assessment of 'Candidatus Liberibacter solanacearum' transmission by the psyllids Bactericera trigonica and B-tremblayi from Apiaceae crops to potato. *Sci Rep.* **2017**, *7*, 1–10.
10. Webster, B. The role of olfaction in aphid host location. *Physiol. Entomol.* **2012**, *37*, 10–18. [CrossRef]
11. Davis, T.S.; Wu, Y.; Eigenbrode, S.D. The effects of bean leafroll virus on life history traits and host selection behavior of specialized pea aphid (Acyrthosiphon pisum, Hemiptera: Aphididae) genotypes. *Environ. Entomol.* **2017**, *46*, 68–74. [PubMed]
12. Fereres, A.; Penaflor, M.F.G.V.; Favaro, C.F.; Azevedo, K.E.X.; Landi, C.H.; Maluta, N.K.P.; Bento, J.M.S.; Lopes, J.R.S. Tomato infection by whitefly-transmitted circulative and non-circulative viruses induce contrasting changes in plant volatiles and vector behaviour. *Viruses-Basel* **2016**, *8*, 225. [CrossRef]
13. Read, D.P.; Feeny, P.P.; Root, R.B. Habitat selection by the aphid parasite Diaeretiella rapae (Hymenoptera: Braconidae) and hyperparasite Charips brassicae (Hymenoptera: Cynipidae). *Can. Entomol.* **1970**, *102*, 1567–1578. [CrossRef]
14. Li, R.; Weldegergis, B.T.; Li, J.; Jung, C.; Qu, J.; Sun, Y.; Qian, H.; Tee, C.; Van Loon, J.J.A.; Dicke, M.; et al. Virulence factors of geminivirus interact with MYC2 to subvert plant resistance and promote vector performance. *Plant Cell.* **2014**, *26*, 4991–5008. [CrossRef] [PubMed]
15. Walsh, J.A.; Sharpe, A.G.; Jenner, C.E.; Lydiate, D.J. Characterisation of resistance to turnip mosaic virus in oilseed rape (*Brassica napus*) and genetic mapping of *TuRB01*. *Theor. Appl. Genet.* **1999**, *99*, 1149–1154. [CrossRef]
16. Li, G.L.; Lv, H.H.; Zhang, S.J.; Zhang, S.F.; Li, F.; Zhang, H.; Qian, W.; Fang, Z.Y.; Sun, R. TuMV management for brassica crops through host resistance: Retrospect and prospects. *Plant Pathol.* **2019**, *68*, 1035–1044. [CrossRef]
17. Chinese Pharmacopoeia Committee. *Pharmacopoeia of the People's Republic of China*; China Medical Science and Technology Press: Beijin, China, 2015; p. 105. (In Chinese)
18. Guo, H.J.; Gu, L.; Liu, F.; Chen, F.; Ge, F.; Sun, Y. Aphid-borne viral spread is enhanced by virus-induced accumulation of plant reactive oxygen species. *Plant Physiol.* **2019**, *179*, 143–155. [CrossRef] [PubMed]
19. Alazem, M.; Lin, N.-S. Roles of plant hormones in the regulation of host-virus interactions. *Mol. Plant Pathol.* **2015**, *16*, 529–540. [CrossRef]
20. Collum, T.D.; Culver, J.N. The impact of phytohormones on virus infection and disease. *Curr. Opin. Virol.* **2016**, *17*, 25–31. [CrossRef] [PubMed]
21. Robert-Seilaniantz, A.; Grant, M.; Jones, J.D.G. Hormone crosstalk in plant disease and defense: More than just JASMONATE-SALICYLATE antagonism. *Annu. Rev. Phytopathol.* **2011**, *49*, 317–343. [CrossRef]
22. Derksen, H.; Rampitsch, C.; Daayf, F. Signaling cross-talk in plant disease resistance. *Plant Sci.* **2013**, *207*, 79–87. [CrossRef]
23. Zarate, S.I.; Kempema, L.A.; Walling, L.L. Silverleaf whitefly induces salicylic acid defenses and suppresses effectual jasmonic acid defenses. *Plant Physiol.* **2007**, *143*, 866–875. [CrossRef] [PubMed]
24. Broin, M.; Cuine, S.; Eymery, F.; Rey, P. The plastidic 2-cysteine peroxiredoxin is a target for a thioredoxin involved in the protection of the photosynthetic apparatus against oxidative damage. *Plant Cell.* **2002**, *14*, 1417–1432. [CrossRef] [PubMed]
25. Schultz, E.S. A transmissible mosaic disease of Chinese cabbage, mustard and Turnip. *Afr. J. Agric. Res.* **1921**, *22*, 173–177.
26. Smith, K.M. A virus disease of cultivated Crucifers. *Ann. Appl. Biol.* **1935**, *22*, 239–242. [CrossRef]
27. Ling, L.; Yang, J.Y. A mosaic disease of Rape and other crucifera in China. *Phytopathology.* **1940**, *30*, 338–342.
28. Walsh, J.A.; Rusholme, R.L.; Hughes, S.L.; Jenner, C.E.; Bambridge, J.M.; Lydiate, D.J.; Green, S.K. Different classes of resistance to turnip mosaic virus in *Brassica rapa*. *Eur. J. Plant Pathol.* **2002**, *108*, 15–20. [CrossRef]
29. D'Alessandro, M.; Turlings, T.C.J. Advances and challenges in the identification of volatiles that mediate interactions among plants and arthropods. *Analyst* **2006**, *131*, 24–32. [CrossRef] [PubMed]
30. Meiners, T.; Hilker, M. Induction of plant synomones by oviposition of a phytophagous insect. *J. Chem. Ecol.* **2000**, *26*, 221–232. [CrossRef]
31. Bleeker, P.M.; Diergaarde, P.J.; Ament, K.; Guerra, J.; Weidner, M.; Schuetz, S.; de Both, M.T.J.; Haring, M.A.; Schuurink, R.C. The role of specific tomato volatiles in tomato-whitefly interaction. *Plant Physiol.* **2009**, *151*, 925–935. [CrossRef] [PubMed]
32. Luan, J.-B.; Yao, D.-M.; Zhang, T.; Walling, L.L.; Yang, M.; Wang, Y.-J.; Liu, S.-S. Suppression of terpenoid synthesis in plants by a virus promotes its mutualism with vectors. *Ecol. Lett.* **2013**, *16*, 390–398. [CrossRef]
33. Eigenbrode, S.D.; Bosque-Perez, N.A.; Davis, T.S. Insect-borne plant pathogens and their vectors: Ecology, evolution, and complex interactions. *Annu. Rev. Entomol.* **2018**, *63*, 169–191. [CrossRef]

34. Mauck, K.E. Variation in virus effects on host plant phenotypes and insect vector behavior: What can it teach us about virus evolution? *Curr. Opin. Virol.* **2016**, *21*, 114–123. [CrossRef]
35. Furlong, M.J.; Pell, J.K. Interactions between the fungal entomopathogen Zoophthora radicans Brefeld (Entomophthorales) and two hymenopteran parasitoids attacking the diamondback moth, *Plutella xylostella* L. *J. Invertebr. Pathol.* **1996**, *68*, 15–21. [CrossRef] [PubMed]
36. Niinemets, U.; Loreto, F.; Reichstein, M. Physiological and physicochemical controls on foliar volatile organic compound emissions. *Trends Plant Sci.* **2004**, *9*, 180–186. [CrossRef] [PubMed]
37. Melotto, M.; Underwood, W.; He, S.Y. Role of stomata in plant innate immunity and foliar bacterial diseases. *Annu. Rev. Phytopathol.* **2008**, *46*, 101–122. [CrossRef]
38. Ameye, M.; Allmann, S.; Verwaeren, J.; Smagghe, G.; Haesaert, G.; Schuurink, R.C.; Audenaert, K. Green leaf volatile production by plants: A meta-analysis. *New Phytol.* **2018**, *220*, 666–683. [CrossRef]
39. Lin, P.A.; Chen, Y.; Ponce, G.; Acevedo, F.E.; Lynch, J.P.; Anderson, C.T.; Ali, J.G.; Felton, G.W. Stomata-mediated interactions between plants, herbivores, and the environment. *Trends Plant Sci.* **2021**. [CrossRef]
40. Seidl-Adams, I.; Richter, A.; Boomer, K.B.; Yoshinaga, N.; Degenhardt, J.; Tumlinson, J.H. Emission of herbivore elicitor-induced sesquiterpenes is regulated by stomatal aperture in maize (*Zea mays*) seedlings. *Plant Cell Environ.* **2015**, *38*, 23–34. [CrossRef] [PubMed]

horticulturae

MDPI

Article

Comparative Transcriptome Identifies Gene Expression Networks Regulating Developmental Pollen Abortion in Ogura Cytoplasmic Male Sterility in Chinese Cabbage (*Brassica rapa* ssp. *pekinensis*)

Lijiao Hu [1,2,†], Xiaowei Zhang [1,†], Yuxiang Yuan [1,†], Zhiyong Wang [1], Shuangjuan Yang [1], Ruina Li [2], Ujjal Kumar Nath [3], Yanyan Zhao [1], Baoming Tian [2], Gongyao Shi [2], Zhengqing Xie [2], Fang Wei [1,2,*] and Xiaochun Wei [1,2,*]

1 Institute of Horticulture, Henan Academy of Agricultural Sciences, Graduate T&R Base of Zhengzhou University, Zhengzhou 450002, China; hu_lijiao@163.com (L.H.); xiaowei5737@163.com (X.Z.); yuxiangyuan126@126.com (Y.Y.); nkywzy@163.com (Z.W.); sjyang_0614@163.com (S.Y.); zhaoyanyan9621@163.com (Y.Z.)

2 Henan International Joint Laboratory of Crop Gene Resources and Improvement, School of Agricultural Sciences, Zhengzhou University, Zhengzhou 450001, China; ruinali0404@163.com (R.L.); tianbm@zzu.edu.cn (B.T.); shigy@zzu.edu.cn (G.S.); zqxie@zzu.edu.cn (Z.X.)

3 Department of Genetics and Plant Breeding, Bangladesh Agricultural University, Mymensingh 2202, Bangladesh; ujjalnath@gmail.com

* Correspondence: fangwei@zzu.edu.cn (F.W.); jweixiaochun@126.com (X.W.); Tel.: +86-371-6778-5055 (F.W.); +86-371-6571-4026 (X.W.)

† Equally contribution.

Citation: Hu, L.; Zhang, X.; Yuan, Y.; Wang, Z.; Yang, S.; Li, R.; Nath, U.K.; Zhao, Y.; Tian, B.; Shi, G.; et al. Comparative Transcriptome Identifies Gene Expression Networks Regulating Developmental Pollen Abortion in Ogura Cytoplasmic Male Sterility in Chinese Cabbage (*Brassica rapa* ssp. *pekinensis*). *Horticulturae* **2021**, *7*, 157. https://doi.org/10.3390/horticulturae7060157

Academic Editors: Xiaowu Wang and Jian Wu

Received: 14 May 2021
Accepted: 18 June 2021
Published: 20 June 2021

Publisher's Note: MDPI stays neutral with regard to jurisdictional claims in published maps and institutional affiliations.

Copyright: © 2021 by the authors. Licensee MDPI, Basel, Switzerland. This article is an open access article distributed under the terms and conditions of the Creative Commons Attribution (CC BY) license (https://creativecommons.org/licenses/by/4.0/).

Abstract: Ogura cytoplasmic male sterility (Ogura CMS), originally identified in wild radish (*Raphanus sativus*), has enabled complete pollen sterility in *Brassica* plants, but the underlying mechanism in Ogura CMS Chinese cabbage (*Brassica rapa* ssp. *pekinensis*) remains unclear. In this study cytological analysis showed that during microsporogenesis the meiosis occurred normally, and the uninucleated pollens subsequently formed, but the development of both binucleated and trinucleated pollens was obviously disrupted due to defects of pollen mitosis in the Ogura CMS line (Tyms) compared with the corresponding maintainer line (231–330). In transcriptome profiling a total of 8052 differentially expressed genes (DEGs) were identified, among which 3890 were up-regulated and 4162 were down-regulated at the pollen abortion stages in an Ogura CMS line. KOG cluster analysis demonstrated that a large number of DEGs were related to the cytoskeleton's dynamics, which may account for the failure of pollen mitosis during development in the Ogura CMS line. The pivotal genes related to the phenylpropane synthesis pathway (*PAL*, *4CL* and *CAD*) were significantly down-regulated, which probably affected the formation and disposition of anther lignin and sporopollenin, and eventually led to abnormality in the pollen exine structure. In addition, several key up-regulated genes (*GPX7*, *G6PD* and *PGD1*) related to the glutathione oxidation-reduction (REDOX) reaction indicated that the accumulation of peroxides in Ogura CMS lines during this period affected the pollen development. Taken together, this cytological and molecular evidence is expected to advance our understanding of pollen abortion induced by Ogura cytoplasmic action in Chinese cabbage.

Keywords: Ogura cytoplasmic male sterility; pollen abortion; transcriptome; phenylpropane synthesis; Chinese cabbage

1. Introduction

Male sterility refers to a failure to develop normal anthers or pollens, but the plants may still develop normal pistils [1]. Both nuclear and cytoplasmic genes are generally involved in regulating male sterility. There are two main types of male sterility: genic male sterility (GMS), governed by nuclear gene(s), and cytoplasmic male sterility (CMS), regulated by the interaction of nuclear and cytoplasmic gene(s) [2]. Specifically, CMS found in many higher plants is governed by maternal inheritance with aborted pollens and normal pistils [3]. CMS plants can manifest pollen abortion depending on the flower apparatus, short filaments and thin anthers in morphology [4]. CMS has been reported in more than 150 plant species to date [5].

In *Brassica*, CMS is determined by the mitochondrial genome and associated with a pollen-sterility phenotype that can be suppressed or counteracted by nuclear genes known as fertility-restorer genes [6]. Ogura CMS was first observed in Japanese radish (*Raphanus sativus*) and is now widely used in breeding *Brassica* crops [7]. The pollen abortion in Ogura CMS cabbage mostly occurred at microspore developmental stages due to the disintegration of the plasma membrane and an abnormal tapetum, which provides nutrition for microspores [8,9]. In microsporogenesis, tapetum cells secrete callose enzymes after the completion of the meiosis of microspore mother cells, which break down the callose wall and release microspores. By contrast, the premature disintegration or delayed degradation of tapetum cells may result in pollen abortion [10].

Ogura CMS is controlled by a mitochondrial *ORF138*, which consists of two co-transcribed open reading frames *ORF138* and *ATP synthase subunit8* (*ATP8*), but the changes in a series of other genetic backgrounds and molecular mechanisms caused by the *orf138* mutation are still unclear [11]. *ATPase* catalyzes the hydrolysis of ATP to provide the energy necessary for pollen development [12]. With the development of pollens, the activity and quantity of *ATPase* theoretically increase, otherwise the pollens will be abortive. Calcium ions act as a second messenger pathway for signal transduction during the development of microspores and gametophytes [13]. In addition, the cytoskeletal organization, programmed cell death, hormone balance and peroxide content may also influence the degrees of pollen fertility [14–16].

Genes involved in carbon metabolism, lipid metabolism, the tricarboxylic acid (TCA) cycle and oxidative phosphorylation are critical for pollen fertility and microspore development in Ogura CMS cabbage lines [17]. These genes could regulate pollen development through changes their expression either at an early stage of pollen development or at the maturity of pollen. In *B. rapa*, Ogura CMS flower buds had lower contents of soluble protein, soluble sugar, free proline, catalase (CAT), peroxidase (POD) and superoxide dismutase (SOD) than maintainer lines [18]. However, the molecular mechanism associated with Ogura CMS, particularly its core molecular mechanism, remains unclear. Therefore, the elucidation of the molecular mechanism of pollen abortion will provide a theoretical basis for the further understanding of the mechanism of male sterility in plants.

In recent years, transcriptome sequencing has been widely used in soybeans [19], cotton [20], rice [21] and other crops as a powerful tool for studying global transcriptional networks to provide high-resolution data, such as those on leaf senescence, leaf color and biological and abiotic stress responses [22]. However, the use of the transcriptome analysis of pollen abortion and male sterility in Chinese cabbage is very limited. In plants, the anther and pollen development process is very complicated, involving gene expression, regulation, metabolism and the activity of many genes. As long as one part of these processes is disturbed, pollen development may be blocked, and pollen abortion may occur [23]. Therefore, high-throughput sequencing methods can help us to better obtain key information regarding pollen development.

In the present study, transcriptome sequencing was performed using stamens during microsporogenesis between Ogura CMS and its maintainer line of Chinese cabbage, to identify critical DEGs related to pollen development. Our findings will help to elucidate the mechanisms controlling sterility at different stages of pollen development in the Ogura CMS system.

2. Results

2.1. The Ogura CMS Chinese Cabbage Displays Complete Male Sterility

Morphologically, the Ogura CMS Chinese cabbage line (Tyms) had a similar flower pattern to its maintainer line (231–330), but the organs (stamens, petals and sepals) were mostly smaller in the Ogura CMS line (Figure S1 and Table S1). At the mature stage, the anthers were completely non-dehiscent and empty in the Ogura CMS line, but the anthers in the maintainer line were naturally split and full of pollen grains (Figure 1). Compared with the maintainer line, the meiosis was normal in the Ogura CMS line (Figure S2), and tetrads were formed after the completion of meiosis. However, the uninucleated pollens could normally form tetrads, but were unable to develop further into binucleated and trinucleated pollens in the Ogura CMS line (Figure 2), which was probably due to the defective mitosis during pollen development.

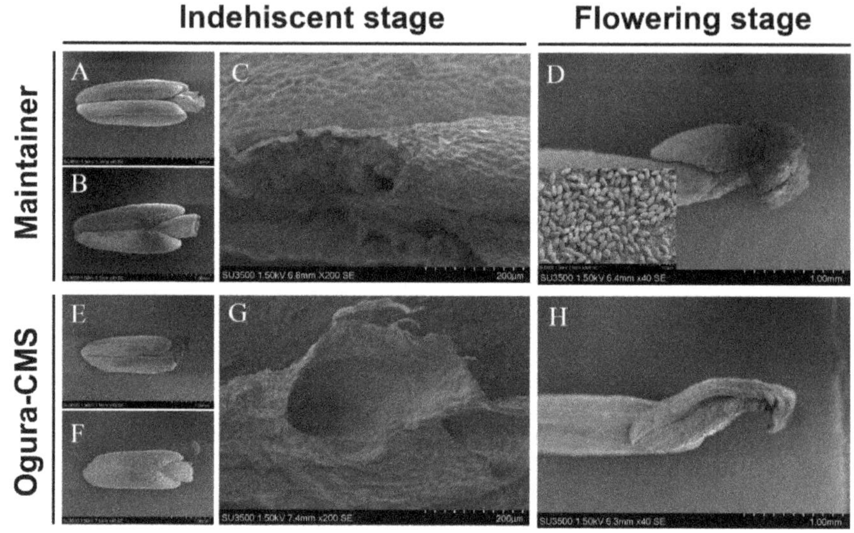

Figure 1. Scanning electron microscopy observation of anthers in the Ogura CMS line and the maintainer line of Chinese cabbage. (**A**,**B**) mature anthers at the indehiscent stage from the maintainer line (231–330); (**C**) local magnification of a mature anther incised at the indehiscent stage in the maintainer line; (**D**) mature anther at the flowering stage with normal oval, plump pollen grains in the maintainer line; (**E**,**F**) Mature anthers at the indehiscent stage from the Ogura CMS line (Tyms); (**G**) local magnification of a mature anther incised at the indehiscent stage in the sterile line, showing empty chambers; (**H**) a mature anther at the flowering stage had no pollen and was collapsed in the sterile line. Bar = 1 mm in (**A**,**B**,**E**,**F**), 200 µm in (**C**,**G**), and 1 mm in (**D**,**H**).

2.2. *Posttranscriptional Regulation, Carbohydrate Metabolism and Cytoskeleton Dynamics Were Probably Associated with Pollen Abortion in Ogura CMS Line*

For comparison with the maintainer line, the young anthers with developing pollens of the Ogura CMS line were collected and subjected to RNA sequencing. Differentially expressed genes (DEGs) were identified with the criteria of fold change (FC) $\geq$2 and false discovery rate (FDR) <0.05. A total of 8052 genes differentially expressed between the maintainer and Ogura CMS lines were identified. Among them, 3890 and 4162 showed up- and down-regulation, respectively (Figure 3). The identified DEGs were classified into 25 categories by using the KOG database (Figure 4A and Table S2). Except for the term "transcription", the category "posttranslational modification, protein turnover, chaperones" was comparatively highlighted, which indicated that posttranscriptional regulation might play important roles in modulating pollen sterility during development in the Ogura CMS line. Additionally, 326 DEGs were also categorized into the term "carbohydrate transport and metabolism" indicating a close relationship between energy metabolism and pollen fertility. In addition, the DEGs were partially classified into two related categories termed "cell cycle control, division, chromosome partitioning" and "cytoskeleton", which likely demonstrates that pollen mitosis and cytoskeletal dynamics were associated with the binucleated and trinucleated pollen development after meiotic tetrad formation. Intact cytoskeleton was essential for pollen growth, and insufficient expression of *ACT* and activity may lead to pollen abortion [24]. For instance, a number of *ACTIN* genes including *ACT1* (*Bra005178* (log_2FC: −0.969); *Bra000010* (log_2FC: −7.701); *Bra017166* (log_2FC: −2.730)), *ACT3* (*Bra014865* (log_2FC: −1.820); *Bra007025* (log_2FC: −1.515)), *ACT4* (*Bra020319* (log_2FC: −10.268)) and *ACT12* (*Bra033819* (log_2FC: −8.022); *Bra019486* (log_2FC: −9.227); *Bra018210* (log_2FC: −9.654)) were obviously down-regulated in the Ogura CMS line (Figure 4B and Table S3). Among them, *ACT4* and *ACT12* genes have been shown to be mainly expressed in Arabidopsis anthers by GUS staining [25]. In addition, the actin depolymerizing factor (ADF), an important player in actin remodeling, increases actin filament treadmilling rates. Furthermore, it has been reported that the actin filaments of *ADF7* and *ADF10* mutants were reduced in the shank and tip, respectively [26]. Another gene, *VLN5*, a typical member of the *VILLIN* family affecting pollen tube growth, were also negatively affected in the Ogura CMS line [27].

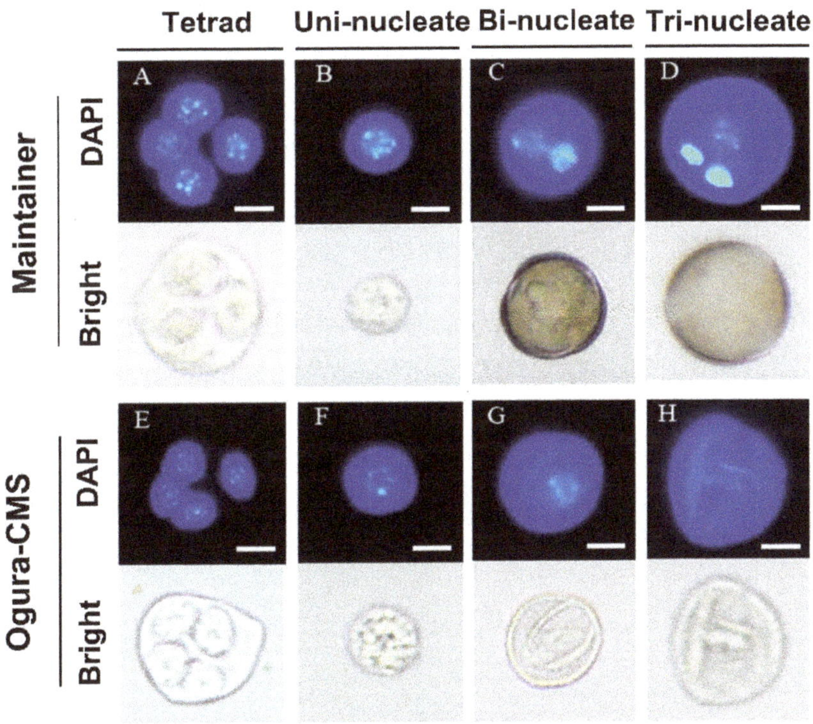

Figure 2. DAPI staining of microspore development process of the Ogura CMS line (Tyms) and the maintainer line (231–330) in Chinese cabbage. (**A**–**D**) DAPI-stained developing spores from the maintainer line (231–330); (**E**–**H**) DAPI-stained developing spores from the Ogura CMS line (Tyms). (**A**,**E**): Tetrad; (**B**,**F**): Uni-nucleate; (**C**,**G**): Bi-nucleate; (**D**,**H**): Tri-nucleate; Bar = 50 μm.

2.3. Up-Regulated Expression of GSH-Oxidation Genes Probably Led to ROS Accumulation and Affected Pollen Fertility in the Ogura CMS Line

In the GO annotations, which include biological process, cellular component and molecular function, the up-regulated DEGs in the top five categories were mostly enriched with "glutathione binding" (GO:0043295) and "glutathione transferase activity" (GO:0004364) (Figure 5A). Glutathione was considered an important metabolite regulator participating in the tricarboxylic acid (TCA) cycle and sugar metabolism and promoting carbohydrate and fat metabolism [28]. In the glutathione oxidation-reduction (REDOX) reaction, the reduced glutathione (GSH) and oxidized glutathione (GSSG) could have transformed each other in the presence of glutathione peroxidase (GPX), glutathione reductase and nicotinamide adenine dinucleotide phosphate (NADPH), accompanied by the decomposition of hydrogen peroxide [29]. Therefore, compared with the maintainer line, the up-regulation of *GPX* genes—*GPX6* (*Bra035211* (log_2FC: 1.083)) and *GPX7 (Bra023978* (log_2FC: 1.568)) and glucose 6-phosphate dehydrogenase genes—*PGD1* (*Bra027685* (log_2FC: 2.108)) and *G6PD2* (*Bra006181* (log_2FC: 2.059); *Bra008855* (log_2FC: 1.110)) during pollen development in the Ogura CMS line suggested a persistent GSH-to-GSSG transformation and accumulation of reactive oxygen species (ROS) such as superoxide, which might eventually affect pollen sterility (Figure 5B).

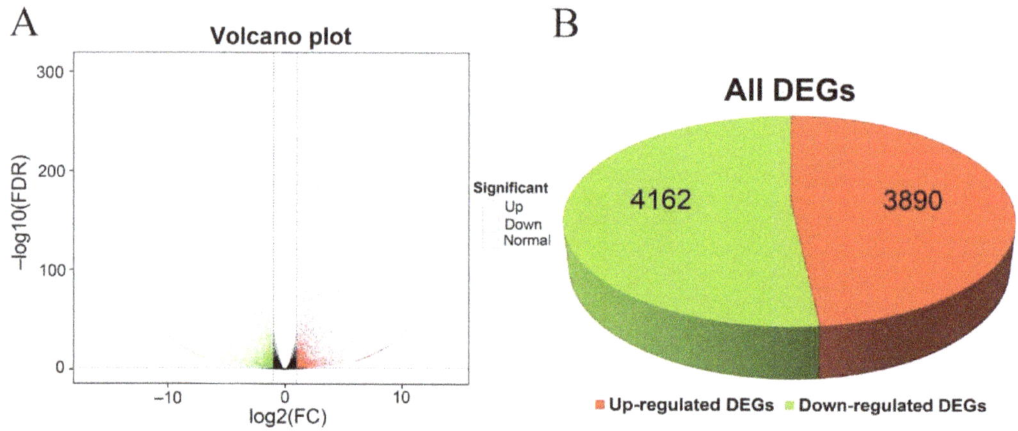

Figure 3. Differentially expressed genes (DEGs) in the Ogura CMS (Tyms) and its maintainer (231–330) lines; (**A**) volcano plot involving DEGs; (**B**) number of DEGs.

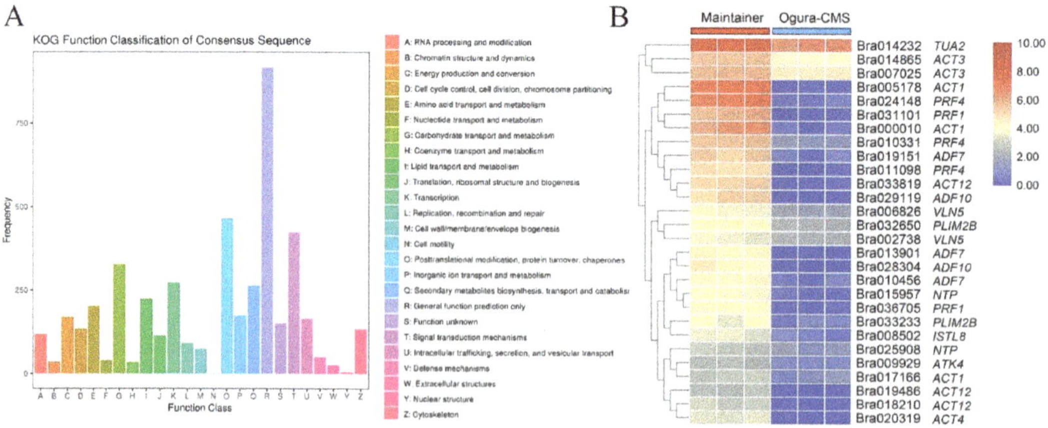

Figure 4. The differentially expressed genes by KOG enrichment in the Ogura CMS (Tyms) and its maintainer (231–330) lines; (**A**) KOG clusters involving DEGs; (**B**) heatmap of DEGs that are related to the cytoskeleton (Z) according to KOG enrichment analysis.

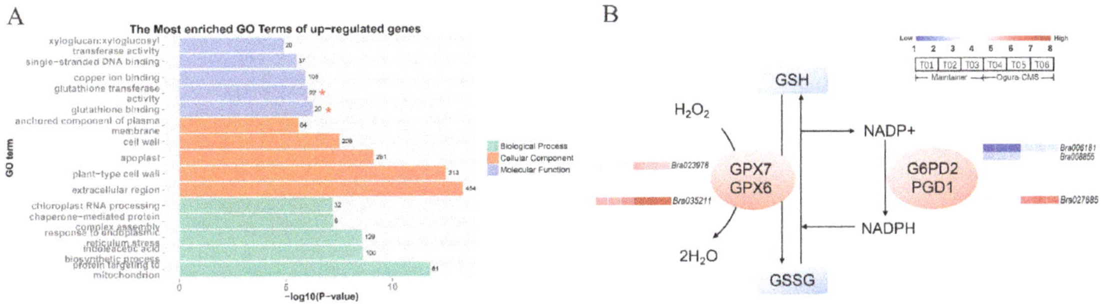

Figure 5. Enrichment analysis of genes up-regulated in the maintainer line vs. Ogura CMS line. (**A**) GO analysis of up-regulated DEGs; (**B**) part of the glutathione metabolic pathway. GSH, glutathione; GSSG, oxidized glutathione; G6PD2, glucose-6-phosphate dehydrogenase 2; PGD1, 6-phosphogluconate dehydrogenase 1; GPX7, glutathione peroxidase 7; GPX6, glutathione peroxidase 6. T01, T02 and T03 represented 3 libraries (the maintainer line, 3 times replication); T04, T05 and T06 represented 3 libraries (the Ogura CMS line, 3 times replication).

2.4. Down-Regulation of DEGs Related to Phenylpropane Synthesis May Affect Sporopollenin Formation during Pollen Development in the Ogura CMS Line

Phenylpropane is an important secondary metabolite, and the pathway of phenylpropane synthesis is related to plant-cell-wall exine during pollen development [30]. Coincidentally, we observed that identified DEGs were classified into the GO term "pollen exine formation" (GO: 0010584) (Figure 6A), and the KEGG pathway analysis showed that these DEGs were preferentially clustered in "phenylpropanoid biosynthesis" (ko00940) and "phenylalanine metabolism" (ko00360) (Figure 6B and Table S4). Phenylpropane synthesis is involved in the formation of sporopollenin [31]. As shown in Figure 6C, two key genes (*PAL1* (*Bra005221* (log_2FC: −1.588)) and *PAL4* (*Bra029831* (log_2FC: −2.376))) encoding phenylalanine ammonia-lyases (PAL) were obviously down-regulated, and in the downstream process of hydroxyphenyl lignin synthesis, the genes encoding 4-coumarate-CoA ligase (*4CL1* (*Bra030429* (log_2FC: −1.348)) and *4CL5* (*Bra001819* (log_2FC: −1.287); *Bra001820* (log_2FC: −1.038)) and cinnamyl alcohol dehydrogenase (*CAD2* (*Bra031216* (log_2FC: −2.334)) and *CAD5* (*Bra011510* (log_2FC: −2.499))) were also down-regulated; however, the gene encoding cinnamyl-CoA reductase gene (*CCR2*) was up-regulated [32]. The changes in the expression levels of these genes probably affected phenylpropane synthesis, thus affecting the accumulation of lignin and sporopollenin, which would modulate the pollen wall composition and eventually affect pollen viability.

2.5. qRT-PCR Validation

The expression of several representative DEGs related to pollen development was validated by qRT-PCR. The results show that the selected genes measured by qRT-PCR had a similar expression pattern to those detected by RNA-seq analysis (Figure 7 and Table S5). The key genes in glutathione metabolism pathway (*GPX7*, *PGD1* and *G6PD2*) were up-regulated in the same trend as the RNA-seq results. Additionally, the key genes in phenylpropane metabolic pathway (*PAL1*, *PAL4*, *4CL1*, *4CL5*, *CAD2* and *CAD5*) were down-regulated by qRT-PCR in accordance with RNA-seq results, suggesting that these key genes played an important role in this process. In addition, the actin family of genes (*ACT1*, *ACT3*, *ACT4* and *ACT12*) showed the same results. The selected gene *Bra006228* (*EXO70C2*), associated with pollen growth, showed higher expression in the maintainer line (231–330) than the Ogura CMS line (Tyms). The gene *(Bra016891 (SPL9))* related to anther development showed up-regulation in the Ogura CMS line (Tyms). *Bra016891* and a gene related to carbohydrate metabolism (*Bra004835 (BGLU15))*, which showed remarkably higher expression in the Ogura CMS than maintainer line (231–330), represent putative candidates for controlling pollen development in Ogura CMS Chinese cabbage.

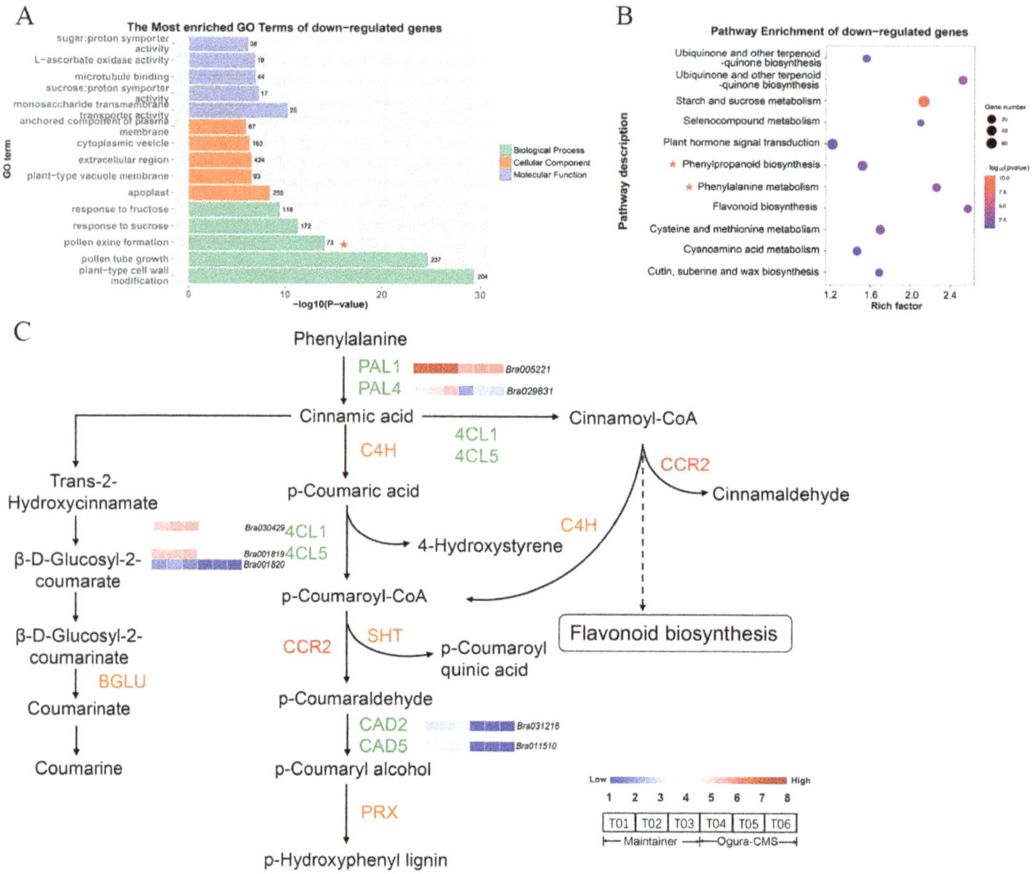

Figure 6. Enrichment analysis of genes down-regulated in the maintainer line vs. the Ogura CMS line. (**A**) GO analysis of down-regulated DEGs; (**B**) KEGG enrichment analysis about pollen exine formation for down-regulated DEGs; (**C**) pathway of phenylpropane biosynthesis (ko00940). PAL1, phenylalanine ammonia-lyase 1; PAL4, phenylalanine ammonia-lyase 4; C4H, cinnamate 4-hydroxylase; 4CL, 4-coumarate-CoA ligase; SHT, spermidine hydroxycinnamoyl transferase; CAD5, cinnamyl alcohol dehydrogenase 5; PRX, peroxidase; CCR2, cinnamyl-CoA reductase; BGLU, β-glucosidase. The red is the up-regulated enzyme gene; the orange represents the enzyme gene with no change in expression; the green represents the down-regulated enzyme gene. T01, T02 and T03 represented 3 libraries (the maintainer line, 3 times replication); T04, T05 and T06 represented 3 libraries (the Ogura CMS line, 3 times replication).

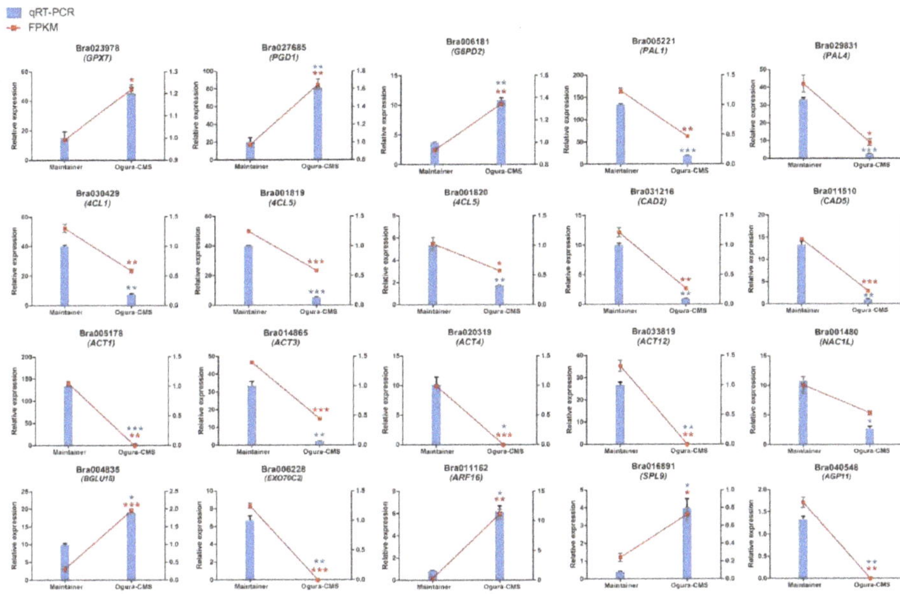

Figure 7. qRT-PCR validation of selected DEGs. The blue columns represent qRT-PCR results and the red lines represent RNA-seq results. *GPX7*: glutathione peroxidase 7; *PGD1*: glucose 6-phosphate dehydrogenase 1; *G6PD2*: glucose 6-phosphate dehydrogenase 2; *PAL1*: phenylalanine ammonia-lyases 1; *PAL4*: phenylalanine ammonia-lyases 4; *4CL1*: 4-coumarate-CoA ligase 1; *4CL5*: 4-coumarate-CoA ligase 5; *CAD2*: cinnamyl alcohol dehydrogenase 2; *CAD5*: cinnamyl alcohol dehydrogenase 5; *ACT1*: actin 1; *ACT3*: actin 3; *ACT4*: actin 4; *ACT12*: actin 12; *NAC1L*: NAC1-Like; *BGLU15*: beta glucosidase 15; *EXO70C2*: exocyst subunit Exo70 family protein C2; *ARF16*: auxin response factor 16; *SPL9*: squamosa promoter binding protein-like 9; *AGP11*: arabinogalactan protein 11. Student's *t*-test was used for statistical analysis of data from the two lines (* $p < 0.05$; ** $p < 0.01$; *** $p < 0.001$).

3. Discussion

3.1. The Accumulation of ROS May Lead to the Premature Degradation of Tapetum, Which Affects the Formation of Pollen Exine

As the outermost layer of anther parietal cells, tapetal cells can provide nutrients such as protein, fat and carbohydrate for the development of microspore by secreting vesicles and self-degradation [33]. Studies have shown that the premature or delayed degradation of tapetum will affect pollen development and lead to pollen abortion [34].

The timely degradation of the tapetum by programmed cell death (PCD) is crucial for microspore development and pollen wall maturation, and reactive oxygen species (ROS) have been shown to be involved in programmed tapetal cell death in plants [35]. Studies have confirmed that mitochondrial protein WA352 can interact with COX11, inhibiting the ROS clearance function of COX11, leading to the increase of ROS, and inducing the occurrence of tapetum PCD in advance, thus leading to male sterility in rice [36]. In addition, tapetum also secretes sporopollenin precursors, which are involved in the construction of the outer wall of pollen [37]. In the way of programmed cell death, the decomposed substances of tapetum are filled into the pollen wall to form a layer of pollen wall composed of lipids, proteins and pigments [29]. The function of tapetum and the formation of pollen envelope are the important processes of pollen development after meiosis.

In our study, *GPX6*, *GPX7*, *G6PD2* and *PGD1* genes were up-regulated in the process of glutathione metabolism, suggesting a large accumulation of ROS in the Ogura CMS lines. The down-regulation of PAL, 4CL and CAD enzymes in the phenylpropane metabolic pathway also suggests the abnormal synthesis of sporopollenin, which may cause the pollen to be less effective in resisting external damage.

In conclusion, according to previous studies, excessive ROS may lead to early degradation of tapetum and affect the uptake of nutrients by pollens. Meanwhile, the abnormal development of tapetum may further hinders the formation of sporopollenin in the pollen exine, which eventually leads to pollen abortion. However, further studies are needed to determine their progressive relationship.

3.2. Cytoskeletal Actin Dynamics Were Probably Involved in Pollen Sterility

Actin is an inclusion in the cytoskeleton that contains a microtubule system and a microfilament system. Limited actin content in the cytoskeleton or disorders of the cytoskeleton system can cause pollen abortion [15]. The protein interaction network depicted in Figure S3 shows the relationship of cytoskeletal genes from the KOG enrichment analysis and the thickness of the straight lines is proportional to the strength of correlation.

We identified *ACT1* (*Bra017166, Bra000010* and *Bra005178*) and *ACT3* (*Bra007025* and *Bra014865*), which are expressed in mature pollen [38]. *ACT4* (*Bra006719* and *Bra020319*) and *ACT12* (*Bra018210, Bra019486* and *Bra033819*) belong to a reproductive actin subclass predominantly expressed in developing reproductive tissues, such as pollen, pollen tubes, ovules and developing seeds, and are expressed at very low levels in vegetative organs [25]. The plant actin bundler *PLIM2s* have been shown to regulate actin bundling in different cells [39]. We found three genes that encode pollen-specific LIM proteins—*PLIM2A* (*Bra000404, Bra004939* and *Bra039313*), *PLIM2B* (*Bra032650* and *Bra033233*) and *PLIM2C* (*Bra014447*)—in the cytoskeleton category of the KOG enrichment analysis. The complete suppression of the *PLIM2s* completely disrupted pollen development, producing abortive pollen grains in transgenic *Arabidopsis thaliana*, whereas the partial suppression of the *PLIM2s* arrested pollen tube growth, resulting in short and swollen pollen tubes [40]. Our transcriptome sequencing results also seem to suggest that one of the causes of pollen abortion in Ogura CMS lines is the down-regulation of *PLIM2s*.

The actin depolymerizing factor (ADF) plays a key role in actin remodeling, which can increase the actin filaments treadmilling rate [26]. Studies have monitored the expression and subcellular localization of ADF7 and ADF10 proteins during male gametophyte development, pollen germination and pollen tube growth. ADF7 is related to the development of the microspore nucleus and vegetative nucleus of mature grains at the stage of low metabolic activity, and ADF10 is associated with the actin filaments in the process of gametophyte development, particularly with the arrays surrounding the apertures of the mature pollen grains [41]. In the study, the down-regulation of *ADF7* (*Bra010456, Bra013901* and *Bra019151*) and *ADF10* (*Bra028304* and *Bra029119*) between the CMS and maintainer lines, which seems to confirm the above results. Another gene, *CAP1*, was found to be involved in cyclase-related proteins, which are important regulators of actin turnover. However, its exact function in regulating actin polymerization, and especially its contribution to actin nucleotide exchange activity, is still not fully understood. In *Arabidopsis*, homozygous *cap1* alleles resulted in short stature and reduced pollen germination efficiency [42]. The *CAP1* gene is located at the core of the protein interaction network; therefore, its expression, even in small quantities, may greatly affect pollen tube growth. DEGs associated with actin, which affect the dynamics of the cytoskeleton by regulating actin motion and play a key role in pollen development, could be explored in connection to the mechanism of male sterility.

4. Materials and Methods

4.1. Morphological Observation

At the blooming stage, flowers of both the Ogura CMS and maintainer lines were collected. The observation of inflorescences was carried out under a Nikon SMZ645 stereoscopic microscope (Nikon Corporation, Tokyo, Japan). For the ultrastructure observation by electron microscopy, the anthers or pollen grains from the newly opened flowers were evenly tiled on the sample table attached to the conductive adhesive and sprayed with an ion-sputtering instrument. Then, a Hitachi SU3500 scanning electron microscope was used to observe and take photos.

4.2. Observation of Meiotic Chromosomal Behaviors

Young inflorescences were collected and fixed overnight in Carnoy's solution (ethanol: glacial acetic acid, 3:1 v/v) at room temperature, and then stored at 4 °C in 70% ethanol for later use. The anthers were removed from the inflorescences with a dissecting needle under a stereomicroscope and soaked in a mixture of cellulase (0.5% w/v) and pectinase (0.5% w/v) to dissolve the microspore cell wall, and then incubated in a citric acid buffer at 37 °C for 4 h. The prepared slides were stained with PI solution (40 g/mg) for 5 min, and then observed under a fluorescence microscope.

4.3. DAPI Staining

The stamens were collected from flower buds of Chinese cabbage at different developmental stages fixed with Carnoy's fixative (ethanol: glacial acetic acid; 3:1). The fixed buds were used for observing different developmental stages of microspores under a fluorescence microscope and photographed after staining with 4,6-diamidino-2-phenylindole (DAPI). A few drops of DAPI dye were added to the glass slides, and the nuclei were stained for 10 min and thereafter flushed with running tap water. Excess water was removed from the slides using filter paper, after which a drop of fluorescent sealing was added and observed under a fluorescence microscope at 360–400 nm wavelengths.

4.4. Transcriptome Sequencing and Identification of DEGs

The stamens of the isogenic Ogura CMS and maintainer lines with different cytoplasmic backgrounds—named as Tyms and 231–330, respectively—of Chinese cabbage were used in this study. The plants were cultivated in experimental plots at the Henan Academy of Agricultural Sciences (Yuanyang, China). Fifty flower buds were trimmed from 10 different plants during microsporogenesis (five buds from each plant), and pooled samples were kept at −80 °C after snap-freezing in liquid nitrogen prior to RNA extraction and sequencing.

A total amount of 1.5 μg RNA per sample was used as input material for rRNA removal using the Ribo-Zero rRNA Removal Kit (Epicentre, Madison, WI, USA). The cDNA library was sequenced via sequencing by synthesis (SBS) technology using Illumina HiSeq 2500 high-throughput Sequencing platform. Raw data (raw reads) of fastq format were firstly processed through in-house perl scripts. In this step, clean data (clean reads) were obtained by removing reads containing adapter, reads containing ploy-N and low quality reads from raw data. Then, the clean data were aligned to the *Brassica rapa* reference genome, v1.5 (http://brassicadb.org/brad/, accessed on 7 August 2019) and the comparative efficiency ranged from 72.73% to 90.22%. StringTie (1.3.1) was used to calculate FPKMs of both mRNAs and coding genes in each sample. Gene FPKMs were computed by summing the FPKMs of transcripts in each gene group.

Differential expression was determined from repeat count data using EdgeR (1.10.1), a Bioconductor software package [43]. Over-dispersion across transcripts was reduced following Poisson's model as well as the empirical Bayesian method for improving the reliability of analysis. A fold change (FC) ≥2 and false discovery rate (FDR) <0.05 were used as criteria for screening the DEGs. All the analyses were performed using software tools on BMK Cloud (https://international.biocloud.net/zh/software/tools/, accessed on 7 August 2019).

4.5. Annotation and Functional Analysis of DEGs

The annotation information for the new gene was obtained through the alignment of DEG sequences against the nonredundant (Nr), BLAST search [44], Swissport [45], Gene Ontology (GO) [46], Clusters of Orthologous Group (COG) [47] and Kyoto Encyclopedia of Genes and Genomes (KEGG) [48] databases. GO enrichment analysis was carried out for the genes with inter-sample differences, whereas Cluster Profiler [49] was used for the genes of biological process, molecular function and cell components. A hypergeometric test in enrichment analysis was performed to find GO terms that were significantly enriched compared to the entire genomic background. The terms obtained from the enrichment results were visualized using the ggplot2 package in R. The euKaryotic Orthologous Group (KOG) diagrams were drawn by using BMKCloud software tools (https://international.biocloud.net/zh/software/tools/, accessed on 7 August 2019).

4.6. *Quantitative Real-Time PCR (qRT-PCR) Validation*

In order to verify the accuracy of the DEG data obtained by RNA-seq, the relative expression levels of the 14 key DEGs and 6 genes selected randomly were analyzed by qRT-PCR. RNA was extracted from 100 mg anthers of Tyms and 231–330 and reverse-transcribed into cDNA as a template for qRT-PCR. The fluorescence quantitative primers for the selected DEGs and the housekeeping gene *GAPDH* (internal control) were designed on NCBI (https://www.ncbi.nlm.nih.gov/, accessed on 8 June 2021) and are listed in Table S6. The specificity and amplification efficiency of the primers were assessed by qRT-PCR. A LightCycler 480 II System (Roche, Basel, Switzerland) and SYBR Premix Ex TaqTM (TaKaRa, Dalian, China) were used for qPCR. The relative expression was calculated according to the $2^{-\Delta\Delta Ct}$ method [50]. A 20 μL reaction-mixture was used in the qPCR; the PCR mixture contained 10 μL of SYBR Premix Ex TaqTM (Tli RNaseH Plus), 0.8 μL of 10 mM concentrations of the forward and reverse primers, 2.0 μL of cDNA (30 ng/μL) and 6.4 μL of dH_2O. Each PCR was repeated three times as technical replicates. The qPCR was performed using the following profile: initial denaturation at 95 °C for 5 min, and then 45 cycles including denaturation for 10 s at 95 °C, annealing for 10 s at 58 °C and extension for 15 s at 72 °C. The results for relative expression according to the qRT-PCR and RNA sequencing were visualized using GraphPad prism 5 software.

5. Conclusions

In this study, cytological observations showed that the anthers that were about to blossom were cavitary and the nucleus gradually degrades in the late uninuclear stage of microspore development until it completely disappears, which causes the sterility rate to reach 100%. Using high throughput RNA-seq technology, a total of 8052 genes were identified as differentially expressed between the maintainer (231–330) and Ogura CMS Chinese cabbage lines. We identified key DEGs clustered in the glutathione oxidation pathway and phenylpropane synthesis pathway, which are probably involved in pollen abortion in the Ogura CMS Chinese cabbage line, due to the accumulation of ROS and abnormal outer wall composition in pollens. In addition, the DEGs related to cytoskeleton and energy metabolism were also enriched, which indicated a possible role in regulating post-meiotic progression during pollen development. The putative candidate DEGs reported in this study pave the way for the exploration of regulatory mechanisms in controlling sterility and fertility in Ogura CMS and its maintainer lines during microsporogenesis.

Supplementary Materials: The following are available online at https://www.mdpi.com/article/10.3390/horticulturae7060157/s1, Figure S1. Morphological features of flower organs and inflorescence in the Ogura CMS line and its maintainer line of Chinese cabbages at anthesis stage. (A–H) a floret at anthesis stage with normal flower organs in the maintainer line; (a–h) a floret at anthesis stage with shorter filaments and anthers without pollen in the Ogura CMS line. Bar = 5 mm. Figure S2. Observation of chromosome behaviors at meiosis of Ogura CMS line and maintainer line in Chinese cabbage. (A–E) chromosome behaviors at meiosis in maintainer line (231–330); (E–H) chromosome behaviors at meiosis in sterile line (Tyms). A, F: Diakinesis; B, G: Metaphase I; C, H: Telophase I; D, I: Metaphase II; E, J: Telophase II; Bar = 10 μm. Figure S3. A protein interaction network was constructed for genes related to cytoskeleton. the thickness of the straight lines is proportionate to the strength of correlation. Table S1. Statistical analysis of morphological properties of maintainers and Ogura sterile lines. The following data are based on the average of 30 samples. Table S2. All the

DEGs were classified into 25 categories by using the KOG database. Table S3. All DEGs' expression level in clusters of "Cytoskeleton". Table S4. The expression levels of all down-regulated genes and their annotations. Table S5. The expression values of qRT-PCR and transcriptome sequencing. Table S6. List of primer sequences for quantitative fluorescence verification involving selected DEGs and internal reference.

Author Contributions: X.W. and F.W. conceived and designed the experiments; Y.Z., S.Y. and Z.W. performed the experiments; G.S., X.Z. and Z.X. analysis the data; L.H., R.L. and B.T. prepared the figures and tables; X.W., Y.Y. and U.K.N. drafted and revised the manuscript critically. All authors have read and agreed to the published version of the manuscript.

Funding: This work was supported by the Zhongyuan Scholar Program (202101510003), National Natural Science Foundation of China (31801874), Programs for Science and Technology Development of Henan Province (212102110124) and Sci-Tech Innovation Team of Henan Academy Agricultural Sciences (2021TD06).

Institutional Review Board Statement: Not applicable.

Informed Consent Statement: Not applicable.

Data Availability Statement: Data is available at NCBI SRA accession PRJNA657160. The data presented in this study are available upon request from the corresponding author.

Acknowledgments: We would like to express our thanks to the anonymous reviewers for their useful comments.

Conflicts of Interest: The authors declared no conflict of interest and absence of any commercial or financial benefits of this research.

References

1. Gang-ping, H.A.O.; Qing, Y. Progress in molecular mechanism of cytoplasmic male sterility and fertility restoration in plant. *Subtrop. Plant Sci.* **2002**, *31*, 78–84.
2. Touzet, P.; Meyer, E.H. Cytoplasmic male sterility and mitochondrial metabolism in plants. *Mitochondrion* **2014**, *19*, 166–171. [CrossRef]
3. Dong, X.; Kim, W.K.; Lim, Y.-P.; Kim, Y.-K.; Hur, Y. Ogura-CMS in Chinese cabbage (*Brassica rapa* ssp. *pekinensis*) causes delayed expression of many nuclear genes. *Plant Sci.* **2013**, *199–200*, 7–17. [CrossRef]
4. Grelon, M.; Budar, F.; Bonhomme, S.; Pelletier, G. Ogura cytoplasmic male-sterility (CMS)-associated *orf138* is translated into a mitochondrial membrane polypeptide in male-sterile Brassica cybrids. *Mol. Gen. Genet.* **1994**, *243*, 540–547. [CrossRef] [PubMed]
5. Brown, G.G.; Formanová, N.; Jin, H.; Wargachuk, R.; Dendy, C.; Patil, P.; Laforest, M.; Zhang, J.; Cheung, W.Y.; Landry, B.S. The radish Rfo restorer gene of Ogura cytoplasmic male sterility encodes a protein with multiple pentatricopeptide repeats. *Plant J.* **2003**, *35*, 262–272. [CrossRef] [PubMed]
6. Chase, C.D. Cytoplasmic male sterility: A window to the world of plant mitochondrial-nuclear interactions. *Trends Genet.* **2007**, *23*, 81–90. [CrossRef]
7. Tanaka, Y.; Tsuda, M.; Yasumoto, K.; Yamagishi, H.; Terachi, T. A complete mitochondrial genome sequence of Ogura-type male-sterile cytoplasm and its comparative analysis with that of normal cytoplasm in radish (*Raphanus sativus* L.). *BMC Genom.* **2012**, *13*, 352. [CrossRef] [PubMed]
8. Du, K.; Liu, Q.; Wu, X.; Jiang, J.; Wu, J.; Fang, Y.; Li, A.; Wang, Y. Morphological structure and transcriptome comparison of the cytoplasmic male sterility line in *Brassica napus* (SaNa-1A) derived from somatic hybridization and its maintainer line saNa-1B. *Front. Plant Sci.* **2016**, *7*, 1313. [CrossRef] [PubMed]
9. Chen, W.; Lv, M.; Wang, Y.; Wang, P.-A.; Cui, Y.; Li, M.; Wang, R.; Gou, X.; Li, J. BES1 is activated by EMS1-TPD1-SERK1/2-mediated signaling to control tapetum development in *Arabidopsis thaliana*. *Nat. Commun.* **2019**, *10*, 4164. [CrossRef] [PubMed]
10. Laurencja, S.; Katarzyna, S.; Wieslawa, P.; Alina, L.; Anna, O.; Katarzyna, K.; Jan, B.; Teresa, C.-T. Development of new restorer lines for CMS Ogura system with the use of resynthesized oilseed rape (*Brassica napus* L.). *Breed. Sci.* **2016**, *66*, 516–521. [CrossRef]
11. Yasumoto, K.; Terachi, T.; Yamagishi, H. A novel *Rf* gene controlling fertility restoration of ogura male sterility by RNA processing of *orf138* found in Japanese wild radish and its STS markers. *Genome* **2009**, *52*, 495–504. [CrossRef]
12. Chen, W.; Jia, P.; Yang, W.; Li, H. Plasma membrane H^+-ATPases-mediated cytosolic proton gradient regulates pollen tube growth. *J. Integr. Plant Biol.* **2020**, *62*, 1817–1822. [CrossRef]
13. Lee, S.; Persson, P.; Mathews, R.D. Role of P-type IIA (ECA) and P-type IIB (ACA) Ca^{2+}-ATPases in plant development and growth Julián. *Rev. Financ. Stud.* **2015**, *29*, 2341–2386. [CrossRef]
14. De Haro, L.A.; Arellano, S.M.; Novák, O.; Feil, R.; Dumón, A.D.; Mattio, M.F.; Tarkowská, D.; Llauger, G.; Strnad, M.; Lunn, J.E.; et al. Mal de Río Cuarto virus infection causes hormone imbalance and sugar accumulation in wheat leaves. *BMC Plant Biol.* **2019**, *19*, 112. [CrossRef]

15. Wang, Y.; Bai, J.; Wang, P.; Duan, W.; Yuan, S.; Zhang, F.; Gao, S.; Liu, L.; Pang, B.; Zhang, L.; et al. Comparative transcriptome analysis identifies genes involved in the regulation of the pollen cytoskeleton in a genic male sterile wheat line. *Plant Growth Regul.* **2018**, *86*, 133–147. [CrossRef]
16. Liu, Z.; Shi, X.; Li, S.; Hu, G.; Zhang, L.; Song, X. Tapetal-delayed programmed cell death (PCD) and oxidative stress-induced male sterility of aegilops uniaristata cytoplasm in wheat. *Int. J. Mol. Sci.* **2018**, *19*, 1708. [CrossRef]
17. Wei, X.; Lv, Y.; Zhao, Y.; Nath, U.K.; Yuan, Y.; Wang, Z.; Yang, S.; Jia, H.; Wei, F.; Zhang, X. Comparative transcriptome analysis in Chinese cabbage (*Brassica rapa* ssp. *pekinesis*) for DEGs of Ogura-, Polima-CMS and their shared maintainer. *Physiol. Mol. Biol. Plants* **2020**, *26*, 719–731. [CrossRef] [PubMed]
18. Duroc, Y.; Hiard, S.; Vrielynck, N.; Ragu, S.; Budar, F. The Ogura sterility-inducing protein forms a large complex without interfering with the oxidative phosphorylation components in rapeseed mitochondria. *Plant Mol. Biol.* **2009**, *70*, 123–137. [CrossRef] [PubMed]
19. Li, J.; Han, S.; Ding, X.; He, T.; Dai, J.; Yang, S.; Gai, J. Comparative transcriptome analysis between the cytoplasmic male sterile line NJCMS1A and its maintainer NJCMS1B in soybean (*Glycine max* L. *Merr.*). *PLoS ONE* **2015**, *10*, e0126771. [CrossRef] [PubMed]
20. Yang, L.; Wu, Y.; Zhang, M.; Zhang, J.; Stewart, J.M.; Xing, C.; Wu, J.; Jin, S. Transcriptome, cytological and biochemical analysis of cytoplasmic male sterility and maintainer line in CMS-D8 cotton. *Plant Mol. Biol.* **2018**, *97*, 537–551. [CrossRef]
21. Chen, L.; Shahid, M.Q.; Wu, J.; Chen, Z.; Wang, L.; Liu, X. Cytological and transcriptome analyses reveal abrupt gene expression for meiosis and saccharide metabolisms that associated with pollen abortion in autotetraploid rice. *Mol. Genet. Genom.* **2018**, *293*, 1407–1420. [CrossRef]
22. Liu, Z.; Li, S.; Li, W.; Liu, Q.; Zhang, L.; Song, X. Comparative transcriptome analysis indicates that a core transcriptional network mediates isonuclear alloplasmic male sterility in wheat (*Triticum aestivum* L.). *BMC Plant Biol.* **2020**, *20*, 10. [CrossRef]
23. Hamid, R.; Marashi, H.; Tomar, R.S.; Shafaroudi, S.M.; Sabara, P.H. Transcriptome analysis identified aberrant gene expression in pollen developmental pathways leading to CGMS in cotton (*Gossypium hirsutum* L.). *PLoS ONE* **2019**, *14*, e0218381. [CrossRef]
24. Li, S.; Dong, H.; Pei, W.; Liu, C.; Zhang, S.; Sun, T.; Xue, X.; Ren, H. LlFH1-mediated interaction between actin fringe and exocytic vesicles is involved in pollen tube tip growth. *New Phytol.* **2017**, *214*, 745–761. [CrossRef]
25. Huang, S.; An, Y.-Q.; McDowell, J.M.; McKinney, E.C.; Meagher, R.B. The *Arabidopsis thaliana ACT4/ACT12* actin gene subclass is strongly expressed throughout pollen development. *Plant J.* **1996**, *10*, 189–202. [CrossRef]
26. Daher, F.B.; Geitmann, A. Actin depolymerizing factors ADF7 and ADF10 play distinct roles during pollen development and pollen tube growth. *Plant Signal. Behav.* **2012**, *7*, 879–881. [CrossRef] [PubMed]
27. Zhang, H.; Qu, X.; Bao, C.; Khurana, P.; Wang, Q.; Xie, Y.; Zheng, Y.; Chen, N.; Blanchoin, L.; Staiger, C.J.; et al. Arabidopsis VILLIN5, an actin filament bundling and severing protein, is necessary for normal pollen tube growth. *Plant Cell* **2010**, *22*, 2749–2767. [CrossRef] [PubMed]
28. Selote, D.S.; Khanna-Chopra, R. Drought-induced spikelet sterility is associated with an inefficient antioxidant defence in rice panicles. *Physiol. Plant.* **2004**, *121*, 462–471. [CrossRef]
29. Zhang, A.; Zhang, J.; Ye, N.; Cao, J.; Tan, M.; Jiang, M. ZmMPK5 is required for the NADPH oxidase-mediated self-propagation of apoplastic H_2O_2 in brassinosteroid-induced antioxidant defence in leaves of maize. *J. Exp. Bot.* **2010**, *61*, 4399–4411. [CrossRef] [PubMed]
30. Li, Y.; Li, D.; Guo, Z.; Shi, Q.; Xiong, S.; Zhang, C.; Zhu, J.; Yang, Z. OsACOS12, an orthologue of Arabidopsis acyl-CoA synthetase5, plays an important role in pollen exine formation and anther development in rice. *BMC Plant Biol.* **2016**, *16*, 256. [CrossRef] [PubMed]
31. Xue, J.-S.; Zhang, B.; Zhan, H.; Lv, Y.-L.; Jia, X.-L.; Wang, T.; Yang, N.-Y.; Lou, Y.-X.; Zhang, Z.-B.; Hu, W.-J.; et al. Phenylpropanoid derivatives are essential components of sporopollenin in vascular plants. *Mol. Plant* **2020**, *13*, 1644–1653. [CrossRef]
32. Wang, N.-D.; Bai, H.; Chen, W.-Q.; Lu, H.; Jiang, X.-N. Identifying a cinnamoyl coenzyme a reductase (CCR) activity with 4-coumaric acid: Coenzyme a ligase (4CL) reaction products in *Populus tomentosa*. *J. Plant Biol.* **2009**, *52*, 482–491. [CrossRef]
33. Han, Y.; Zhou, S.-D.; Fan, J.-J.; Zhou, L.; Shi, Q.-S.; Zhang, Y.-F.; Liu, X.-L.; Chen, X.; Zhu, J.; Yang, Z.-N. OsMS188 Is a key regulator of tapetum development and sporopollenin synthesis in rice. *Rice* **2021**, *14*, 4. [CrossRef] [PubMed]
34. Hu, L.; Liang, W.; Yin, C.; Cui, X.; Zong, J.; Wang, X.; Hu, J.; Zhang, D. Rice MADS3 regulates ROS homeostasis during late anther development. *Plant Cell* **2011**, *23*, 515–533. [CrossRef]
35. Solis, M.-T.; Chakrabarti, N.; Corredor, E.; Cortés-Eslava, J.; Rodriguez-Serrano, M.; Biggiogera, M.; Risueno, M.C.; Testillano, P.S. Epigenetic changes accompany developmental programmed cell death in tapetum cells. *Plant Cell Physiol.* **2014**, *55*, 16–29. [CrossRef]
36. Luo, D.; Xu, H.; Liu, Z.; Guo, J.; Li, H.; Chen, L.; Fang, C.; Zhang, Q.; Bai, M.; Yao, N.; et al. A detrimental mitochondrial-nuclear interaction causes cytoplasmic male sterility in rice. *Nat. Genet.* **2013**, *45*, 573–577. [CrossRef]
37. Parish, R.W.; Li, S.F. Death of a tapetum: A programme of developmental altruism. *Plant Sci.* **2010**, *178*, 73–89. [CrossRef]
38. An, Y.Q.; Huang, S.; McDowell, J.M.; McKinney, E.C.; Meagher, R.B. Conserved expression of the Arabidopsis ACT1 and ACT3 actin subclass in organ primordia and mature pollen. *Plant Cell* **1996**, *8*, 15–30. [CrossRef] [PubMed]
39. Sudo, K.; Park, J.-I.; Sakazono, S.; Masuko-Suzuki, H.; Osaka, M.; Kawagishi, M.; Fujita, K.; Maruoka, M.; Nanjo, H.; Suzuki, G.; et al. Demonstration in vivo of the role of Arabidopsis PLIM2 action-binding proteins during pollination. *Genes Genet. Syst.* **2013**, *88*, 279–287. [CrossRef]

40. Ye, J.; Xu, M. Actin bundler PLIM2s are involved in the regulation of pollen development and tube growth in Arabidopsis. *J. Plant Physiol.* **2012**, *169*, 516–522. [CrossRef]
41. Daher, F.B.; Van Oostende, C.; Geitmann, A. Spatial and temporal expression of actin depolymerizing factors ADF7 and ADF10 during male gametophyte development in *Arabidopsis thaliana*. *Plant Cell Physiol.* **2011**, *52*, 1177–1192. [CrossRef] [PubMed]
42. Deeks, M.J.; Rodrigues, C.; Dimmock, S.; Ketelaar, T.; Maciver, S.K.; Malhó, R.; Hussey, P.J. Arabidopsis CAP1-A key regulator of actin organisation and development. *J. Cell Sci.* **2007**, *120*, 2609–2618. [CrossRef]
43. Love, M.I.; Huber, W.; Anders, S. Moderated estimation of fold change and dispersion for RNA-seq data with DESeq2. *Genome Biol.* **2014**, *15*, 550. [CrossRef] [PubMed]
44. Altschul, S.F.; Madden, T.L.; Schäffer, A.A.; Zhang, J.; Zhang, Z.; Miller, W.; Lipman, D.J. Gapped BLAST and PSI-BLAST: A new generation of protein database search programs. *World J. Microbiol. Biotechnol.* **2007**, *25*, 3389–3402. [CrossRef] [PubMed]
45. Boeckmann, B.; Bairoch, A.; Apweiler, R.; Blatter, M.-C.; Estreicher, A.; Gasteiger, E.; Martin, M.J.; Michoud, K.; O'Donovan, C.; Phan, I.; et al. The SWISS-PROT protein knowledgebase and its supplement TrEMBL in 2003. *Nucleic Acids Res.* **2003**, *31*, 365–370. [CrossRef]
46. Harris, M.A.; Clark, J.; Ireland, A.; Lomax, J.; Ashburner, M.; Foulger, R.; Eilbeck, K.; Lewis, S.; Marshall, B.; Mungall, C.; et al. The Gene Oncology (GO) database and informatics resource. *Nucleic Acids Res.* **2004**, *32*, 258–261. [CrossRef]
47. Tatusov, R.L.; Fedorova, N.D.; Jackson, J.D.; Jacobs, A.R.; Kiryutin, B.; Koonin, E.V.; Natale, D.A.; Vasudevan, S.; Wolf, Y.I.; Yin, J.J.; et al. The COG database: An updated version includes eukaryotes. *BMC Bioinform.* **2003**, *4*, 41. [CrossRef]
48. Kanehisa, M.; Goto, S. KEGG: Kyoto Encyclopedia of Genes and Genomes. *Oncol. Lett.* **2000**, *19*, 3316–3332. [CrossRef]
49. Yu, G. clusterProfiler: Universal enrichment tool for functional and comparative study. *bioRxiv* **2018**. [CrossRef]
50. Livak, K.J.; Schmittgen, T.D. Analysis of relative gene expression data using real-time quantitative PCR and the $2^{-\Delta\Delta CT}$ method. *Methods* **2001**, *25*, 402–408. [CrossRef]

horticulturae

Article

Genome-Wide Analysis of Purple Acid Phosphatase Genes in *Brassica rapa* and Their Association with Pollen Development and Phosphorus Deprivation Stress

Yongfang Cai, Jiao Qi, Chun Li, Kehui Miao, Baixue Jiang, Xiaoshuang Yang, Wenyu Han, Yang Wang, Jing Gao * and Xiangshu Dong *

School of Agriculture, Yunnan University, Kunming 650091, China; yongfangcai@mail.ynu.edu.cn (Y.C.); qijiao@mail.yun.edu.cn (J.Q.); lichun1177@mail.ynu.edu.cn (C.L.); 631196202@mail.ynu.edu.cn (K.M.); jiangbaixue@mail.ynu.edu.cn (B.J.); yangxiaoshuang@mail.ynu.edu.cn (X.Y.); hanwenyu@mail.ynu.edu.cn (W.H.); wangyang@ynu.edu.cn (Y.W.)

* Correspondence: jinggao@ynu.edu.cn (J.G.); dongxiangshu@ynu.edu.cn (X.D.); Tel.: +86-871-6503-1539 (X.D.)

Abstract: PAPs (purple acid phosphatases) belong to the metallo-phosphoesterase superfamily and play important roles in developmental processes, phosphorus foraging, and recycling. However, the specific functions of BrPAPs in *Brassica rapa* are poorly understood. In this study, 39 *BrPAPs* were identified and divided into three major clades and nine subgroups. In 8 of the 39 BrPAPs, some invariant amino acid residues were lost or shifted. Based on an expression profiling analysis, *BrPAP11, 14, 20, 24, 29*, and *34* were specifically expressed in fertile floral buds, indicating their critical roles during pollen development. A total of 21 *BrPAPs* responded to Pi deprivation in either shoots or roots. Of these, *BrPAP4, 5, 19*, and *21* were upregulated in roots under Pi depravation conditions, while *BrPAP12* was upregulated in the roots in normal conditions. *BrPAP28* was upregulated in shoots under Pi depravation conditions, indicating its function shifted compared with its *Arabidopsis* homolog, *AtPAP26*. The present work contributes to further investigation of *BrPAPs* as candidate genes for genetic improvement studies of low phosphorus tolerance as well as for creating male sterile lines based on gene editing methods in *Brassica rapa*.

Keywords: *Brassica rapa*; purple acid phosphatases; *BrPAP*; Pi-deprivation; pollen development

Citation: Cai, Y.; Qi, J.; Li, C.; Miao, K.; Jiang, B.; Yang, X.; Han, W.; Wang, Y.; Gao, J.; Dong, X. Genome-Wide Analysis of Purple Acid Phosphatase Genes in *Brassica rapa* and Their Association with Pollen Development and Phosphorus Deprivation Stress. *Horticulturae* **2021**, *7*, 363. https://doi.org/10.3390/horticulturae7100363

Academic Editors: Xiaowu Wang, Jian Wu and Xu Cai

Received: 31 August 2021
Accepted: 30 September 2021
Published: 5 October 2021

Publisher's Note: MDPI stays neutral with regard to jurisdictional claims in published maps and institutional affiliations.

Copyright: © 2021 by the authors. Licensee MDPI, Basel, Switzerland. This article is an open access article distributed under the terms and conditions of the Creative Commons Attribution (CC BY) license (https://creativecommons.org/licenses/by/4.0/).

1. Introduction

Purple acid phosphatases (PAPs) belong to the metallo-phosphoesterase superfamily and are a type of acid phosphatase (APase) comprised of a binuclear metal center binding Fe (III)-M (II) complex (where M = Fe, Zn, or Mn) at the active site [1,2]. Due to the existence of a tyrosine residue ligated to a ferric iron, this group of acid phosphatases has a purple color [3]. They have been isolated from most eukaryotic organisms and some bacteria [3,4]. APases can hydrolyze a wide range of anhydrides and phosphate esters [5]. Therefore, improvement of APase activity is an important avenue for improving the efficiency of phosphorus utilization of plants in a low Pi (inorganic phosphate) environment. This enables plants to utilize extracellular and intracellular organic phosphorus (Po) [5–7].

With the aid of five conserved motif blocks (**D**XG/G**D**XX**Y**/G**N**H(D/E)/VXX**H**/G**H**X**H**, where bold and underlined letters indicate metal ligating residues that are required for metal coordination) and the availability of plant genome sequences, many *PAPs* have been identified and annotated from plant species at the genomic level [2,8,9]. For example, there are 29 *PAPs* in *Arabidopsis thaliana*, 38 in *Glycine max*, 33 in *Zea mays* ssp. *mays* var. B73, 19 in *Camellia sinensis*, 25 in *Jatropha curcas*, 26 in *Oryza sativa*, and 25 in *Cicer arietinum* [2,8–15]. Based on protein mass and structure, the PAPs isolated from plants can be divided into two forms: low molecular-mass monomeric PAPs (LMMs, with a mass of approximately 35 kDa) and high molecular-mass oligomeric PAPs (HMMs, with a mass around 55–60 kDa) [2,8,15].

LMM PAPs usually contain only one catalytic domain, and HMM PAPs are homodimeric or heterodimeric proteins, where each subunit has a C-terminal domain with an active site and an N-terminal domain of unknown function [16].

Due to the APase activity of PAPs, phosphorus acquisition and metabolism are the focus of most studies relating to PAP function. In *Arabidopsis*, *AtPAP10* was induced by Pi starvation and was shown to play a role in plant tolerance to Pi limitation [17]. *AtPAP15* was related to the mobilization of phosphorus reserves during seed and pollen germination [18]. *AtPAP12* and *AtPAP26* appear to be the predominant PAP isozymes responsive to Pi starvation and they can utilize Po [19,20]. In rice, APase activity and Po degradation are significantly increased by the overexpression of *OsPAP10a*, *OsPAP10c*, and *OsPAP21b*, and the expression levels of these can be induced by Pi starvation [21–24]. In soybean, *GmPAP14* was induced in response to Po and utilized external Po to promote plant growth and development [7].

In addition to phosphorus acquisition and metabolism, the biological functions of *PAP* genes include other biological processes. Heterologous expression of *AtPAP2* in potato can lead to a higher photosynthesis rate, faster growth, higher tuber number, and tuber starch content [25]. In tobacco (*Nicotiana tabacum*), beta-glucan synthesis and cellulose deposition were increased in *NtPAP12* overexpressing cells during cell wall biosynthesis [26]. In soybean, *GmPAP3* is related to the adaptation of soybean to NaCl stress through the involvement of ROS (reactive oxygen species) metabolism processes [27]. In *Arabidopsis*, seven *AtPAPs* (*AtPAP6, 11, 14, 19, 23, 24,* and *25*) are predominantly expressed in flowers [4], and *AtPAP11* and *AtPAP5* are involved in the biological process of pollen tube growth [28]. This implies their potential functions during flower, pollen, and pollen tube development processes. Taken together, evidence suggests that *PAPs* have diverse functions during plant growth and development under normal or environmental stress conditions.

Brassica rapa is an important vegetable crop in China, Korea, and other Asian countries. PAP genes in *B. rapa* have been isolated due to the importance of *PAPs* for plant growth and development [9]. The detailed characteristics and features of *B. rapa* purple acid phosphatase genes (*BrPAPs*) are limited. In this study, *BrPAPs* were systematically identified and characterized. In silico and semi-RT-PCR analyses indicated that *BrPAP11, 14, 20, 24, 29,* and *34* might be related to pollen development, and their potential functions were highlighted by co-expression analyses. The responses of *BrPAPs* to Pi deprivation conditions were also investigated. The results from this study provide a basis for further understanding of the functions of *BrPAPs* during pollen development and *B. rapa* responses to Pi deprivation.

2. Materials and Methods

2.1. Plant Growth and Pi Deprivation Treatment

Brassica rapa L. *ssp. pekinensis* 'Chiifu' seeds were germinated in Petri dishes at 23 ± 1 °C in darkness. The germinated seeds were sown into plastic pots (7 cm × 7 cm) filled with vermiculite and grown for 20 d at 23 ± 2 °C with a light intensity of 6000–7000 Lux on a 16:8 h (L:D) photoperiod. During this growth phase, the plants were watered every two days with Hoagland solution (in mmol/L: KNO_3 6, $Ca(NO)_3 \bullet 4H_2O$ 4, KH_2PO_4 1, $MgSO_4 \bullet 7H_2O$ 1, H_3BO_3 0.046, $MnSO_4 \bullet H_2O$ 0.014, $ZnSO_4 \bullet 7H_2O$ 1.36×10^{-3}, $CuSO_4 \bullet 5H_2O$ 4.8×10^{-4}, $(NH_4)_6MoO_{24} \bullet 4H_2O$ 1.62×10^{-5}, $FeSO_4 \bullet 7H_2O$ 2×10^{-5}, and $C_{10}H_{14}N_2Na_2O_8$ 2×10^{-5}) [29]. The pH of the solution was adjusted to 6.0 using HCl and NaOH. For Pi deprivation treatment, the KH_2PO_4 in the Hoagland solution was replaced with KCl. After 20 d of cultivation, 15 to 20 plants were selected randomly. Five plants were used for root length, root weight, fresh weight, leaf area, and phosphorus content measurements. The roots and shoots of the remaining plants were separated and sampled. After sampling, the roots and shoots were frozen immediately in liquid nitrogen and stored at −80 °C until use. All Pi deprivation treatments were carried out with three independent biological replicates.

For the fertile and sterile plants of *B. rapa*, seeds were germinated in Petri dishes at 23 ± 1 °C in darkness, then the germinated seeds were transferred to a 4 °C growth chamber in darkness for 30 d to induce vernalization. After vernalization, the seeds were sown into pots (15 × 15 × 18 cm) containing potting soil and grown in a greenhouse at 23 ± 1 °C with a light intensity of 6000–7000 Lux and a 16:8 (L:D) photoperiod. The floral buds were then collected from five plants using three biological replicates, as previously described [30]. Root and shoot tissues were collected from three-week-old seedlings without vernalization. Stems and stem leaf tissues were sampled from these plants seven days after bolting. All tissues were stored at −80 °C until use.

2.2. Leaf Area and Phosphorus Content

For leaf area determination, the outermost leaf of Chiifu seedlings was dissected, and the leaf area was determined with a Yaxin-1241 leaf meter (Beijing Yaxinliyi, China) following manufacturer's instructions. Total phosphorus concentration was determined as described previously [31]. In brief, 0.2 g of leaves was ground and extracted using a laboratory-scale high-performance microwave digestion system (Milestone Ethos UP, Italy, Sorisole) with 8 mL HNO_3-H_2O_2 (3:1) reagent. Afterward, the mixtures were reacted with molybdenum blue reagent and measured 20 min later at 700 nm by using a spectrophotometer (UV-1200, Instrument Co., Ltd., Shanghai, China).

2.3. Identification of BrPAP Genes in Brassica rapa

To identify the PAP genes from *B. rapa*, all putative protein sequences of *B. rapa* (version 3.0) were downloaded from BRAD (http://brassicadb.agridata.cn/brad/ accessed on 28 January 2021) [32]. All of the protein sequences of the 29 *AtPAPs* (*A. thaliana purple acid phosphatase*) were retrieved from TAIR (https://www.arabidopsis.org/index.jsp; Araport11 accessed on 28 January 2021) based on a previous publication [2] and used as queries to search against all putative protein sequences of *B. rapa* using BLASTP. A total of 43 putative BrPAP proteins were obtained based on two criteria: an E-value below 10^{-5} and sequence identity above 20%. Next, the amino acid sequences of the 43 putative BrPAP proteins were analyzed to determine the presence of conserved sequence motifs (**D**XG/G**D**XX**Y**/G**N**H(E/D)/VX2**H**/G**H**X**H**) defined previously for BrPAPs [2], and BraA02g012470.3C, BraA04g002700.3C, BraA06g026740.3C, and BraA10g003360.3C were removed. As a result, 39 proteins containing these conserved sequence motifs were identified. To find all potential PAPs in *B. rapa*, each of the 39 BrPAPs were used as queries for BLAST searching at NCBI and Phytozome 13 (https://phytozome-next.jgi.doe.gov/phytomine/begin.do accessed on 28 January 2021). However, no additional predicted BrPAPs were found. All BrPAPs were named BrPAP1 to BrPAP39 based on their genomic locations.

2.4. Phylogenetic Tree and Bioinformatics Analysis of BrPAPs

To carry out phylogenetic tree analysis, the protein sequences of PAPs from *Arabidopsis* [2] and *B. rapa* were aligned using MUSCLE with default parameters [33]. An unrooted phylogenetic tree was constructed using MEGA6 with the Neighbor-joining method and parameters set as follows: Jones–Taylor–Thornton (JTT) model, pairwise deletion, and a bootstrap test of 1000 replications [34]. The position of each BrPAP on the *B. rapa* chromosomes was isolated from BRAD (http://brassicadb.agridata.cn/brad/ accessed on 28 January 2021) and visualized with a custom Python script. The isoelectric point (PI) and molecular weight (M.W.) of BrPAPs were determined using the ProtParam tool (https://web.expasy.org/protparam/ accessed on 28 January 2021) [35]. Conserved motifs in BrPAPs were identified using the MEME suite 5.1.1 (http://meme-suite.org/ accessed on 28 January 2021) [36]. The predictions of signal peptides and N-glycosylation sites were carried out using SignalP-5.0 (http://www.cbs.dtu.dk/services/SignalP/ accessed on 28 January 2021) [37] and NetNGlyc 1.0 (http://www.cbs.dtu.dk/services/NetNGlyc/ accessed on 28 January 2021) with default settings, respectively. Gene structures of *BrPAPs* were drawn using the Gene Structure Display Server (GSDS, version 2.0, http://gsds.cbi.pku.edu.cn/ accessed on 28 January 2021) [38].

2.5. Expression and Co-Expression Analysis of BrPAPs

All RNA-seq reads from shoots (vernalized and non-vernalized), leaves, stems, roots, flowers, differential developmental seeds, and male sterile lines of *B. rapa* [39–44] were downloaded from the SRA (sequence read archive) on NCBI (https://www.ncbi.nlm.nih.gov/sra accessed on 28 January 2021) from accession IDs PRJNA339187, PRJNA185152, PRJNA310313, and PRJNA579431. The quality of these RNA-seq reads was analyzed using fastQC (https://www.bioinformatics.babraham.ac.uk/projects/fastqc/ accessed on 28 January 2021), and the adapter sequences and low-quality reads were removed using the FASTX-Toolkit (Version 0.0.13) [45]. For gene expression, the groomed reads were mapped to genes using Bowtie2 [46], and fragments per kilobases of exons per million mapped reads (FPKM) values for transcripts were calculated using Cufflinks [47]. Heatmaps were generated using Multiple Experiment Viewer (T-MEV) with log transformed FPKM values [48].

For co-expression analysis, RNA-seq data from male sterility lines of *B. rapa* were reassembled [40,41,43,44], and the genes with no differences between fertile and sterile floral buds were removed. Pearson's correlation coefficient (PPC) values were calculated in Excel using *BrPAP* candidates as queries. The threshold values for PPCs were set between 0.60 and −0.60. GO (Gene Ontology) enrichment analysis was carried out using agriGO (http://systemsbiology.cau.edu.cn/agriGOv2/ accessed on 28 January 2021) [49] and the clusterProfiler package in R [50].

2.6. RNA Extraction, Semi-RT-PCR, and qRT-PCR

Total RNA was isolated using the RNAiso Plus Reagent (Takara Biomedical Technology Co., Ltd., Beijing, China), based on manufacturer protocols. The PrimeScript™ RT reagent Kit with gDNA Eraser (Takara Biomedical Technology Co., Ltd., Beijing, China) was used to synthesize 1st strand cDNA with 1 μg of total RNA. After synthesis, this cDNA was diluted to 10 ng/μL for PCR analysis. Semi-RT-PCR was carried out using 20 ng/μL cDNA with the following program: denaturation at 94 °C for 5 min, followed by 28 cycles of 94 °C for 30 s, 55 °C for 30 s, and 72 °C for 60 s. For qRT-PCR analysis, 10 ng/μL of cDNA was used and performed with a 30 s pre-denaturation at 95 °C, followed by 40 cycles of 95 °C for 5 s, 55 °C for 30 s, and 72 °C for 60 s. All of the primers used are listed in Table S1. The *B. rapa* actin gene was used as an internal control. The products of semi-RT-PCR were separated on 1.5% agarose gels and stained with GeneGreen (Tiangen Technology Co., Ltd., Beijing, China). The qRT-PCR results were analyzed using the $2^{-\Delta Ct}$ method with three biological replicates. After removing the genes with Ct value above 35 and less than two-fold change, the results are represented by a heatmap using R scripts with row normalization.

3. Results

3.1. Brassica rapa PAP Genes

After BLASTP search and conserved motif determination, 39 *PAP* genes were identified from the *B. rapa* genome and designated as *BrPAP1* to *BrPAP39* according to their positions on chromosomes (Figure 1). Among them, 31 BrPAP proteins contained the seven conserved metal-binding residues (D, D, Y, N, H, H, and H) in five consensus blocks (Table 1). Another three BrPAPs (BrPAP14, BrPAP17, and BrPAP21) lacked the first conserved block. The fourth block was lost in BrPAP9. For BrPAP7, one conserved residue in the fourth block was changed from H to S and the fifth block was missing. In the proteins from BrPAP22, BrPAP32, and BrPAP39, one conserved residue in the second block changed from Y to F, which has been reported in *Arabidopsis* [2], soybean [51], rice [11], maize [12], chickpea [13], and tea [15].

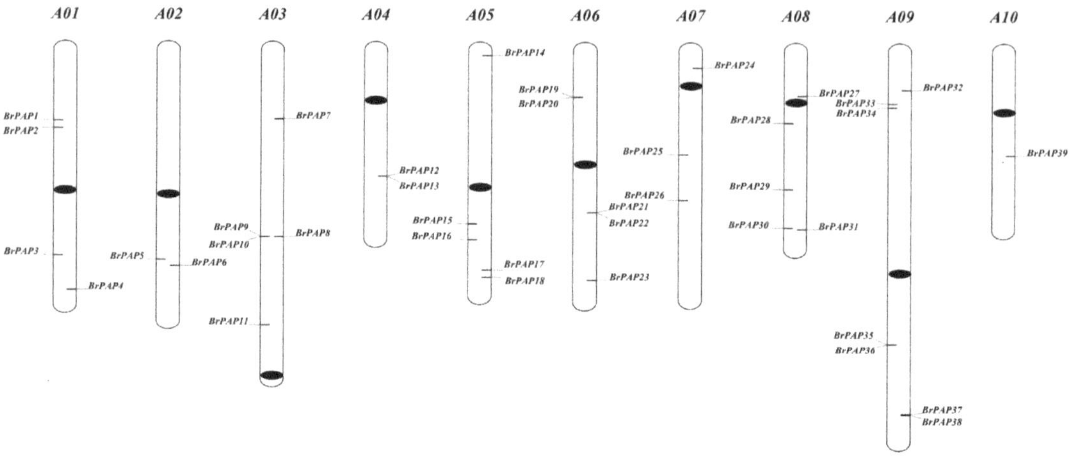

Figure 1. Chromosomal locations of the 39 *BrPAP* genes identified in this study. Chromosome number is indicated above each chromosome. Black ovals on each chromosome represent centromeric regions.

Table 1. Conserved metal ligating amino acid motifs present in BrPAP proteins. Conserved residues are underlined.

Gene ID	Gene Name	PAP Conserved Motifs				
		GDXG	GDXXY	GNH(D/E)	VXXH	GHXH
BraA01g015600.3C	*BrPAP1*	GDMG	GDITY	GNHE	FIAH	GHVH
BraA01g017190.3C	*BrPAP2*	GDLG	GDLSY	GNHE	VLNH	GHVH
BraA01g034870.3C	*BrPAP3*	GDWG	GDNFY	GNHD	VVGH	GHDH
BraA01g041630.3C	*BrPAP4*	GDTG	GDVSY	GNHE	VTWH	GHVH
BraA02g035210.3C	*BrPAP5*	GDWG	GDNFY	GNHD	VVGH	GHDH
BraA02g036110.3C	*BrPAP6*	GDMG	GDISY	GNHE	VQGH	GHVH
BraA03g017330.3C	*BrPAP7*	GDLG	CDSCY	GEHE	ATWS	
BraA03g043110.3C	*BrPAP8*	GDLG	GDFSY	GNHE	VLMH	GHVH
BraA03g043120.3C	*BrPAP9*	GDLE	YDTKY	GNCD		
BraA03g043130.3C	*BrPAP10*	GDLG	GDLSY	GNHE	VLNH	GHVH
BraA03g059950.3C	*BrPAP11*	GDLG	GDLSY	GNHE	VIVH	GHVH
BraA04g019300.3C	*BrPAP12*	GDLG	GDLSY	GNHE	VLVH	GHVH
BraA04g019310.3C	*BrPAP13*	GDLG	GDLSY	GNHE	VLVH	GHVH
BraA05g001000.3C	*BrPAP14*		SDMHY	GNHD	VYVH	GHDH
BraA05g026730.3C	*BrPAP15*	GDLG	GDLSY	GNHE	ALFH	GHVH
BraA05g029790.3C	*BrPAP16*	GDWG	GDNFY	GNHD	VVGH	GHDH
BraA05g036550.3C	*BrPAP17*		GDVVT	GNHD	IFWH	GHNH
BraA05g038510.3C	*BrPAP18*	GDLG	GDVSY	GNHE	VSWH	GHVH
BraA06g009980.3C	*BrPAP19*	GDMG	GDICY	GNHE	FLAH	GHAH
BraA06g010040.3C	*BrPAP20*	GDMG	GDISY	GNHE	VQGH	GHVH
BraA06g026750.3C	*BrPAP21*		GDNIF	GNHD	AYFH	GHDH
BraA06g026760.3C	*BrPAP22*	ADMH	GDNIF	GNHD	AYFH	GHDH
BraA06g040030.3C	*BrPAP23*	GDWG	GDNFY	GNHD	VVGH	GHDH
BraA07g002610.3C	*BrPAP24*	GDLG	GDLSY	GNHE	VLVH	GHVH
BraA07g012620.3C	*BrPAP25*	GDWG	GDNIY	GNHD	VVGH	GHDH
BraA07g020920.3C	*BrPAP26*	GDLG	GDLSY	GNHE	AVMH	GHIH
BraA08g006360.3C	*BrPAP27*	GDLG	GDFTY	GNHE	ATMH	GHVH
BraA08g009450.3C	*BrPAP28*	GDLG	GDLSY	GNHE	VLMH	GHVH
BraA08g021060.3C	*BrPAP29*	GDLG	GDLAY	GNHE	VMVH	GHVH
BraA08g030080.3C	*BrPAP30*	GDWG	GDNFY	GNHD	VVGH	GHDH

Table 1. *Cont.*

Gene ID	Gene Name	GDXG	GDXXY	GNH(D/E)	VXXH	GHXH
		PAP Conserved Motifs				
BraA08g030430.3C	*BrPAP31*	GDMG	GDISY	GNHE	VQGH	GHVH
BraA09g007730.3C	*BrPAP32*	ADMH	GDNIF	GNHD	AYFH	GHDH
BraA09g010460.3C	*BrPAP33*	GDLG	GDFSY	GNHE	VLMH	GHVH
BraA09g011240.3C	*BrPAP34*	GDLG	GDLSY	GNHE	VLVH	GHVH
BraA09g043870.3C	*BrPAP35*	GDLG	GDLSY	GNHE	AVIH	GHVH
BraA09g043900.3C	*BrPAP36*	GDLG	GDLSY	GNHE	VLLH	GHVH
BraA09g059830.3C	*BrPAP37*	GDMG	GDISY	GNHE	VQGH	GHVH
BraA09g059970.3C	*BrPAP38*	GDMG	GDICY	GNHE	FLAH	GHAH
BraA10g015040.3C	*BrPAP39*	ADMH	GDNIF	GNHD	AFFH	GHDH

General information for *BrPAPs*, including gene size, molecular weight, *Arabidopsis* homologs, and potential N-linked glycosylation sites, were also analyzed (Table 2). The sizes of the genomic DNA sequences for *BrPAPs* varied from 1196 bp (*BrPAP16*) to 6766 bp (*BrPAP23*), with an average length of 2443 bp. Molecular weight of identified BrPAPs ranged from 34.8 kDa (BrPAP21) to 75.79 kDa (BrPAP20). Based on predictions from SignalP, signal peptides were found in 34 BrPAPs except for BrPAP2, BrPAP9, BrPAP21, BrPAP25, and BrPAP39. N-glycosylation site prediction analysis indicated that BrPAPs exhibited different numbers of glycosylation sites, from 0 to 6 (Table 2).

3.2. *Phylogenetic and Gene Structure Analyses of BrPAPs*

A phylogenetic analysis was carried out using amino acid sequences from BrPAPs and AtPAPs with the neighbor-joining method. All PAP proteins were classified into three major clades (I to III), and a further classification of these three major clades produced nine subgroups (Figure 2). This analysis was similar to the findings of a previous study in *Arabidopsis* [2], with one exception. AtPAP10, 12, and 26 were classified into the same group in *Arabidopsis* [2], whereas AtPAP26 was separated from AtPAP10 and 12 and classified into a new subgroup (I a-3) in this study. Previously, a single gene was identified as a potential ortholog of AtPAP26 in maize and rice, indicating its conserved functions in the Pi deprivation response, which was also observed in *B. rapa* and was grouped into I a-3 (Figure 2). Based on our phylogenetic analysis, one AtPAP was related to one to three BrPAP homologs, which may have been due to whole-genome polyploidization and a gene loss event of *B. rapa* after separation from *Arabidopsis*, although this was not the case for AtPAP10 (Table 2 and Figure 2). For AtPAP10, five homologs (BrPAP2, BrPAP8, BrPAP9, BrPAP10, and BrPAP33) were identified in *B. rapa*, which indicated the expansion of AtPAP10 in *B. rapa* over evolutionary time.

During the evolution of a multigene family, gene structure usually diversifies. We therefore wanted to study *BrPAPs* related to their functional diversification and evolution. The intron size and number of *BrPAPs* were highly variable, showing 11 distinct exon–intron organization patterns. Seven exons separated by six introns was the most common organizational pattern and was present in 10 *BrPAPs* (Table 2 and Figure 3). Except for *BrPAP6*, most *BrPAPs* contained more than one intron, indicating the possible existence of alternative splicing during gene expression. For Ia-1, IIb, and IIIb, genes from the same subgroups showed similar gene structures, indicating that they might have conserved functions. The diversity of gene structures of genes from IIIa implied their diverse functions (Figure 3).

Table 2. General information for BrPAP genes and their deduced proteins. M.W., molecular weight; PI, isoelectric point.

Gene Name	Chromosome Location	No. of Exons	Gene Length(bp)	M.W. (kDa)	Protein Length (a.a.)	*PI*	Signal Peptide (Length, Cleave Site)	Number of N-Glycosylation Sites	Best Hit to *Arabidopsis*		
									*At*_ID	*At*_Name	*E*-Value
BrPAP1	A01	10	2778	72.67	642	6.09	24, GHA-SN	1	*AT4G24890.1*	*AtPAP24*	0
BrPAP2	A01	8	2942	45.42	393	5.81	—	4	*AT2G16430.1*	*AtPAP10*	4.05×10^{-171}
BrPAP3	A01	3	1240	37.80	334	5.28	26, TNG-EL	1	*AT3G17790.1*	*AtPAP17*	0
BrPAP4	A01	6	2607	60.35	541	5.23	32, SDA-IP	6	*AT3G07130.1*	*AtPAP15*	0
BrPAP5	A02	7	4923	37.60	331	6.3	24, SFS-KL	2	*AT2G01880.1*	*AtPAP7*	0
BrPAP6	A02	1	1929	72.89	643	6.07	18, VHS-TP	4	*AT2G03450.1*	*AtPAP9*	0
BrPAP7	A03	6	2705	60.27	536	4.97	21, VDA-FP	6	*AT2G32770.3*	*AtPAP13*	0
BrPAP8	A03	7	2372	56.76	489	7.16	24, CHG-GT	1	*AT2G16430.2*	*AtPAP10*	0
BrPAP9	A03	4	1726	34.34	309	9.37	—	1	*AT2G16430.2*	*AtPAP10*	2.48×10^{-51}
BrPAP10	A03	7	2070	53.73	467	7.76	25, CNG-GI	5	*AT2G16430.2*	*AtPAP10*	0
BrPAP11	A03	7	1894	52.19	461	6.22	22, ING-GM	3	*AT1G56360.1*	*AtPAP6*	0
BrPAP12	A04	8	2354	55.49	482	6.16	28, CDG-GI	4	*AT2G27190.1*	*AtPAP12*	0
BrPAP13	A04	7	2057	54.08	470	6.25	28, CDG-GI	3	*AT2G27190.1*	*AtPAP12*	0
BrPAP14	A05	5	1682	44.56	394	8.47	24, VDA-YG	1	*AT2G46880.1*	*AtPAP14*	0
BrPAP15	A05	4	1947	47.48	414	5.68	19, VAA-DD	0	*AT3G20500.1*	*AtPAP18*	0
BrPAP16	A05	3	1196	38.33	338	5.57	—	1	*AT3G17790.1*	*AtPAP17*	0
BrPAP17	A05	4	1399	42.91	385	5.86	23, AVG-WE	1	*AT3G10150.2*	*AtPAP16*	0
BrPAP18	A05	5	2307	61.43	541	5.64	26, SSA-DY	4	*AT3G07130.1*	*AtPAP15*	0
BrPAP19	A06	9	2755	69.52	621	7.9	22, GGA-IQ	2	*AT1G13750.1*	*AtPAP1*	0
BrPAP20	A06	2	2599	75.79	680	5.6	22, ANA-EA	3	*AT1G13900.1*	*AtPAP2*	0
BrPAP21	A06	3	1378	34.08	308	6.24	—	0	*AT5G63140.1*	*AtPAP29*	0
BrPAP22	A06	3	1460	42.19	383	9.01	31, ASA-QG	1	*AT5G63140.1*	*AtPAP29*	0
BrPAP23	A06	14	6766	69.10	607	5.51	29, SRA-EL	3	*AT2G01890.1*	*AtPAP8*	0
BrPAP24	A07	7	2127	52.10	460	5.86	24, SHA-GV	4	*AT4G36350.1*	*AtPAP25*	0
BrPAP25	A07	8	2551	52.34	468	7.17	—	2	*AT1G25230.3*	*AtPAP4*	0
BrPAP26	A07	6	4130	50.56	437	8.25	22, SQA-YN	1	*AT3G52810.1*	*AtPAP21*	0
BrPAP27	A08	7	2273	61.20	548	5.02	31, AGG-ES	4	*AT4G13700.1*	*AtPAP23*	0
BrPAP28	A08	9	2313	55.40	481	7.35	27, GEG-GI	1	*AT5G34850.1*	*AtPAP26*	0
BrPAP29	A08	7	1887	52.42	460	6.6	19, ING-GI	1	*AT1G56360.1*	*AtPAP6*	0
BrPAP30	A08	7	2595	38.03	335	5.81	29, STA-EL	1	*AT1G14700.1*	*AtPAP3*	0
BrPAP31	A08	2	2540	73.13	652	5.95	2, ANA-KA	6	*AT1G13900.1*	*AtPAP2*	0
BrPAP32	A09	3	1746	42.92	386	9.2	32, TSS-HR	2	*AT5G63140.1*	*AtPAP29*	0
BrPAP33	A09	8	2490	54.12	469	6.77	29, CHG-GR	3	*AT2G16430.2*	*AtPAP10*	0
BrPAP34	A09	7	1977	52.82	468	5.49	23, SHA-GV	3	*AT2G18130.1*	*AtPAP11*	0
BrPAP35	A09	6	2974	46.99	415	6.01	21, VSS-YD	1	*AT3G52780.1*	*AtPAP20*	0
BrPAP36	A09	6	3163	51.08	448	5.79	35, SQA-DV	2	*AT3G52820.1*	*AtPAP22*	0
BrPAP37	A09	2	2436	73.36	653	6.18	21, ANA-KA	4	*AT1G13900.1*	*AtPAP2*	0
BrPAP38	A09	9	3035	69.22	622	6.16	24, ALG-GR	4	*AT1G13750.1*	*AtPAP1*	0
BrPAP39	A10	4	1966	45.34	404	6.83	—	3	*AT5G57140.1*	*AtPAP28*	0

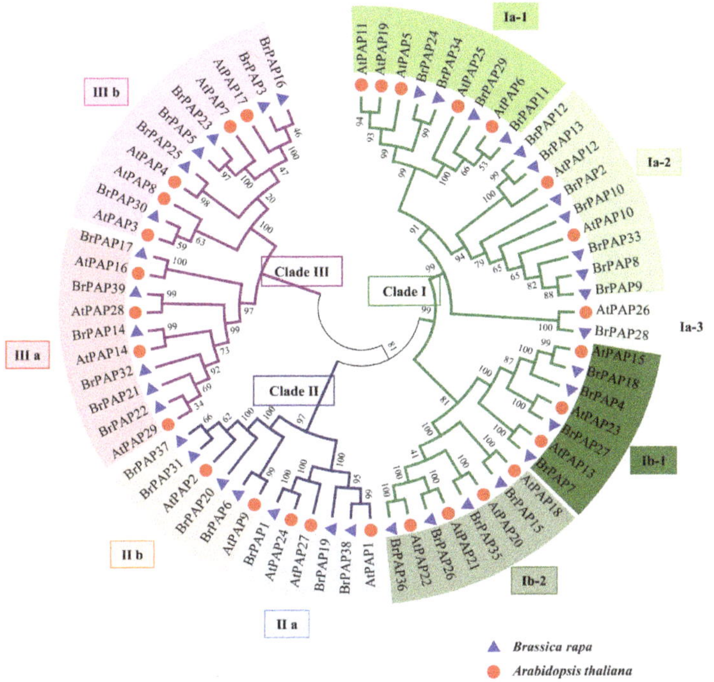

Figure 2. Phylogenetic relationships of BrPAP and AtPAP genes. Multiple protein sequence alignment was performed using ClustalW implemented in MEGA6. A neighbor-joining (NJ) tree was constructed in MEGA6 using the Jones–Taylor–Thornton (JTT) model with the following options: homogeneous pattern, pairwise deletion, and 1000-replicate bootstrap. Different clades and branches (subclades) are indicated by the colors of the branched lines and background, respectively.

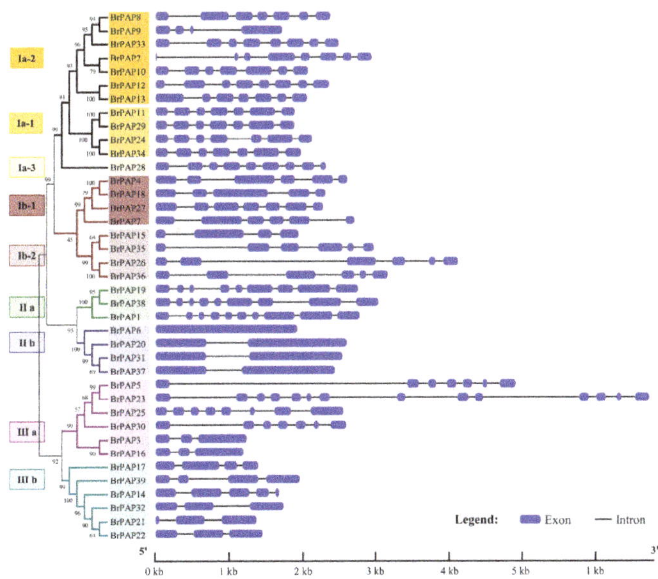

Figure 3. Gene structures of *BrPAP* genes.

3.3. Potential Functions of BrPAPs during Pollen Development

To study the functional divergence of *BrPAPs*, the spatial and temporal expression patterns of *BrPAPs* in *B. rapa* were analyzed. RNA-seq data derived from shoots (vernalized and non-vernalized), leaves, stems, roots, flowers, differential developmental stages of seeds, and male sterility lines were explored [22,39,41–44]. Based on this RNA-seq data, *BrPAP* transcripts accumulated in all tissues examined (Figure 4), as previously reported for *AtPAPs* under standard growth conditions [4]. This indicated that *BrPAPs* might play diversified functions during development. Interestingly, *BrPAP11, 24, 29,* and *34* from subgroup Ia-1 showed specific expression in fertile buds in *B. rapa* (Figure 4). In addition to these four genes, *BrPAP14* and *20* also showed high expression in fertile buds in *B. rapa*. This indicated that these genes specifically expressed in fertile buds might be related to pollen development. To confirm the expression patterns of these six genes, different developmental stages of floral buds from genetic male sterility (GMS) lines were collected based on a previous study (Figure 5A) [30]. According to our semi-RT-PCR results, *BrPAP11* and *24* were first detected in F2 floral buds (with tetrad pollen), and these *BrPAPs* were highly expressed in F3 floral buds (with pollen grain after tetrad formation but before maturation) (Figure 5B). *BrPAP14* and *29* were highly expressed in both F2 and F3 floral buds. *BrPAP20* and *BrPAP34* were only expressed in F4 (with pollen grain after tetrad formation but before maturation) and F3 (with mature pollen grains) floral buds, respectively (Figure 5B). To narrow down and predict the functions of *BrPAP11, 14, 20, 24, 29,* and *34* during pollen development, we carried out co-expression analysis using previously published male-sterility RNA-seq data in *B. rapa* [22,41,43,44]. When the threshold value for PPC (Pearson's correlation coefficient) was set between 0.60 and −0.60, 1511, 979, 2355, 1593, 1886, and 1667 genes were found to be co-expressed with *BrPAP11, 14, 20, 24, 29,* and *34*, respectively (Table S2 and Figure 6A). GO enrichment analysis was carried out to highlight the biological processes influenced by targeted *BrPAPs* (Figure 6B). The pollen development process was overrepresented by the co-expressed genes of *BrPAP11, 20, 24,* and *34*. The pollination process was only highlighted by the co-expressed genes of *BrPAP20, 24, 29,* and *34*. Lipid metabolic and fatty acid biosynthetic processes were enriched by co-expressed genes of *BrPAP14*. Carbohydrate transport, pollen sperm cell differentiation, pectin catabolic, polysaccharide catabolic, cell wall organization, pollen germination, and carbohydrate metabolic processes were only overrepresented by the co-expressed genes of *BrPAP20*.

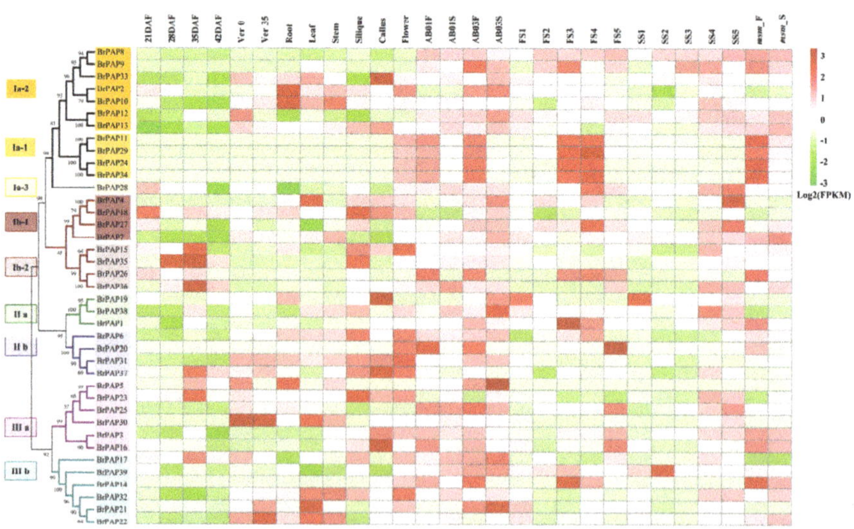

Figure 4. Heatmap representation of expression patterns for *BrPAPs* with various tissues and male sterility lines. DAF,

days after flowering; Ver0, 0 day after vernalization; Ver35, 35 days after vernalization; AB01_F and AB01_S, indicate fertile sterile floral buds of GMS Chinese cabbage line 'AB01', respectively; AB03_F and AB03_S represent fertile sterile floral buds of GMS Chinese cabbage line 'AB03', respectively; FS1–FS5, fertile flower buds stage 1 to 5 of GMS Chinese cabbage 'Bcajh97-01A/B'; SS1-SS5, sterile flower buds stage 1 to 5 of GMS Chinese cabbage 'Bcajh97-01A/B; *msm*_F and *msm*_S, floral buds from GMS Chinese cabbage line '*msm*'.

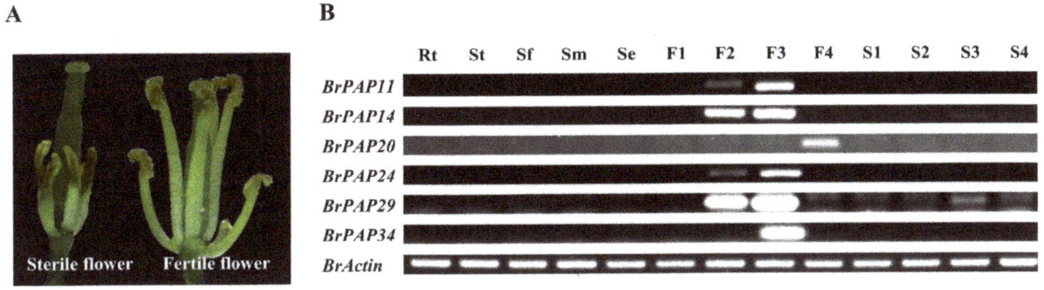

Figure 5. Confirmation of the expression patterns of *BrPAPs* in fertile buds using semi-RT-PCR. (**A**) Flower structure of fertile and sterile lines of *Brassica rapa*. In order to focus on the stamen, the petals and sepals were removed. (**B**) Analysis of the expression patterns of fertile bud specifically expressed genes using semi-RT-PCR. Rt, root; St, shoot; Sf, stem leaf; Sm, stem; Se, silique; F1–F4 indicated the floral buds from fertile plants. F1, before the tetrad stage. F2, at the tetrad stage. F3, after the tetrad stage but before containing mature pollen. F4, containing mature pollen. S1–S4, floral buds from sterile plants. S1, before the tetrad stage. S2, at the tetrad stage. S3 and S4, after the tetrad stage with aborted pollen grains.

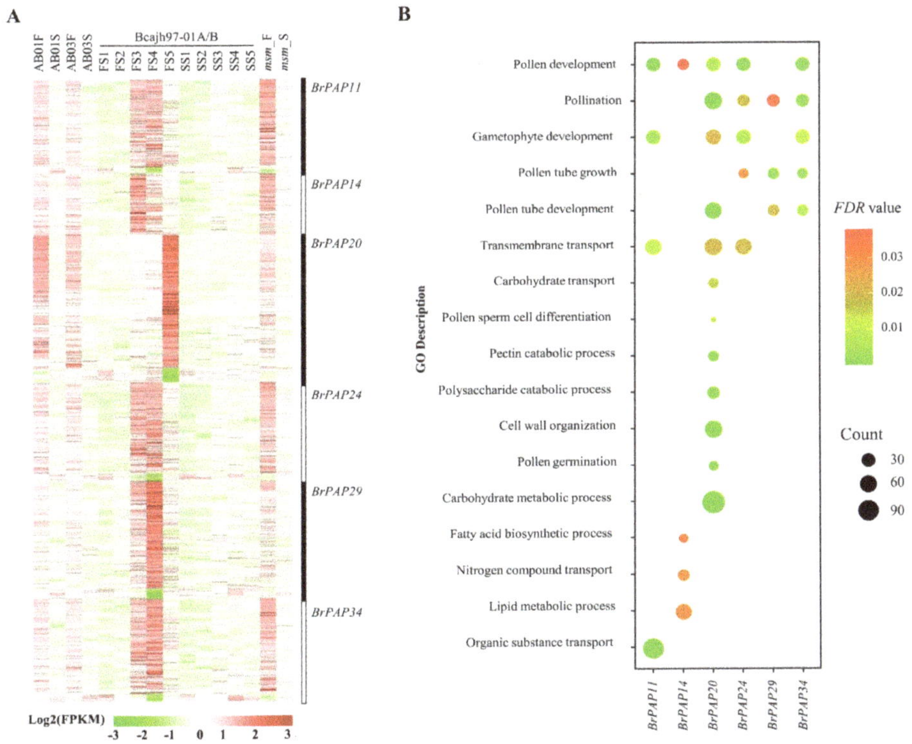

Figure 6. Analysis of co-expression genes of *BrPAP11, 14, 20, 24, 29,* and *34*. (**A**), Heatmap representation for expression patterns of co-expressed genes of *BrPAP11, 14, 20, 24, 29,* and *34*. (**B**), GO enrichment analysis of co-expression genes of *BrPAP11, 14, 20, 24, 29,* and *34*.

3.4. Expression of BrPAP Genes in Response to Pi Deprivation in *Brassica rapa*

PAPs are considered to play an essential role in the processes of phosphorus foraging and recycling [5,14]. To identify the potential functions of *BrPAPs* related to phosphorus acquisition and utilization, germinated *B. rapa* seeds were sown in pots containing vermiculite and watered with Hoagland solution [28]. For Pi deprivation treatments, the KH_2PO_4 in Hoagland solution was replaced with KCl. After treatments, the expression patterns of *BrPAPs* and physiological characteristics were measured. With Pi deprivation treatment, the plant height, fresh weight, leaf area, and Pi-content were all significantly decreased, while the root–shoot ratio of biomass increased compared with the normal growth condition (Figure 7). The shoot and root parts of seedlings were sampled separately for qRT-PCR analysis. Using a Ct value below 35 and fold change above two as thresholds, 21 *BrPAPs* responded to Pi deprivation either in shoots or roots (Figure 8). Four genes were mainly upregulated in roots under Pi deprivation (Figure 8A). Three genes were upregulated both in shoots and roots under Pi deprivation (Figure 8B). One gene, *BrPAP12*, was preferentially upregulated by Pi deprivation (Figure 8C). Ten genes were induced by Pi deprivation in shoots (Figure 8D). Three genes were consistently upregulated in shoots (Figure 8E).

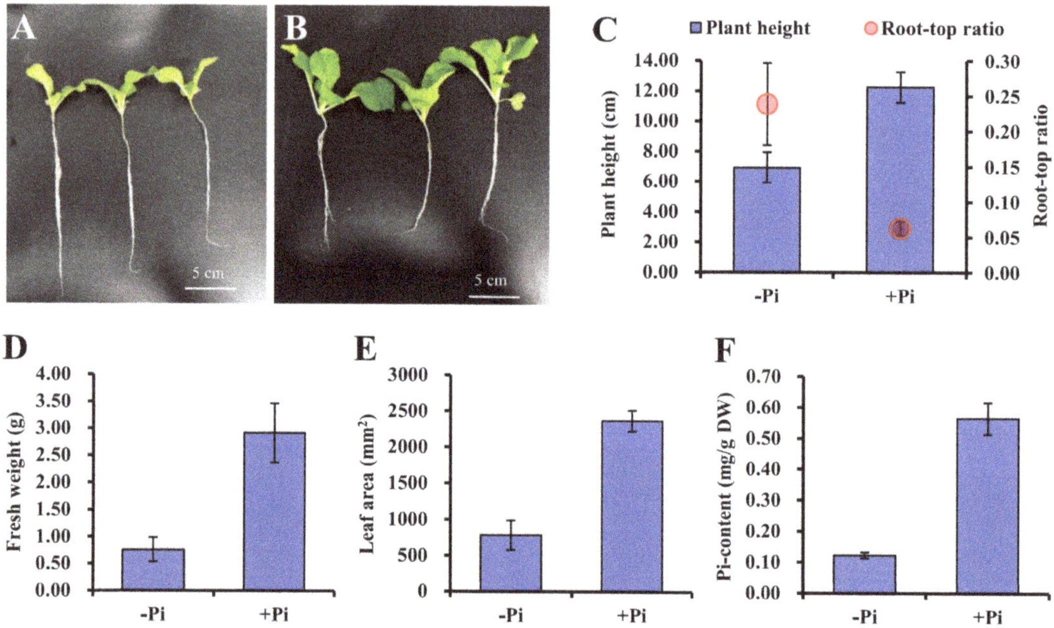

Figure 7. Phenotypical characteristics of Chiifu seedlings after 20 days of Pi deprivation treatment. (**A**,**B**) Picture representing Chiifu seedlings under –Pi or +Pi growth conditions. (**C**–**F**) Plant height, root–shoot ratio of biomass, fresh weight, leaf area, and Pi-content differences of Chiifu seedlings in –Pi or +Pi growth conditions.

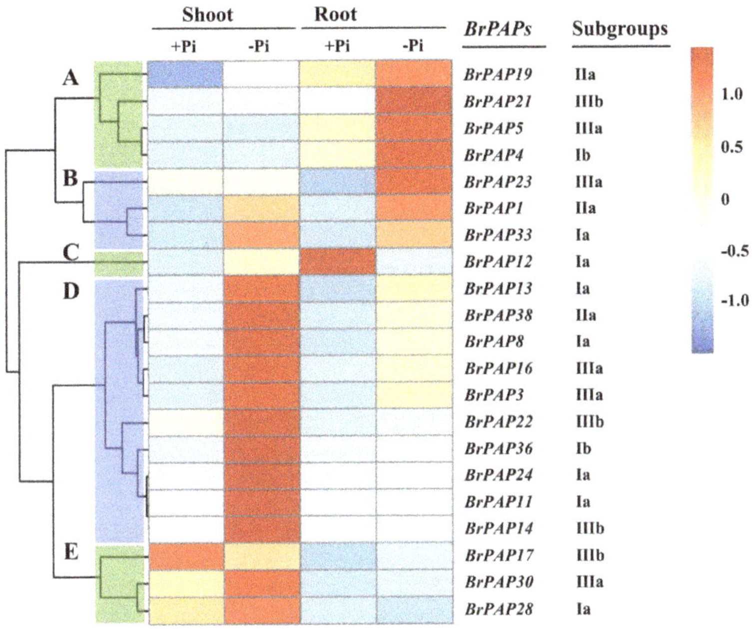

Figure 8. qRT-PCR analysis of expression patterns of *BrPAP* genes in response to Pi deprivation. (**A**) Genes upregulated in roots under Pi deprivation. (**B**) Genes upregulated both in shoots and roots in response to Pi deprivation. (**C**) Genes upregulated in roots under normal growth conditions. (**D**) Genes upregulated in shoots under Pi deprivation. (**E**) Genes upregulated in shoots both under normal and Pi deprivation conditions. The expression of *BrPAPs* was calculated using the $2^{-\Delta Ct}$ method and plotted using R with row normalization.

4. Discussion

4.1. Identification and Analysis of BrPAPs

Based on conserved sequences and motifs, genome-wide analysis of *PAPs* has been performed on many plants, including *Arabidopsis* [2], soybean [51], maize [12], tea [15], rice [11], *Brassica napus* [10], and ten vegetable species [9]. In this study, 39 BrPAPs were identified and classified into three different clades with nine subgroups based on a phylogenetic analysis using PAPs from *Arabidopsis* and *B. rapa* (Figure 2). Previously, 37 *BrPAPs* were identified in *B. rapa* [9], which could be explained by the fact that a higher genome version was used [32], leading to significantly higher accuracy of identification of *BrPAPs* [9]. The classification results in our study are mostly consistent with previous results from *Arabidopsis* and rice [2,11]. AtPAP26 from Ia-2 clustered into Ia-3 with BrPAP28 (Figure 2) [2]. Due to the whole-genome duplication events and gene loss events that have occurred in *B. rapa* over evolutionary time, one *Arabidopsis* gene usually has one to three *B. rapa* homologs [32], and the results from this study also demonstrated this observation except for AtPAP10. Based on our phylogenetic analysis, AtPAP10 corresponded to five BrPAPs (BrPAP2, BrPAP8, BrPAP9, BrPAP10, and BrPAP33), indicating the functional expansion of AtPAP10 (Figure 2). Consistent with studies in maize and rice, only one single gene, BrPAP28, was identified to be the ortholog of AtPAP26 in subgroup Ia-3, indicating its highly conserved function during evolution (Figure 2) [12].

Based on the results from *Arabidopsis*, rice, maize and other plants, some PAPs lack the invariant amino acid residues involved in the coordination of the di-metal nuclear center of known PAPs [2,11,12]. In *B. rapa*, 31 of 39 BrPAPs contained seven invariant amino acid residues (Table 1). The first block was absent in BrPAP14, BrPAP17, and BrPAP21. BrPAP9 did not possess the fourth and fifth blocks, and the fifth block was also lost in BrPAP7.

The third invariant amino acid residue Y shifted to T or F in BrPAP17, BrPAP21, BrPAP22, BrPAP32, and BrPAP39. In BrPAP7, the fourth and fifth conserved residues N and H were replaced by E and S, respectively. Compared with *Arabidopsis*, all BrPAPs had *Arabidopsis* homologs except for BrPAP9, and there was no conserved block absence in *Arabidopsis* (Tables 1 and 2) [2]. This indicates that these atypical BrPAPs might not be biochemically active PAPs and their true functions need further study.

4.2. BrPAPs and Pollen Development

Previous studies have suggested that *PAPs* play roles in the processes of carbon metabolism [25], cell wall synthesis [26], ROS metabolism, and response to salt stress [27]. Little is known about their functions related to pollen development, even for *AtPAP6, 11, 14, 19, 23, 24,* and *25*, which are predominantly expressed in flowers [4]. In this study, *BrPAP11, 14, 20, 24, 29,* and *34* were specifically expressed in fertile floral buds in *B. rapa* (Figures 4 and 5). Combining the differences between fertile and sterile floral buds in the current and previous studies (Figure 5A) [30], in addition to our semi-RT-PCR results, we concluded that these *BrPAPs* might function and play different roles during the process of pollen development (Figure 5B). In support of this hypothesis, *BrPAP11* was initially expressed in F2 floral buds and highly expressed in F3 floral buds, and its co-expressed genes were involved in the biological processes of organic substance transport, transmembrane transport, gametophyte development, and pollen development. This suggested that *BrPAP11* might be associated with nutrition transport during pollen development (Figure 5B). *BrPAP14* was specifically expressed in F2 and F3 floral buds, and the biological processes of lipid metabolic, nitrogen compound transport, fatty acid biosynthetic, and pollen development were enriched in its co-expressed genes (Figures 5B and 6B). This implies that *BrPAP14* might be involved in secondary metabolism during pollen development. The expression levels of *BrPAP20* were high in F4 floral buds (containing mature pollen grains), and carbohydrate metabolic, polysaccharide catabolic, and pectin catabolic were enriched in its co-expressed genes (Figures 5B and 6B), suggesting that this gene might have functions during pollen maturation. The normal processing of cell wall degradation and reorganization are required for the development and functions of pollen grains [30,41]. In tobacco, *NtPAP12* were found binding to the cell wall and enhanced the activities of cellulose and callose synthases [26]. The highlighted cell wall organization process by *BrPAP20* and its co-expressed genes indicate that the functions of *BrPAP20* might have similar functions during pollen development to *NtPAP12* (Figure 6B). Even *BrPAP24* had similar expression patterns to those of *BrPAP11*, and pollination and pollen tube growth were only enriched by *BrPAP24* and its co-expressed genes. The processes of transmembrane transport, gametophyte development, and pollen development were enriched by the co-expressed genes of *BrPAP11* and *BrPAP24*. This observation suggests that the functions of *BrAPA11* and *BrPAP24* partially overlap. A similar phenomenon was observed for *BrPAP14* and *BrPAP29*. For *BrPAP34*, the biological processes of pollen development, pollination, pollen tube growth, and pollen tube development were revealed by its co-expressed genes based on GO enrichment analysis, highlighting its possible roles during pollen germination and pollen development. Previously, *AtPAP5* and *11* were highly expressed in anthers and involved in pollen tube growth processes [28]. In this study, *BrPAP11, BrPAP24, BrPAP34, AtPAP11,* and *AtPAP5* were classified into subgroup Ia-1, indicating the importance of subgroup Ia-1 during pollen tube growth. Phytic acid is one of the most important components of pollen, *AtPAP15* and *AtPAP23* were expressed in pollen or floral buds with phytase activity in *Arabidopsis* [4,18], indicating the specifically expressed *BrPAP11, 14, 20, 24, 29,* and *34* in fertile floral buds might have similar functions. Collectively, these analyses suggest that *BrPAP11, 14, 20, 24, 29,* and *34* have multiple functions during pollen development with some overlap. These findings will assist future molecular breeding for GMS in *B. rapa*.

4.3. Responses of Brassica rapa and BrPAPs under Pi Deprivation Conditions

Phosphorus is one of most important essential nutrients for plant growth and is involved in many metabolic processes such as respiration and photosynthesis [12]. However, the content of inorganic phosphorus (Pi) is lower than that required for plant growth in most soils, and most phosphorus is present in the Po (organic phosphorus) form or is fixed by iron, calcium, and other elements. This means that plants cannot directly utilize it [52]. APases can enable plants to use intracellular and extracellular organic phosphorus to survive in low Pi conditions [5]. Among the plant APases, PAPs are one of the most important classes and they can be induced by low Pi conditions [53]. In *Arabidopsis*, several *AtPAPs* can be induced by Pi deprivation such as *AtPAP10, 11, 12, 17,* and *26* [2,17,54]. In rice, the overexpression of *OsPAP10a* and *OsPAP10c* can improve the utilization of external organic phosphorus, and both of these can be induced by Pi starvation [21,22,24]. In the current study, 21 of the 39 *BrPAPs* were found to be responsive to Pi deprivation and these can be divided into five groups (Figure 8). The first group showed a high expression in roots under Pi deprivation, including *BrPAP4, 5, 19,* and *21* (Figure 8A). *BrPAP1, 23,* and *33* were clustered into a second group and their expression patterns were higher in both shoots and roots under Pi deprivation compared with normal growth conditions (Figure 8B). The third group contained only one gene, *BrPAP12*, which was upregulated in roots in normal conditions and shoots with Pi deprivation (Figure 8C). Most of the Pi deprivation-responsive genes were clustered into the fourth group, which included ten *BrPAPs* (*BrPAP3, 8, 11, 13, 14, 16, 22, 24, 36,* and *38*); these genes were predominantly upregulated in shoots under Pi deprivation (Figure 8D). *BrPAP17, 28* and *30* were classified into a fifth group and were upregulated in shoots in both normal and Pi deprivation conditions (Figure 8D). In previous studies, *AtPAP10* was specifically induced by Pi limitation on the root surface and was considered to be one of the major Pi starvation-induced APases in *Arabidopsis* [17,55]. Among *BrPAPs*, five homologs of *BrPAP10* were found and only two of them, *BrPAP8* and *BrPAP33*, were Pi deprivation-responsive genes (Figures 2 and 8), indicating the functional expansion of the *AtPAP10* group during evolution. The expression levels of *AtPAP12* increased in low phosphate conditions in *Arabidopsis*, and its homologs, *BrPAP12* and *BrPAP13*, were also induced in shoots under Pi-deprivation (Figures 2 and 8) [2]. *AtPAP26* functions as a dual-targeted PAP and is the predominant intracellular APase as well as a major secreted APase in Pi limited conditions [19]. *BrPAP28*, the homolog of *AtPAP26*, was only upregulated in shoots in Pi deprivation, indicating that its function had shifted compared with *AtPAP26* (Figures 2 and 8E). Future studies exploring BrPAP responses to Pi deprivation will provide insights into the functions of different *BrPAPs* and novel gene sources for engineering crops with increased tolerance to Pi deprivation.

5. Conclusions

We systematically identified and analyzed the *BrPAP* genes from the *B. rapa* genome at the genome-wide level. A total of 39 *BrPAPs* were identified and divided into three major clades with nine subgroups. Expression profiles revealed that all *BrPAPs* were expressed in all examined tissues, thus indicating their various functions under standard conditions. Interestingly, *BrPAP11, 14, 20, 24, 29,* and *34* were specifically expressed in fertile floral buds, indicating their critical roles during pollen development. Their potential functions were further highlighted by co-expression analysis. Twenty-one *BrPAPs* responded to Pi deprivation in either shoots or roots. Further studies are required to determine the functions of *BrPAPs*. The information presented here provides a foundation for future investigations to gain a deeper understanding of the roles of each *BrPAP* as well as the molecular mechanisms underlying pollen development and low Pi tolerance.

Supplementary Materials: The following are available online at https://www.mdpi.com/article/10.3390/horticulturae7100363/s1, Table S1: List of primers used in this study, Table S2: List of genes co-expressed with *BrPAP11, 14, 20, 24, 29,* and *34*, with a Pearson's correlation coefficient (PPC) threshold value set between 0.60 and −0.60.

Author Contributions: Conceptualization, Y.C.; data curation, Y.W., J.G. and X.D.; formal analysis, Y.C.; funding acquisition, X.D.; project administration, X.D.; supervision, X.D.; validation, J.Q., C.L., K.M., B.J., X.Y. and W.H.; writing—original draft, Y.C.; writing—review and editing, J.G. and X.D. All authors have read and agreed to the published version of the manuscript.

Funding: This research was funded by the National Science Foundation of China (31601771), the Applied Basic Research Project of Yunnan (202001BB050017 and 202101AT070016), and the Yunnan University Innovation Fund for Graduate Students (2019175 and 2020324).

Acknowledgments: We thank LetPub (www.letpub.com accessed on 2 July 2021) for the linguistic assistance and pre-submission expert review.

Conflicts of Interest: The authors declare no conflict of interest. The funders had no role in the design of the study; in the collection, analyses, or interpretation of data; in the writing of the manuscript; or in the decision to publish the results.

References

1. Barford, D.; Das, A.K.; Egloff, M.-P. The structure and mechanism of protein phosphatases: Insights into catalysis and regulation. *Annu. Rev. Biophys. Biomol. Struct.* **1998**, *27*, 133–164. [CrossRef] [PubMed]
2. Li, D.; Zhu, H.; Liu, K.; Liu, X.; Leggewie, G.; Udvardi, M.; Wang, D. Purple acid phosphatases of *Arabidopsis thaliana*. Comparative analysis and differential regulation by phosphate deprivation. *J. Biol. Chem.* **2002**, *277*, 27772–27781. [CrossRef] [PubMed]
3. Schenk, G.; Korsinczky, M.L.; Hume, D.; Hamilton, S.; DeJersey, J. Purple acid phosphatases from bacteria: Similarities to mammalian and plant enzymes. *Gene* **2000**, *255*, 419–424. [CrossRef]
4. Zhu, H.; Qian, W.; Lu, X.; Li, D.; Liu, X.; Liu, K.; Wang, D. Expression patterns of purple acid phosphatase genes in *Arabidopsis* organs and functional analysis of *AtPAP23* predominantly transcribed in flower. *Plant Mol. Biol.* **2005**, *59*, 581–594. [CrossRef] [PubMed]
5. Tran, H.T.; Hurley, B.A.; Plaxton, W.C. Feeding hungry plants: The role of purple acid phosphatases in phosphate nutrition. *Plant Sci.* **2010**, *179*, 14–27. [CrossRef]
6. Xiao, K.; Katagi, H.; Harrison, M.; Wang, Z.-Y. Improved phosphorus acquisition and biomass production in *Arabidopsis* by transgenic expression of a purple acid phosphatase gene from *M. truncatula*. *Plant Sci.* **2006**, *170*, 191–202. [CrossRef]
7. Kong, Y.; Li, X.; Wang, B.; Li, W.; Du, H.; Zhang, C. The soybean purple acid phosphatase *GmPAP14* predominantly enhances external phytate utilization in plants. *Front. Plant Sci.* **2018**, *9*, 292. [CrossRef] [PubMed]
8. Zhu, S.; Chen, M.; Liang, C.; Xue, Y.; Lin, S.; Tian, J. Characterization of purple acid phosphatase family and functional analysis of *GmPAP7a/7b* involved in extracellular ATP utilization in soybean. *Front. Plant Sci.* **2020**, *11*, 661. [CrossRef]
9. Xie, L.; Shang, Q. Genome-wide analysis of purple acid phosphatase structure and expression in ten vegetable species. *BMC Genom.* **2018**, *19*, 646. [CrossRef]
10. Lu, K.; Chai, Y.R.; Zhang, K.; Wang, R.; Chen, L.; Lei, B.; Lu, J.; Xu, X.F.; Li, J.N. Cloning and characterization of phosphorus starvation inducible *Brassica napus* PURPLE ACID PHOSPHATASE 12 gene family, and imprinting of a recently evolved MITE-minisatellite twin structure. *Theor. Appl. Genet.* **2008**, *117*, 963–975. [CrossRef]
11. Zhang, Q.; Wang, C.; Tian, J.; Li, K.; Shou, H. Identification of rice purple acid phosphatases related to posphate starvation signalling. *Plant Biol.* **2010**, *13*, 7–15. [CrossRef]
12. Gonzalez-Munoz, E.; Avendano-Vazquez, A.O.; Montes, R.A.; De Folter, S.; Andres-Hernandez, L.; Abreu-Goodger, C.; Sawers, R.J. The maize (Zea mays ssp. mays var. B73) genome encodes 33 members of the purple acid phosphatase family. *Front. Plant Sci.* **2015**, *6*, 341. [CrossRef]
13. Bhadouria, J.; Singh, A.P.; Mehra, P.; Verma, L.; Srivastawa, R.; Parida, S.K.; Giri, J. Identification of purple acid phosphatases in chickpea and potential roles of *CaPAP7* in seed phytate accumulation. *Sci. Rep.* **2017**, *7*, 11012. [CrossRef]
14. Venkidasamy, B.; Selvaraj, D.; Ramalingam, S. Genome-wide analysis of purple acid phosphatase (PAP) family proteins in *Jatropha curcas* L. *Int. J. Biol. Macromol.* **2019**, *123*, 648–656. [CrossRef]
15. Yin, C.; Wang, F.; Fan, H.; Fang, Y.; Li, W. Identification of tea plant purple acid phosphatase genes and their expression responses to excess iron. *Int. J. Mol. Sci.* **2019**, *20*, 1954. [CrossRef]
16. Schenk, G.; Mitić, N.; Hanson, G.R.; Comba, P. Purple acid phosphatase: A journey into the function and mechanism of a colorful enzyme. *Coord. Chem. Rev.* **2013**, *257*, 473–482. [CrossRef]
17. Wang, L.; Li, Z.; Qian, W.; Guo, W.; Gao, X.; Huang, L.; Wang, H.; Zhu, H.; Wu, J.W.; Wang, D.; et al. The *Arabidopsis* purple acid phosphatase *AtPAP10* is predominantly associated with the root surface and plays an important role in plant tolerance to phosphate limitation. *Plant Physiol.* **2011**, *157*, 1283–1299. [CrossRef] [PubMed]
18. Kuang, R.; Chan, K.H.; Yeung, E.; Lim, B.L. Molecular and biochemical characterization of AtPAP15, a purple acid phosphatase with phytase activity, in *Arabidopsis*. *Plant Physiol.* **2009**, *151*, 199–209. [CrossRef] [PubMed]
19. Tran, H.T.; Qian, W.; Hurley, B.A.; She, Y.M.; Wang, D.; Plaxton, W.C. Biochemical and molecular characterization of *AtPAP12* and *AtPAP26*: The predominant purple acid phosphatase isozymes secreted by phosphate-starved *Arabidopsis thaliana*. *Plant Cell Environ.* **2010**, *33*, 1789–1803. [CrossRef] [PubMed]

20. Ghahremani, M.; Park, J.; Anderson, E.M.; Marty-Howard, N.J.; Mullen, R.T.; Plaxton, W.C. Lectin *AtGAL1* interacts with high-mannose glycoform of the purple acid phosphatase *AtPAP26* secreted by phosphate-starved *Arabidopsis*. *Plant Cell Environ.* **2019**, *42*, 1158–1166. [CrossRef]
21. Tian, J.; Wang, C.; Zhang, Q.; He, X.; Whelan, J.; Shou, H. Overexpression of *OsPAP10a*, a root-associated acid phosphatase, increased extracellular organic phosphorus utilization in rice. *J. Integr. Plant Biol.* **2012**, *54*, 631–639. [CrossRef] [PubMed]
22. Lu, L.; Qiu, W.; Gao, W.; Tyerman, S.D.; Shou, H.; Wang, C. *OsPAP10c*, a novel secreted acid phosphatase in rice, plays an important role in the utilization of external organic phosphorus. *Plant Cell Environ.* **2016**, *39*, 2247–2259. [CrossRef] [PubMed]
23. Mehra, P.; Pandey, B.K.; Giri, J. Improvement in phosphate acquisition and utilization by a secretory purple acid phosphatase (*OsPAP21b*) in rice. *Plant Biotechnol. J.* **2017**, *15*, 1054–1067. [CrossRef] [PubMed]
24. Deng, S.; Lu, L.; Li, J.; Du, Z.; Liu, T.; Li, W.; Xu, F.; Shi, L.; Shou, H.; Wang, C. Purple acid phosphatase 10c encodes a major acid phosphatase that regulates plant growth under phosphate-deficient conditions in rice. *J. Exp. Bot.* **2020**, *71*, 4321–4332. [CrossRef] [PubMed]
25. Zhang, Y.; Sun, F.; Fettke, J.; Schottler, M.A.; Ramsden, L.; Fernie, A.R.; Lim, B.L. Heterologous expression of *AtPAP2* in transgenic potato influences carbon metabolism and tuber development. *FEBS Lett.* **2014**, *588*, 3726–3731. [CrossRef] [PubMed]
26. Kaida, R.; Satoh, Y.; Bulone, V.; Yamada, Y.; Kaku, T.; Hayashi, T.; Kaneko, T.S. Activation of beta-glucan synthases by wall-bound purple acid phosphatase in tobacco cells. *Plant Physiol.* **2009**, *150*, 1822–1830. [CrossRef] [PubMed]
27. Liao, H.; Wong, F.-L.; Phang, T.-H.; Cheung, M.-Y.; Li, W.-Y.F.; Shao, G.; Yan, X.; Lam, H.-M. *GmPAP3*, a novel purple acid phosphatase-like gene in soybean induced by NaCl stress but not phosphorus deficiency. *Gene* **2003**, *318*, 103–111. [CrossRef]
28. Lan, P.; Li, W.; Schmidt, W. Genome-wide co-expression analysis predicts protein kinases as important regulators of phosphate deficiency-induced root hair remodeling in *Arabidopsis*. *BMC Genom.* **2013**, *14*, 210. [CrossRef]
29. Hoagland, D.R.; Arnon, D.I. The water-culture method for growing plants without soil. *Circular. Calif. Agric. Exp. Stn.* **1950**, *347*, 32.
30. Dong, X.; Feng, H.; Xu, M.; Lee, J.; Kim, Y.K.; Lim, Y.P.; Piao, Z.; Park, Y.D.; Ma, H.; Hur, Y. Comprehensive analysis of genic male sterility-related genes in *Brassica rapa* using a newly developed Br300K oligomeric chip. *PLoS ONE* **2013**, *8*, e72178. [CrossRef]
31. Miao, X.; Miao, Y.; Gong, H.; Tao, S.; Chen, Y.; Chen, Z.; Liao, W. Digestion methods for determining phosphorus content in plants. *Chin. Agric. Sci. Bull.* **2019**, *35*, 132–137.
32. Zhang, L.; Cai, X.; Wu, J.; Liu, M.; Grob, S.; Cheng, F.; Liang, J.; Cai, C.; Liu, Z.; Liu, B.; et al. Improved *Brassica rapa* reference genome by single-molecule sequencing and chromosome conformation capture technologies. *Hortic. Res.* **2018**, *5*, 50. [CrossRef]
33. Edgar, R.C. MUSCLE: Multiple sequence alignment with high accuracy and high throughput. *Nucleic Acids Res.* **2004**, *32*, 1792–1797. [CrossRef]
34. Tamura, K.; Stecher, G.; Peterson, D.; Filipski, A.; Kumar, S. MEGA6: Molecular evolutionary genetics analysis version 6.0. *Mol. Biol. Evol.* **2013**, *30*, 2725–2729. [CrossRef] [PubMed]
35. Wilkins, M.R.; Gasteiger, E.; Bairoch, A.; Sanchez, J.C.; Williams, K.L.; Appel, R.D.; Hochstrasser, D.F. Protein identification and analysis tools in the ExPASy server. *Methods Mol. Biol.* **1999**, *112*, 531–552. [CrossRef] [PubMed]
36. Bailey, T.L.; Boden, M.; Buske, F.A.; Frith, M.; Grant, C.E.; Clementi, L.; Ren, J.; Li, W.W.; Noble, W.S. MEME SUITE: Tools for motif discovery and searching. *Nucleic Acids Res.* **2009**, *37*, w202–w208. [CrossRef] [PubMed]
37. Armenteros, J.J.A.; Tsirigos, K.D.; Sonderby, C.K.; Petersen, T.N.; Winther, O.; Brunak, S.; Von Heijne, G.; Nielsen, H.B. SignalP 5.0 improves signal peptide predictions using deep neural networks. *Nat. Biotechnol.* **2019**, *37*, 420–423. [CrossRef] [PubMed]
38. Hu, B.; Jin, J.; Guo, A.Y.; Zhang, H.; Luo, J.; Gao, G. GSDS 2.0: An upgraded gene feature visualization server. *Bioinformatics* **2015**, *31*, 1296–1297. [CrossRef] [PubMed]
39. Tong, C.; Wang, X.; Yu, J.; Wu, J.; Li, W.; Huang, J.; Dong, C.; Hua, W.; Liu, S. Comprehensive analysis of RNA-seq data reveals the complexity of the transcriptome in *Brassica rapa*. *BMC Genom.* **2013**, *14*, 689. [CrossRef]
40. Liu, C.; Liu, Z.; Li, C.; Zhang, Y.; Feng, H. Comparative transcriptome analysis of fertile and sterile buds from a genetically male sterile line of Chinese cabbage. *Vitr. Cell. Dev. Biol. Plant* **2016**, *52*, 130–139. [CrossRef]
41. Zhou, X.; Liu, Z.; Ji, R.; Feng, H. Comparative transcript profiling of fertile and sterile flower buds from multiple-allele-inherited male sterility in Chinese cabbage (*Brassica campestris* L. ssp. pekinensis). *Mol. Genet. Genom.* **2017**, *292*, 967–990. [CrossRef]
42. Lee, Y.H.; Kim, K.S.; Lee, J.E.; Cha, Y.L.; Moon, Y.H.; Song, Y.S.; Jeong, E.G.; Ahn, S.J.; Park, W. Comprehensive transcriptome profiling in relation to seed storage compounds in tetralocular *Brassica rapa*. *J. Plant Growth Regul.* **2018**, *37*, 867–882. [CrossRef]
43. Shen, X.; Xu, L.; Liu, Y.; Dong, H.; Zhou, D.; Zhang, Y.; Lin, S.; Cao, J.; Huang, L. Comparative transcriptome analysis and ChIP-sequencing reveals stage-specific gene expression and regulation profiles associated with pollen wall formation in *Brassica rapa*. *BMC Genom.* **2019**, *20*, 264. [CrossRef] [PubMed]
44. Huang, S.; Peng, S.; Liu, Z.; Li, C.; Tan, C.; Yao, R.; Li, D.; Li, X.; Hou, L.; Feng, H. Investigation of the genes associated with a male sterility mutant (msm) in Chinese cabbage (*Brassica campestris* ssp. pekinensis) using RNA-Seq. *Mol. Genet. Genom.* **2020**, *295*, 233–249. [CrossRef] [PubMed]
45. Blankenberg, D.; Gordon, A.; Von Kuster, G.; Coraor, N.; Taylor, J.; Nekrutenko, A.; Galaxy, T. Manipulation of FASTQ data with Galaxy. *Bioinformatics* **2010**, *26*, 1783–1785. [CrossRef]
46. Langmead, B.; Salzberg, S.L. Fast gapped-read alignment with Bowtie 2. *Nat. Methods* **2012**, *9*, 357–359. [CrossRef] [PubMed]
47. Trapnell, C.; Williams, B.A.; Pertea, G.; Mortazavi, A.; Kwan, G.; Van Baren, M.J.; Salzberg, S.L.; Wold, B.J.; Pachter, L. Transcript assembly and quantification by RNA-Seq reveals unannotated transcripts and isoform switching during cell differentiation. *Nat. Biotechnol.* **2010**, *28*, 511–515. [CrossRef] [PubMed]

48. Saeed, A.I.; Bhagabati, N.K.; Braisted, J.C.; Liang, W.; Sharov, V.; Howe, E.A.; Li, J.; Thiagarajan, M.; White, J.A.; Quackenbush, J. TM4 microarray software suite. *Methods Enzymol.* **2006**, *411*, 134–193. [CrossRef]
49. Tian, T.; Liu, Y.; Yan, H.; You, Q.; Yi, X.; Du, Z.; Xu, W.; Su, Z. agriGO v2.0: A GO analysis toolkit for the agricultural community, 2017 update. *Nucleic Acids Res.* **2017**, *45*, W122–W129. [CrossRef]
50. Yu, G.; Wang, L.G.; Han, Y.; He, Q.Y. clusterProfiler: An R package for comparing biological themes among gene clusters. *OMICS* **2012**, *16*, 284–287. [CrossRef]
51. Li, C.; Gui, S.; Yang, T.; Walk, T.; Wang, X.; Liao, H. Identification of soybean purple acid phosphatase genes and their expression responses to phosphorus availability and symbiosis. *Ann. Bot.* **2011**, *109*, 275–285. [CrossRef] [PubMed]
52. George, T.; Richardson, A. Potential and limitations to improving crops for enhanced phosphorus utilization. In *The Ecophysiology of Plant-Phosphorus Interactions*; Springer: Dordrecht, The Netherlands, 2008; pp. 247–270.
53. Sun, F.; Liang, C.; Whelan, J.; Yang, J.; Zhang, P.; Lim, B.L. Global transcriptome analysis of *AtPAP2*-overexpressing *Arabidopsis thaliana* with elevated ATP. *BMC Genom.* **2013**, *14*, 752. [CrossRef] [PubMed]
54. Hurley, B.A.; Tran, H.T.; Marty, N.J.; Park, J.; Snedden, W.A.; Mullen, R.T.; Plaxton, W.C. The dual-targeted purple acid phosphatase isozyme *AtPAP26* is essential for efficient acclimation of *Arabidopsis* to nutritional phosphate deprivation. *Plant Physiol.* **2010**, *153*, 1112–1122. [CrossRef]
55. Zhang, Y.; Wang, X.; Lu, S.; Liu, D. A major root-associated acid phosphatase in *Arabidopsis*, *AtPAP10*, is regulated by both local and systemic signals under phosphate starvation. *J. Exp. Bot.* **2014**, *65*, 6577–6588. [CrossRef] [PubMed]

horticulturae

MDPI

Article

Marker-Assisted Pyramiding of Genes for Multilocular Ovaries, Self-Compatibility, and Clubroot Resistance in Chinese Cabbage (*Brassica rapa* L. ssp. *pekinensis*)

Jingyi Zheng [1], Huicai Zhao [1], Yingmei Ma [1], Mingliang Jiang [2], Zongxiang Zhan [1], Xiaonan Li [1,*] and Zhongyun Piao [1,*]

[1] College of Horticulture, Shenyang Agricultural University, Shenyang 110866, China; amberz0301@163.com (J.Z.); zjxsbd0126@163.com (H.Z.); mayingmei0126@163.com (Y.M.); zhanzxiang@126.com (Z.Z.)
[2] School of Agriculture, Jilin Agricultural Science and Technology College, Jilin 132101, China; jml005@163.com
* Correspondence: Gracesleexn@163.com (X.L.); zypiao@syau.edu.cn (Z.P.); Tel.: +86-42-8848-7143 (Z.P.)

Abstract: Molecular marker-assisted gene pyramiding combined with backcrossing has been widely applied for crop variety improvement. Molecular marker identification could be used in the early stage of breeding to achieve the rapid and effective pyramiding of multiple genes. To create high-quality germplasm for Chinese cabbage breeding, multi-gene pyramiding for self-compatibility, multilocular, and clubroot resistance was performed through molecular marker-assisted selection. The results showed that self-compatibility and multilocular traits were controlled by a pair of recessive genes. Two flanking markers, sau_um190 and cun_246a, and marker Teo-1, based on the gene sequence related to multilocular ovaries, were used for multilocular ovary trait selection. Two flanking markers, SCF-6 and SC-12, and marker Sal-SLGI /PK1+PK4, based on the gene sequence, were used for self-compatibility selection. Two flanking markers, TCR74 and TCR79, closely linked to clubroot resistance gene *CRb*, were used as foreground selection markers. Based on Chinese cabbage genomic information, 111 SSR markers covering 10 chromosomes were applied for background selection. After multiple generations of selection, a multi-gene pyramided line from a BC_4F_2 population with self-compatibility, multilocular ovaries, and clubroot resistance was obtained with a high genomic background recovery rate. The improved pyramided line is expected to be utilized as a potential material in further breeding programs.

Keywords: *Brassica rapa*; gene pyramiding; marker-assisted selection

Citation: Zheng, J.; Zhao, H.; Ma, Y.; Jiang, M.; Zhan, Z.; Li, X.; Piao, Z. Marker-Assisted Pyramiding of Genes for Multilocular Ovaries, Self-Compatibility, and Clubroot Resistance in Chinese Cabbage (*Brassica rapa* L. ssp. *pekinensis*). *Horticulturae* **2022**, *8*, 139. https://doi.org/10.3390/horticulturae8020139

Academic Editor: Luigi De Bellis

Received: 20 December 2021
Accepted: 30 January 2022
Published: 6 February 2022

Publisher's Note: MDPI stays neutral with regard to jurisdictional claims in published maps and institutional affiliations.

Copyright: © 2022 by the authors. Licensee MDPI, Basel, Switzerland. This article is an open access article distributed under the terms and conditions of the Creative Commons Attribution (CC BY) license (https://creativecommons.org/licenses/by/4.0/).

1. Introduction

Chinese cabbage (*Brassica rapa* L. ssp. *pekinensis*), belonging to the *Brassica* subspecies of the Brassicaceae family, is widely cultivated in China, Korea, and Japan. At present, most of the common commercial cultivars on the market are first-generation hybrids. However, owing to self-incompatibility, it is difficult to reproduce the parent lines, and the cost of artificial seed production is high. The seed yield of Chinese cabbage is limited, which is influenced by silique-related traits. Additionally, clubroot, as one of the main diseases of cruciferous crops, has severely threatened the production of Chinese cabbage. Breeding clubroot-resistant (CR) varieties is the most effective method for clubroot prevention since physical and biological control methods have limited effects. Therefore, cultivar pyramiding with multilocular silique, self-compatibility, and clubroot resistance would be an ideal resource for Chinese cabbage breeding.

Cruciferous plants, such as Brassica cultivars included in the U's triangle [1], are mostly bilocular plants, including *Brassica rapa* (AA, 2n = 20), *Brassica nigra* (BB, 2n = 16), *Brassica oleracea* (CC, 2n = 18), *Brassica napus* (AACC, 2n = 38), *Brassica juncea* (AABB, 2n = 36), and *Brassica carinata* (BBCC, 2n = 34). However, with the collection and sorting of germplasm resources, some silique variation types, such as multilocular rapeseed, have been discovered. Among them, yellow sarson, an oil-type *B. rapa*, has the characteristics of multiple ovaries. Its number of silique ventricles is four, and the seed number is significantly higher than that of Chinese cabbage. Zhao et al. [2] used AFLP markers to identify the developmental relationship of 161 materials from different *B. rapa* subspecies and demonstrated that yellow sarson is an independent group that has a relatively distant relationship with other subspecies. Other studies have shown that multilocular traits can increase the yield of seeds, and the width of the silique is positively correlated with thousand-seed weight [3]. Yadava et al. [4] found that the multilocular trait and wider silique width of yellow sarson may be due to the *Bra034340* gene mutation in Chinese cabbage that is homologous to *Arabidopsis CLAVATA3*.

In the process of plant evolution, the self-incompatibility mechanism evolved [5]. Self-incompatibility is divided into sporophytic self-incompatibility (SSI) and gametophytic self-incompatibility (GSI) [6]. Brassica plants have sporophytic self-incompatibility, which is genetically controlled by an S-locus with multiple alleles called the S-haplotype [7,8]. There are three linked genes at the S-locus, namely S-locus glycoprotein (*SLG*), S-locus receptor kinase (*SRK*) [9–11], and S-locus cysteine-rich protein (*SCR*) or the S-locus protein 11 (*SP11*) [12,13]. The stigma localized SRK interact with *SP11/SCR* in the S-locus of Brassica plants to recognize self-pollen and cross-pollen. This interaction is based on the specificity of the S haplotype; that is, only *SRK* and *SP11/SCR* of the same haplotype can interact with each other, which causes self-incompatibility [14–16]. Owing to this feature, Chinese cabbage seed production requires breaking buds and selfing, which wastes a lot of manpower and increases the cost. Thus, the introduction of self-compatibility genes into Chinese cabbage would be of great significance for self-fertilization breeding and preservation of important germplasm resources.

Clubroot disease is an obligate parasitic soil-borne disease caused by *Plasmodiophora brassicae* Woron. It mainly harms cruciferous crops and is distributed in most countries and regions around the world [17]. It is estimated that 3 to 4 million hectares of farmland in China are threatened by this disease every year, causing 20–30% yield losses [18,19]. However, it is difficult to prevent and control the disease through physical and chemical methods [20,21] because the resting spores of *P. brassicae* can survive in the soil for many years [22]. At present, breeding-resistant varieties are the most effective and economical prevention methods. To date, several CR loci have been identified by genetic mapping/quantitative trait locus (QTL) mapping in *B. rapa* using different resistance resources, including *Crr1, Crr2, Crr3, Crr4, CRa, CRb, CRc, CRd, CRk, PbBa3.1, PbBa3.2. PbBa8.1, Rcr1, Rcr4*, and *Rcr9* [23–33]. These CR loci and their corresponding linkage markers enable the introgression of a single CR gene or pyramiding of multiple CR genes into one variety through marker-assisted selection (MAS).

The marker-assisted pyramiding strategy that introgresses multiple QTL/genes for one or multiple traits has been widely used in several crops such as soybean (multiple *Rpp* genes for Asian soybean rust resistance) [34] and rice (*Bph27* (*t*) and *Bph3* for brown planthopper resistance [35]; *R* genes for blast-resistance [36]). In Brassica crops, Shah et al. [37] developed a new gene pyramided *Brassica napus* line by combining two clubroot resistant genes, *CRb* and *PbBa8.1*, which showed strong resistant to *P. brassicae* field isolates. Matsumoto et al. [38] accumulated three clubroot resistant genes (*CRa, CRk*, and *CRc*) through MAS in Chinese cabbage. Accumulation of multiple CR-QTL in *B. oleracea* conferred broad-spectrum clubroot resistance against six *P. brassicae* isolates [39]. The above reports were mostly based on the pyramiding of different genes for one trait. In common wheat, eight genes for seven different traits were introgressed into one cultivar. The improved pyramided lines exhibited resistance against three rusts and excellent quality [40]. However, at present, there are fewer reports on gene pyramiding in Chinese cabbage. In this study, we pyramided genes for multilocular, self-compatibility, and clubroot resistance into one material using MAS. The result provides new insights into the creation and breeding of new Chinese cabbage resources.

2. Materials and Methods

2.1. Plant Material

Two *B. rapa* inbred lines, namely, 'CR BJN3-2' and 'KYS,' were used in this study. CR BJN3-2, a near-isogenic line of Chinese cabbage 'BJN3-2,' which is susceptible to clubroot disease and carries the resistance allele *CRb* [28], was used as a recurrent parent [41]. It was a bilocular and self-incompatible material. The donor, parental line KYS (*B. rapa* L. var. yellow sarson), is an oil-type *B. rapa* with tetralocular ovaries, self-compatibility, yellow seed coat, and clubroot susceptibility.

2.2. Marker Development, Validation, and Phenotype Genetic Analysis

Yadava et al. [4] identified a major QTL *tet-o* for tetralocular ovary for *B. rapa* and predicted *BrCLAVATA3* (*Bra034340*) as the candidate gene. Simple sequence repeats (SSRs) were searched based on the *BrCLAVATA3* gene sequence using SSRHunter 1.3, and two SSR markers were designed for polymorphism screening between two parental lines (Table S1). The PCR reaction and amplification conditions for genotyping with SSR markers has been previously described by Li et al. [42] and Ge et al. [43].

For self-compatibility-related genes, we used multiplex PCR to screen co-dominant marker combinations based on multiple reported primers designed based on *SRK*, *SLG*, and *SP11* genes and their genomic regions (Table S2). PCR amplification of genomic DNA using a 20 μL reaction volume containing 3 μL of genomic DNA, 1 μL of each of the forward and reverse primer, 10 μL of 10 × PCR buffer, and 5 μL of ddH2O to complete the reaction volume. PCR products were detected using 2% agarose gel electrophoresis. Additionally, four SSR and one InDel marker reported by Zhang et al. [44] were used for polymorphism screening of the *CRb* gene. All polymorphic markers were then tested for linkage in the F_2 population of CR BJN3-2 and KYS. Chi-square fitness was used to detect the separation of phenotype and genotype.

2.3. Crossing Program and MAS for Gene Pyramiding

To introduce the multilocular and self-compatibility genes carried by KYS into the clubroot disease-resistant line CR BJN3-2, we designed a set of backcross breeding programs. The two paternal lines were crossed to produce the F_1 hybrid, which was then backcrossed to produce BC_4F_1, using CR BJN3-2 as the recurrent parent, accompanied by MAS in each generation. Subsequently, BC_4F_2 was generated by selfing the selected BC_4F_1 lines (Figure 1). Finally, the new pyramided line (BC_4F_3) with high genetic background recovery of the recurrent parent CR BJN3-2, high rate of tetralocular, self-compatibility, and clubroot resistance was obtained.

The screening method for the BC_nF_n generation is detailed in Figure 1, which mainly included the foreground marker selection of the target genes, background marker selection of the genome, and phenotypic screening. The marker information used for target gene foreground selection is shown in Table 1. For genetic background selection, SSR markers developed previously [45,46] and distributed on 10 chromosomes of the Chinese cabbage genome were used to determine polymorphisms. The polymorphic markers were used in subsequent genomic background screening.

In BC_1F_1, low-density markers were used for background selection, and plants with higher recovery rates were selected for backcrossing. In the BC_3F_1 generation, we increased the background selection marker density to have a clearer understanding of the genetic composition of the background and select individuals with a higher genomic background recovery rate.

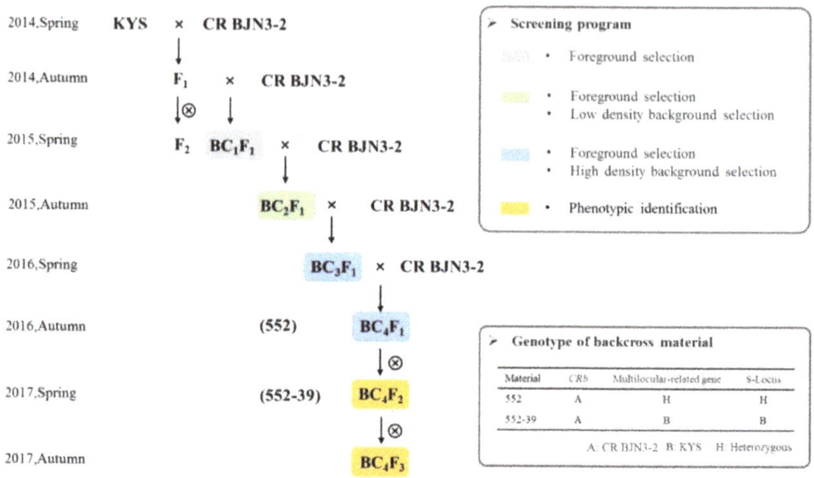

Figure 1. Schematic illustration of marker-assisted gene pyramiding in this study. CR BJN3-2: recurrent parent; KYS: donor parent; 552: selected BC_4F_1 plant; 552-39: selected pyramided line. A: genotype of CR BJN3-2; B: genotype of KYS; H: heterozygous genotype.

Table 1. Molecular marker information for foreground selection of three targeted loci.

Gene	Chromosomal Location	Markers	Type of Marker	Character
Bra034340	A4	sau_um190	Based on PCR	Co-dominant
		cun_146a	Based on PCR	Co-dominant
		Teo-1	SSR	Co-dominant
S-Locus	A7	PK1+PK4/Sal-SLGI	Based on PCR	Co-dominant
		SCF-6	SSR	Co-dominant
		SC-12	SSR	Co-dominant
CRb	A3	TCR74	SSR	Co-dominant
		TCR79	SSR	Co-dominant

2.4. Phenotypic Characterization

P. brassicae pathotype 4, defined by Williams' system [47] in a previous study [44], was used to test clubroot resistance in the parental and pyramided lines. The isolate was propagated in the susceptible Chinese cabbage lines as described by Piao et al. [28]. The swollen infected roots, which were soaked in water, were blended into solution and then filtrated with gauze. The filtrate was collected in a 50-mL centrifuge tube, dissolved in sterile water, and finally stored in a refrigerator at 4 °C. Resting spores were adjusted to the concentration of 10^7 spores/mL for inoculation by injecting the soil around thirty-day-old plants. The disease resistance of the plants was investigated at 45 days after infection. If the plant grew normally and there were no visible clubs in the main roots and fibrous roots, it was recorded as disease resistant and scored as level 0. A few small swollen clubs on the lateral roots indicated level 1. If the lateral roots or main root had larger clubs, it was scored as level 2, and significant taproot swelling was classed as level 3.

The self-compatibility of plants was determined by the self-compatibility index (SCI), which was evaluated as the ratio of the total number of seeds to the number of pods in self-pollinated flowers. The self-incompatibility type was defined as SCI less than 1, and the self-compatibility type was defined as SCI greater than 1.

The agronomic traits of Chinese cabbage during the harvest period, including plant height, plant width, plant weight, number of outer leaves, leaf length, leaf width, petiole length, petiole width, petiole color, head weight, head shape, head solidity, head length, head width, head color, stem length, and stem width, as described by Ge et al. [43], were investigated in pyramided and recurrent parental lines. The t-test analysis was performed between pyramided line and recurrent parental line at $p \leq 0.05$ significant level.

3. Results

3.1. Marker Development and Genetic Analysis for Multilocular and Self-Compatibility Traits

Two SSR markers, Teo-1 and Teo-2, based on the candidate gene *BrCLAVATA3* for multilocular, were designed (Table S1). Both markers showed polymorphism between CR BJN3-2 and KYS. Teo-1 with more motif repeats was selected as a linked marker for further analysis.

Among eight pairs of marker combinations designed for self-compatibility, as shown in Table S2, PCR bands were amplified only in KYS using PK1+PK4 and in CR BJN3-2 using Sal-SRKI and Sal-SLGI. PCR products were not detected using PS3+PS21, Sal-SLGII, Sal-SRKII, and SP11II between the two parents (Figure S1). To develop a codominant marker for self-compatibility, we conducted multiplex PCR by combining the PK1+PK4 primer with Sal-SRKI and Sal-SLGI. As shown in Figure 2, the PK1+PK4/Sal-SLGI primer combination could be used as a co-dominant marker to identify self-compatibility.

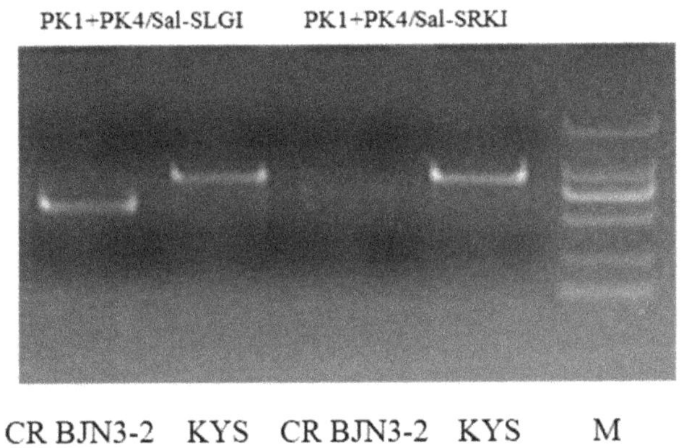

Figure 2. Multiplex PCR showing polymorphism of two pairs of primer combinations (PK1+PK4/Sal-SLGI; PK1+PK4/Sal-SRKI) between two parents (CR BJN3-2 and KYS). M: The DNA ladder 2000.

To perform genetic analysis and verify the molecular markers for locular and self-compatibility phenotypes, the F_1 population was generated by reciprocal crossing of CR BJN3-2 and KYS, and the F_2 population was constructed by selfing the F_1 generation. The F_1 generations showed bilocular and self-incompatibility, which indicated that the inheritance of multilocular and self-compatibility traits was not affected by a reciprocal cross. Both multilocular and self-incompatibility traits had no cytoplasmic effect and were genetically controlled by recessive nuclear genes.

Furthermore, a total of 204 F_2 individual plant phenotypes were investigated. Ventricular traits were segregated among the F_2 population. Of the individuals, 164 had bilocular ovaries and 40 had multilocular ovary (trilocular and tetralocular). The separation ratio of bilocular to multilocular was 3:1 according to the chi-square test (χ^2 = 2.8824), indicating that the ventricular traits are genetic traits controlled by a pair of genes. Moreover, the genotype of the F_2 population using the Teo-1 marker was consistent with the ventricular phenotype (Figure 3), which could be used as a linkage marker for further MAS.

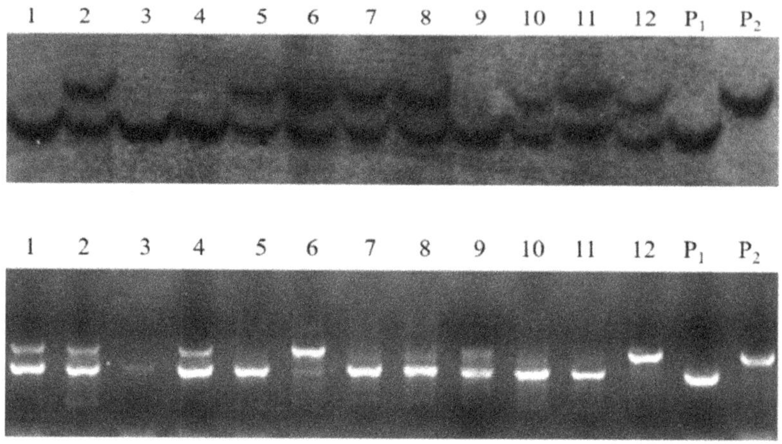

Figure 3. Genotype of partial individuals of F_2 population for locular number (up) and self-compatibility (down). P_1: CR BJN3-2; P_2: KYS; 1-12: F_2 individual plant number.

In addition, among 204 F_2 plants, 149 individual plants showed self-incompatibility and 55 individuals showed self-compatibility. The ratio of self-compatible to self-incompatible plants was close to 1:3 by the chi-square fitness test (χ^2 = 0.3203), indicating that self-compatibility is controlled by a pair of genes. The genotype of 204 F_2 plants detected by the newly developed marker PK1+PK4/Sal-SLGI was also identical to this phenotype (Figure 3).

3.2. Polymorphic Marker Screening for Foreground Selection

For clubroot resistance, five *CRb* gene linked markers developed by Zhang et al. [44] were used for polymorphism detection, as shown in Table S1. TCR74 and TCR79, which were closely linked with *CRb*, showed polymorphism between two parental lines, which were selected as foreground markers. For self-compatibility, 12 SSR markers were developed in the chromosome region of SI-related genes SRK, SLG, and SP11 (Table S1). Among them, SCF-6 and SC-12 showed polymorphism between CR BJN3-2 and KYS. Thus, SCF-6, SC-12, and gene-based PK1+PK4/Sal-SLGI (Figure 2) were used as foreground selection markers, co-segregating with self-compatibility. Two flanking SSR markers of *BrCLAVATA3*, sau_um190, and cun_146a, showing polymorphism between CR BJN3-2, KYS, and Teo-1, were used as selection markers for the multilocular trait.

All markers used in this study were co-dominant markers that could clearly distinguish heterozygous and homozygous plants. Detailed information for these markers is shown in Table 1.

3.3. Polymorphic Markers Screening for Background Selection of the Whole Genome

Among 1140 genomic SSR markers [45,46], 319 pairs showed polymorphism between two parents, and the polymorphism ratio was28.0%. For background selection of each generation, 111 pairs of markers were chosen. The numbers of these 111 background selection markers distributed on 10 *B. rapa* chromosomes were 10, 9, 11, 12, 11, 11, 12, 10, 13, and 12, respectively. The average interval of these markers was 2.25 Mb, and the total physical distance covered was 249.27 Mb (Table S3). The distribution of these markers on the 10 chromosomes is shown in Figure 4.

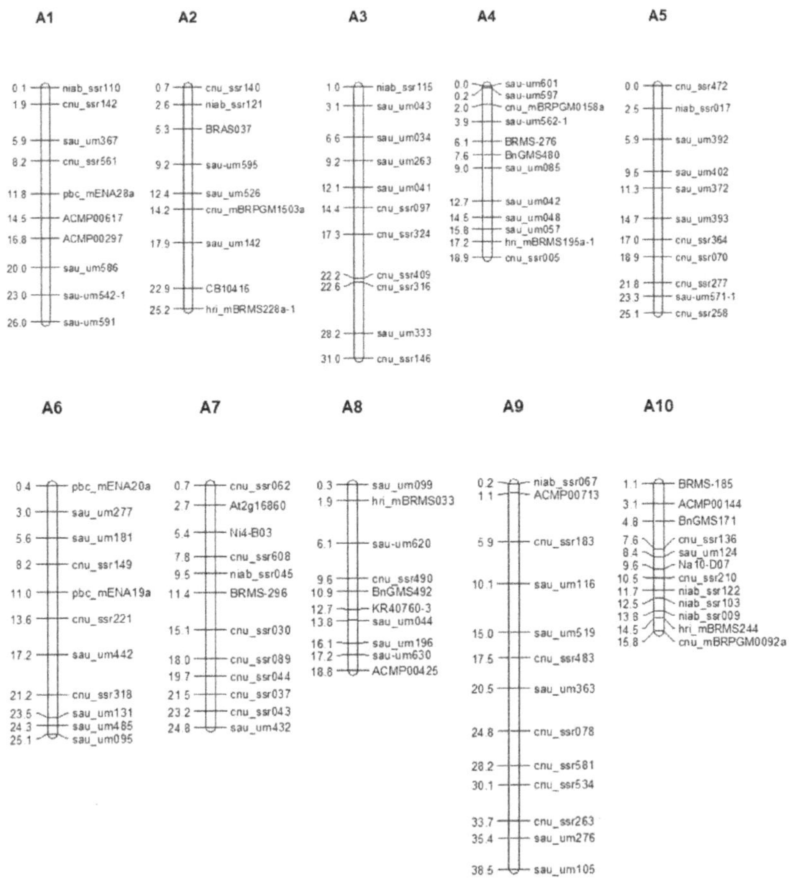

Figure 4. Distribution of 111 background selection markers on 10 *B. rapa* chromosomes (A1–A10).

3.4. Marker-Assisted Pyramiding of Multilocular, Self-Compatibility, and CRb Genes

Using the co-segregation markers for multilocular, self-compatibility, and *CRb* genes, foreground selection was exercised in each generation of backcross hybrids from BC_1F_1 to BC_4F_1 (Figure 1). Among 100 BC_1F_1, 9 plants that carried all 3 loci in heterozygous condition were selected to perform background genomic screening in low marker density (37 SSR markers). Four plants with an average of 51% genomic recovery were used to raise BC_2F_1. Ten plants among 276 BC_2F_1, which were "positive" in three loci, were used for background selection using 21 markers that were not recovered in recurrent parental genotypes. Four plants with an average genomic recovery of 74.12% were backcrossed to generate the BC_3F_1 progeny. In the BC_3F_1 generation, 10 of the 300 plants were found to be heterozygous for three targeted loci. A high marker density background assay revealed that the genomic recovery of these 10 plants ranged from 78.10 to 86.67%, and these plants were backcrossed to produce BC_4F_1. In BC_4F_1, we found that 10 of 154 plants were triple heterozygous, and their background recovery ranged from 88.18 to 94.55% (Figure 5). Four well-growing plants with high recovery were then self-pollinated to generate the BC_4F_2 population. Ten homozygous BC_4F_2 plants for three target loci were propagated for BC_4F_3. One best line with 95% recovery, namely, '552-39,' was finally selected for the phenotypic assay.

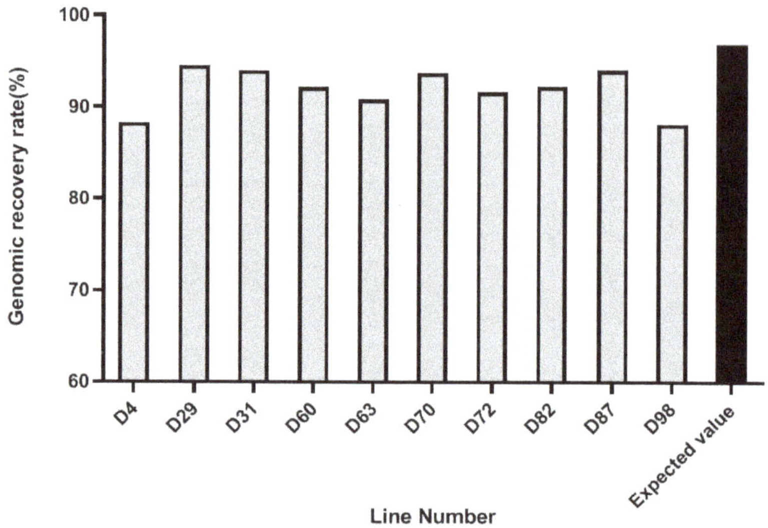

Figure 5. Genomic background recovery rate of ten BC_4F_1 individual plants (D4–D98). Black bar chart represented the expected value of genomic recovery rate of BC_4F_1 generation.

3.5. Phenotypic Evaluation of a Pyramided Line

The clubroot-resistant parent CR BJN3-2, possessing the *CRb* gene, exhibited high resistance to *P. brassicae* pathotype 4, and the negative control, BJN3-2, was susceptible. The pyramided line 552-39 showed high resistance with a low disease incidence rate and disease index value (Figure 6, Table 2).

CR BJN3-2 BJN3-2 552-39

Figure 6. Phenotypic investigation of clubroot resistance for the pyramided line and two parental lines. CR BJN3-2: recurrent parent; BJN3-2: susceptible line as control; 552-39: selected pyramided line.

Table 2. Phenotypic investigation of clubroot resistance.

Material	Level 0	Level 1	Level 2	Level 3	Disease INCIDENCE (%)	Disease Index (DI)
CR BJN3-2	30	3	6	5	31.8	23.5
BJN3-2				50	100	100
552-39	48			2	4	4

Note: The table shows the number of plants at different levels of clubroot disease.

Ten homozygous pyramided plants were evaluated for multi-ventricular and self-compatibility traits after two weeks of selfing. The 10 individuals were all multilocular, as shown in Figure 7. The SCI for these 10 plants was greater than 1, which was expressed as self-compatibility (Table 3).

552-39 CR BJN3-2

Figure 7. Phenotypic investigation of locule number for the pyramided line and two parental lines. CR BJN3-2: recurrent parent; 552-39: selected pyramided line.

Table 3. Self-compatibility index (SCI) of the pyramided line 552-39 and two parental lines (CR BJN3-2 and KYS).

Material	SCI
CR BJN3-2	0.48 ± 0.02
KYS	1.37 ± 0.29
552-39	1.15 ± 0.18

Moreover, 50 improved pyramided plants of the 552-39 line were grown in the open field together with recurrent parent CR BJN3-2. Agronomy-related traits were investigated in their heading stage. The results showed that the pyramided line was not significantly different from the recurrent parent in terms of heading-related and yield traits (Table 4).

Table 4. Agronomic trait evaluation of recurrent parent and pyramided line.

Agronomic Traits	552-39	CR BJN3-2
Plant height (cm)	35.3 ± 0.58	34.5 ± 4.95
Plant width (cm)	49.0 ± 7.00	48.0 ± 0.00
Plant weight (kg)	2.23 ± 0.29	2.17 ± 0.15
Number of outer leaves	25.7 ± 3.06	19.5 ± 2.12
Leaf length (cm)	32.0 ± 1.73	34.0 ± 2.12
Leaf width (cm)	19.8 ± 2.75	21.0 ± 0.00

Table 4. *Cont.*

Agronomic Traits	552-39	CR BJN3-2
Petiole length (cm)	18.5 ± 1.00	17.8 ± 1.06
Petiole width (cm)	5.3 ± 0.58	5.5 ± 0.71
Petiole color	Green	Green
Head weight (kg)	1.16 ± 0.17	1.09 ± 0.08
Head shape	Folded	Folded
Head solidity	Compaction	Compaction
Head length (cm)	27.7 ± 2.84	23.9 ± 0.21
Head width (cm)	12.8 ± 2.47	10.7 ± 0.35
Head color	Pale Yellow	Pale Yellow
Stem length (cm)	2.83 ± 1.26	3.25 ± 0.35
Stem width (cm)	2.30 ± 0.75	2.90 ± 0.14

Significance level at $p \leq 0.05$.

4. Discussion

The large number of gene/QTL mapping studies for diverse crops has provided an abundance of molecular markers associated with traits [48]. The use of MAS combined with the backcross strategy allows the introgression of multiple target genes into one elite cultivar or line, which enhances the precision and efficiency of the breeding program.

Self-compatibility in *B. rapa* crops is a characteristic with a complex genetic mechanism [49]. To improve this trait, conventional breeding is difficult and time-consuming. In our study, a co-dominant marker combination Sal-SLGI/PK1+PK4 based on well-known self-compatibility genes was developed and verified in the F_2 population. It also showed a better foreground selection efficiency in the pyramiding process.

Genetic analysis using segregation populations of CR BJN3-2 and KYS demonstrated that the bilocular silique phenotype was dominant over the tetralocular phenotype, which was coincident with previous studies [3,4]. We found that both trilocular and tetralocular types appeared on a single plant in F_2 individuals and backcross progeny, even though their genotypes were the same as KYS detected by linkage marker Teo-1. Xu et al. [50] reported that two independent recessive genes controlled the trait of a trilocular silique. We predicted that other loci/genes probably control a trilocular silique or interact with the major QTL tet-o for a multilocular silique in *B. rapa*. Therefore, we finally selected 552-39, which had the highest tetralocular ratio, as the best pyramided line.

The recurrent parent CR BJN3-2, harboring the *CRb* gene, is resistant to *P. brassica* pathotype 4 [29,44]. Phenotypic investigation showed that the pyramided line had strong clubroot resistance, indicating the successful MAS strategy. Owing to the isolate-specific association between CR genes and *P. brassicae*, multiple CR gene pyramiding challenged us for resistant breeding. Shah et al. [37] combined *CRb* and *PbBa8.1* genes and developed a pyramided homozygous inbred line of *B. napus* that was resistant to various *P. brassicae* isolates. In rice, many studies have focused on developing disease-resistant lines that introgress multiple resistance genes [51,52]. Several studies found that pyramiding three or more bacterial blight (BB) resistance genes exhibited higher resistance than the lines with one or two genes [53,54]. The developed pyramided line, 553-39, could be used as a recurrent parent for further gene pyramiding of other CR genes. Multiple CR gene pyramiding would facilitate not only the creation of resistance resources but also CR gene interaction studies in the future.

The percentages of recurrent parent genomes referring to %RPG in the backcross population reflect the degree of recovery of the genetic material of a single plant to the recurrent parent [55]. Many studies have shown that increasing the genomic distance between background markers could improve the efficiency of foreground selection, thereby reducing the population size required, but the reduction of marker density increased the length of donor fragments, which was prone to linkage drag. Therefore, it is not possible to rely solely on molecular markers for selection; foreground marker selection combined with phenotype identification was necessary to achieve the desired goal.

In our study, the background selection marker of the previous generation, which was restored to the recurrent parent genotype, was not used in the next generation. Before the BC_2F_1 generation, the average genetic background recovery rate of each generation was significantly lower than the expected value of the genetic background recovery rate. Although the BC_4F_1 generation recovered, it still did not meet the expected value. This result was also found in previous gene pyramiding studies in other crops [54,56]. The reduction in the recovery rate of the recurrent parent's genomic background was influenced by the backcross population size, linkage drag, number of background selection markers, and purity of the recurrent parents [40,48].

In conclusion, we introgressed multilocular and self-compatibility trait in the background of CR BJN3-2, which carries the *CRb* gene to improve seed yield and clubroot resistance. The improved pyramided line recovered the agro-morphological phenotype of recurrent parents and could be released as a potential resource for Chinese cabbage breeding.

Supplementary Materials: The following supporting information can be downloaded at: https://www.mdpi.com/article/10.3390/horticulturae8020139/s1, Figure S1: Screening results of polymorphic markers; Table S1: Molecular markers designed for foreground selection of three targeted loci; Table S2: Self-compatibility primers designed based on the sequences of the *SLG*, *SRK*, and *SP11* genes; Table S3: Background selection marker distribution on ten chromosomes of *Brassica rapa*.

Author Contributions: J.Z. performed experiments and data analysis as well as manuscript drafting. H.Z. and Y.M. performed the marker development and backcross strategies. M.J. and Z.Z. helped with phenotype investigation and manuscript drafting. X.L. and Z.P. conceived the project and designed the research. All authors have read and agreed to the published version of the manuscript.

Funding: This work was supported by LiaoNing Revitalization Talents Program (XLYC2002034) and China Agriculture Research System of MOF and MARA (CARS-12). This work was also supported by the National Natural Science Foundation of China (No. 31772326).

Institutional Review Board Statement: Not applicable.

Informed Consent Statement: Not applicable.

Data Availability Statement: The study did not report any data.

Acknowledgments: We thank LetPub (www.letpub.com) for its linguistic assistance during the preparation of this manuscript.

Conflicts of Interest: The authors declare that they have no conflict of interest.

References

1. Nagaharu, U. Genome analysis in *Brassica carinata* with special reference to the experimental formation of *Brassica napus*, a peculiar mode of fertilization. *Jpn. J. Bot.* **1935**, *7*, 389–452.
2. Zhao, J.; Wang, X.; Deng, B.; Lou, P.; Bonnema, G. Genetic relationships within *Brassica rapa* as inferred from AFLP fingerprints. *Theor. Appl. Genet.* **2005**, *110*, 1301–1314. [CrossRef] [PubMed]
3. Somnath, R.; Sinhamahapatra, S.P. Relationship between seed yield and yield components in bilocular and tetralocular yellow sarson (*Brassica rapa*). *Indian J. Agric. Sci.* **2011**, *81*, 643–647.
4. Yadava, S.K.; Paritosh, K.; Panjabi-Massand, P.; Gupta, V.; Chandra, A.; Sodhi, Y.S.; Pradhan, A.K.; Pental, D. Tetralocular ovary and high silique width in yellow sarson lines of *Brassica rapa* (subspecies *trilocularis*) are due to a mutation in *Bra034340* gene, a homologue of *CLAVATA3* in Arabidopsis. *Theor. Appl. Genet.* **2014**, *127*, 2359–2369. [CrossRef]
5. Takayama, S.; Isogai, A. Self-incompatibility in plants. *Annu. Rev. Plant Biol.* **2005**, *56*, 467–489. [CrossRef]
6. Yamamoto, M.; Nishio, T. Commonalities and differences between Brassica and Arabidopsis self-incompatibility. *Hortic. Res.* **2014**, *1*, 14054. [CrossRef]
7. Bateman, A.J. Self-incompatibility systems in angiosperms: III. *Crucif. Hered.* **1955**, *9*, 53–68. [CrossRef]
8. Nasrallah, J.B.; Nishio, T.; Nasrallah, M.E. The Self-incompatibility genes of Brassica: Expression and use in genetic ablation of floral tissues. *Plant Mol. Biol.* **1991**, *42*, 393–422. [CrossRef]
9. Watanabe, M.; Takasaki, T.; Toriyama, K.; Yamakawa, S.; Isogai, A.; Suzuki, A.; Hinata, K. A high degree of homology exists between the protein encoded by *SLG* and the S receptor domain encoded by *SRK* in self-incompatible *Brassica campestris* L. *Plant Cell Physiol.* **1994**, *35*, 1221–1229. [CrossRef]
10. Stein, J.C.; Howlett, B.; Boyes, D.C.; Nasrallah, M.E.; Nasrallah, J.B. Molecular cloning of a putative receptor protein kinase gene encoded at the self-incompatibility locus of *Brassica oleracea*. *Proc. Natl. Acad. Sci. USA* **1991**, *88*, 8816–8820. [CrossRef]

11. Hatakeyama, K.; Takasaki, T.; Watanabe, M.; Hinata, K. Molecular characterization of S locus genes, *SLG* and *SRK*, in a pollen-recessive self-incompatibility haplotype of *Brassica rapa* L. *Genetics* **1998**, *149*, 1587–1597. [CrossRef]
12. Schopfer, C.R.; Nasrallah, M.E.; Nasrallah, J.B. The male determinant of self-incompatibility in Brassica. *Science* **1999**, *286*, 1697–1700. [CrossRef]
13. Suzuki, G.; Kai, N.; Hirose, T.; Fukui, K.; Nishio, T.; Takayama, S.; Isogai, A.; Watanabe, M.; Hinata, K. Genomic organization of the S locus: Identification and characterization of genes in *SLG/SRK* region of S(9) haplotype of *Brassica campestris* (syn. *rapa*). *Genetics* **1999**, *153*, 391–400. [CrossRef]
14. Kachroo, A.; Schopfer, C.R.; Nasrallah, M.E.; Nasrallah, J.B. Allele-specific receptor-ligand interactions in Brassica self-incompatibility. *Science* **2001**, *293*, 1824–1826. [CrossRef] [PubMed]
15. Takayama, S.; Shimosato, H.; Shiba, H.; Funato, M.; Che, F.S.; Watanabe, M.; Iwano, M.; Isogai, A. Direct ligand-receptor complex interaction controls Brassica self-incompatibility. *Nature* **2001**, *413*, 534–538. [CrossRef]
16. Hiroko, S.; Yokota, N.; Shiba, H.; Iwano, M.; Entani, T.; Che, F.S.; Watanabe, M.; Isogai, A.; Takayama, S. Characterization of the SP11/SCR high-affinity binding site involved in self/nonself recognition in Brassica self-incompatibility. *Plant Cell* **2007**, *19*, 107–117.
17. Dixon, G.R. The occurrence and economic impact of *plasmodiophora brassicae* and clubroot disease. *J. Plant Growth Regul.* **2009**, *28*, 194–202. [CrossRef]
18. Jing, W.; Yun, H.; Li, X.; Li, H. Research progress in clubroot of crucifers. *Plant Prot.* **2011**, *37*, 153–158.
19. Chai, A.L.; Xie, X.W.; Shi, Y.X.; Li, B.J. Special Issue: Research status of clubroot (*Plasmodiophora brassicae*) on cruciferous crops in China. *Can. J. Plant Pathol.* **2014**, *36*, 142–153. [CrossRef]
20. Donald, C.; Porter, I. Integrated Control of Clubroot. *J. Plant Growth Regul.* **2009**, *28*, 289–303. [CrossRef]
21. Kageyama, K.; Asano, T. Life Cycle of *Plasmodiophora brassicae*. *J. Plant Growth Regul.* **2009**, *28*, 203–211. [CrossRef]
22. Mehraj, H.; Akter, A.; Miyaji, N.; Miyazaki, J.; Shea, D.J.; Fujimoto, R.; Doullah, M.A.-U. Genetics of clubroot and fusarium wilt disease resistance in Brassica vegetables: The Application of marker assisted breeding for disease resistance. *Plants* **2020**, *9*, 726. [CrossRef] [PubMed]
23. Matsumoto, E.; Yasui, C.; Ohi, M.; Tsukada, M. Linkage analysis of RFLP markers for clubroot resistance and pigmentation in Chinese cabbage. *Euphytica* **1998**, *104*, 79–86. [CrossRef]
24. Suwabe, K.; Tsukazaki, H.; Iketani, H.; Hatakeyama, K.; Fujimura, M.; Nunome, T.; Fukuoka, H.; Matsumoto, S.; Hirai, M. Identification of two loci for resistance to clubroot (*Plasmodiophora brassicae Woronin*) in *Brassica rapa* L. *Theor. Appl. Genet.* **2003**, *107*, 997–1002. [CrossRef]
25. Suwabe, K.; Tsukazaki, H.; Iketani, H.; Hatakeyama, K.; Kondo, M.; Fujimura, M.; Nunome, T.; Fukuoka, H.; Hirai, M.; Matsumoto, S. Simple sequence repeat-based comparative genomics between *Brassica rapa* and *Arabidopsis thaliana*: The genetic origin of clubroot resistance. *Genetics* **2006**, *173*, 309–319. [CrossRef]
26. Hirai, M.; Harada, T.; Kubo, N.; Tsukada, M.; Suwabe, K.; Matsumoto, S. A novel locus for clubroot resistance in *Brassica rapa* and its linkage markers. *Theor. Appl. Genet.* **2004**, *108*, 639–643. [CrossRef]
27. Saito, M.; Kubo, N.; Matsumoto, S.; Suwabe, K.; Tsukada, M.; Hirai, M. Fine mapping of the clubroot resistance gene, *Crr3*, in *Brassica rapa*. *Theor. Appl. Genet.* **2006**, *114*, 81–91. [CrossRef]
28. Piao, Z.Y.; Deng, Y.Q.; Choi, S.R.; Park, Y.J.; Lim, Y.P. SCAR and CAPS mapping of *CRb*, a gene conferring resistance to *Plasmodiophora brassicae* in Chinese cabbage (*Brassica rapa* ssp. *pekinensis*). *Theor. Appl. Genet.* **2004**, *108*, 1458–1465. [CrossRef]
29. Chen, J.; Jing, J.; Zhan, Z.; Zhang, T.; Zhang, C.; Piao, Z. Identification of novel QTLs for isolate-specific partial resistance to *plasmodiophora brassicae* in *Brassica rapa*. *PLoS ONE* **2013**, *8*, e85307. [CrossRef]
30. Pang, W.; Fu, P.; Li, X.; Zhan, Z.; Yu, S.; Piao, Z. Identification and mapping of the clubroot resistance gene *CRd* in Chinese cabbage (*Brassica rapa* ssp. *pekinensis*). *Front. Plant Sci.* **2018**, *9*, 653. [CrossRef]
31. Yu, F.; Zhang, X.; Huang, Z.; Chu, M.; Song, T.; Falk, K.C.; Deora, A.; Chen, Q.; Yan, Z.; Mcgregor, L. Identification of genome-wide variants and discovery of variants associated with *Brassica rapa* clubroot resistance gene *Rcr1* through bulked segregant RNA sequencing. *PLoS ONE* **2016**, *11*, e0153218. [CrossRef] [PubMed]
32. Yu, F.; Zhang, X.; Peng, G.; Falk, K.C.; Strelkov, S.E.; Gossen, B.D. Genotyping-by-sequencing reveals three QTL for clubroot resistance to six pathotypes of *Plasmodiophora brassicae* in *Brassica rapa*. *Sci. Rep.* **2017**, *7*, 4516. [CrossRef] [PubMed]
33. Zhen, H.; Peng, G.; Liu, X.; Abhinandan, D.; Falk, K.C.; Gossen, B.D.; Mcdonald, M.R.; Yu, F. Fine mapping of a clubroot resistance gene in Chinese Cabbage using SNP markers identified from bulked segregant RNA sequencing. *Front. Plant Sci.* **2017**, *8*, 1448.
34. Yamanaka, N.; Hossain Md, M. Pyramiding three rust-resistance genes confers a high level of resistance in soybean (Glycine max). *Plant Breed.* **2019**, *138*, 686–695. [CrossRef]
35. Liu, Y.; Chen, L.; Liu, Y.; Dai, H.; He, J.; Kang, H.; Pan, G.; Huang, J.; Qiu, Z.; Wang, Q.; et al. Marker assisted pyramiding of two brown planthopper resistance genes, *Bph3* and *Bph27* (t), into elite rice cultivars. *Rice* **2016**, *9*, 27. [CrossRef]
36. Xiao, W.; Yang, Q.; Huang, M.; Guo, T.; Liu, Y.; Wang, J.; Yang, G.; Zhou, J.; Yang, J.; Zhu, X.; et al. Improvement of rice blast resistance by developing monogenic lines, two-gene pyramids and three-gene pyramid through MAS. *Rice* **2019**, *12*, 78. [CrossRef]
37. Shah, N.; Li, Q.; Xu, Q.; Liu, J.; Huang, F.; Zhan, Z.; Qin, P.; Zhou, X.; Yu, W.; Zhu, L.; et al. *CRb* and *PbBa8.1* synergically increases resistant genes expression upon infection of *Plamodiophora brassicae* in *Brassica napus*. *Genes* **2020**, *11*, 202. [CrossRef] [PubMed]
38. Matsumoto, E.; Ueno, H.; Aruga, D.; Sakamoto, K.; Hayashida, N. Accumulation of three clubroot resistance genes through marker-assisted selection in Chinese cabbage (*Brassica rapa* ssp. *pekinensis*). *J. Jpn. Soc. Hortic. Sci.* **2012**, *81*, 184–190. [CrossRef]

39. Tomita, H.; Shimizu, M.; Doullah, M.A.U.; Fujimoto, R.; Okazaki, K. Accumulation of quantitative trait loci conferring broad-spectrum clubroot resistance in *Brassica oleracea*. *Mol. Breed.* **2013**, *32*, 889–900. [CrossRef]
40. Tyagi, S.; Mir, R.R.; Kaur, H.; Chhuneja, P.; Ramesh, B.; Balyan, H.S.; Gupta, P.K. Marker-assisted pyramiding of eight QTLs/genes for seven different traits in common wheat (*Triticum aestivum* L.). *Mol. Breed.* **2014**, *34*, 167–175. [CrossRef]
41. Chen, J.; Pang, W.; Chen, B.; Zhang, C.; Piao, Z. Transcriptome analysis of *Brassica rapa* near-isogenic lines carrying clubroot-resistant and –susceptible alleles in response to *Plasmodiophora brassicae* during early infection. *Front. Plant Sci.* **2016**, *6*, 1183. [CrossRef] [PubMed]
42. Li, X.; Ramchiary, N.; Choi, S.R.; VanNguyen, D.; Hossain, M.J.; Yang, H.; Lim, Y.P. Development of a high density integrated reference genetic linkage map for the multinational *Brassica rapa* Genome Sequencing Project. *Genome* **2010**, *53*, 939–947. [CrossRef] [PubMed]
43. Ge, Y.; Ramchiary, N.; Wang, T.; Liang, C.; Wang, N.; Wang, Z.; Choi, S.R.; Lim, Y.P.; Piao, Z.Y. Mapping quantitative trait loci for leaf and heading-related traits in Chinese cabbage (*Brassica rapa* L.ssp. *pekinensis*). *Hortic. Environ. Biotechnol.* **2011**, *52*, 494–501. [CrossRef]
44. Zhang, T.; Zhao, Z.; Zhang, C.; Pang, W.; Choi, S.R.; Lim, Y.P.; Piao, Z. Fine genetic and physical mapping of the *CRb* gene conferring resistance to clubroot disease in *Brassica rapa*. *Mol. Breed.* **2014**, *34*, 1173–1183. [CrossRef]
45. Li, X.; Ramchiary, N.; Dhandapani, V.; Choi, S.R.; Hur, Y.; Nou, I.-S.; Yoon, M.K.; Lim, Y.P. Quantitative trait loci mapping in *Brassica rapa* revealed the structural and functional conservation of genetic loci governing morphological and yield component traits in the A, B, and C subgenomes of Brassica species. *DNA Res.* **2013**, *20*, 1–16. [CrossRef]
46. Ge, Y.; Ramchiary, N.; Wang, T.; Liang, C.; Wang, N.; Wang, Z.; Choi, S.R.; Lim, Y.P.; Piao, Z.Y. Development and linkage mapping of unigene-derived microsatellite markers in *Brassica rapa* L. *Breed. Sci.* **2011**, *61*, 160–167. [CrossRef]
47. Williams, P.H. A system for the determination of races of *Plasmodiophora brassicae* that infect Cabbage and Rutabaga. *Phytopathology* **1966**, *56*, 624–626.
48. Collard, B.C.Y.; Mackill, D.J. Marker-assisted selection: An approach for precision plant breeding in the twenty-first century. *Philos. Trans. R. Soc. B Biol. Sci.* **2008**, *363*, 557–572. [CrossRef]
49. Nasrallah, J.B. Plant mating systems: Self-incompatibility and evolutionary transitions to self-fertility in the mustard family. *Curr. Opin. Genet. Dev.* **2017**, *47*, 54–60. [CrossRef]
50. Xu, P.; Lv, Z.; Zhang, X.; Wang, X.; Pu, Y.; Wang, H.; Yi, B.; Wen, J.; Ma, C.; Tu, J.; et al. Identification of molecular markers linked to trilocular gene (*mc1*) in *Brassica juncea* L. *Mol. Breed.* **2014**, *33*, 425–434. [CrossRef]
51. Singh, A.K.; Singh, V.K.; Singh, S.P.; Pandian, R.T.P.; Ellur, R.K.; Singh, D. Molecular breeding for the development of multiple disease resistance in Basmati rice. *AoB Plants* **2012**, *2012*, pls029. [CrossRef]
52. Das, G.; Rao, G.J.N. Molecular marker assisted gene stacking for biotic and abiotic stress resistance genes in an elite rice cultivar. *Front. Plant Sci.* **2015**, *6*, 698. [CrossRef] [PubMed]
53. Ramalingam, J.; Raveendra, C.; Savitha, P.; Vidya, V.; Chaithra, T.L.; Velprabakaran, S.; Saraswathi, R.; Ramanathan, A.; Pillai, M.P.A.; Arumugachamy, S.; et al. Gene Pyramiding for achieving enhanced resistance to bacterial blight, blast, and sheath blight diseases in rice. *Front. Plant Sci.* **2020**, *11*, 591457. [CrossRef] [PubMed]
54. Hsu, Y.C.; Chiu, C.H.; Yap, R.; Tseng, Y.C.; Wu, Y.P. Pyramiding bacterial blight resistance genes in Tainung82 for broad-spectrum resistance using marker-assisted selection. *Int. J. Mol. Sci.* **2020**, *21*, 1281. [CrossRef] [PubMed]
55. Cobb, J.N.; Biswas, P.S.; Platten, J.D. Back to the future: Revisiting MAS as a tool for modern plant breeding. *Theor. Appl. Genet.* **2019**, *132*, 647–667. [CrossRef] [PubMed]
56. Balachiranjeevi, C.; Bhaskar, N.S.; Abhilash, V.; Akanksha, S.; Viraktamath, B.C.; Madhav, M.S.; Hariprasad, A.S.; Laha, G.S.; Prasad, M.S.; Balachandran, S.M.; et al. Marker-assisted introgression of bacterial blight and blast resistance into DRR17B, an elite, fine-grain type maintainer line of rice. *Mol. Breed.* **2015**, *35*, 151. [CrossRef]

horticulturae MDPI

Article

miR398 Attenuates Heat-Induced Leaf Cell Death via Its Target CSD1 in Chinese Cabbage

Biting Cao [1,2,†], Jianxia Jiang [3,†], Jinjuan Bai [2], Xuan Wang [2], Yajie Li [2], Wenna Shao [4], Shengwu Hu [1,*], Yuke He [2,*] and Xiang Yu [4,*]

1 State Key Laboratory of Crop Stress Biology in Arid Areas and College of Agronomy, Northwest A&F University, Xianyang 712100, China; caobis@163.com
2 National Key Laboratory of Plant Molecular Genetics, Center for Excellence in Molecular Plant Science, Shanghai Institute of Plant Physiology and Ecology, Chinese Academy of Sciences, Fenglin Road 300, Shanghai 200032, China; jjbai@cemps.ac.cn (J.B.); doublexzz@outlook.com (X.W.); yjli02@sjtu.edu.cn (Y.L.)
3 Crop Breeding and Cultivation Research Institute, Shanghai Academy of Agricultural Sciences, Shanghai 201403, China; xiajianjiang321@163.com
4 Joint International Research Laboratory of Metabolic & Developmental Sciences, School of Life Sciences and Biotechnaology, Shanghai Jiao Tong University, Shanghai 200240, China; swn_9511@sjtu.edu.cn
* Correspondence: swhu83251@nwafu.edu.cn (S.H.); heyk@sippe.ac.cn (Y.H.); yuxiang2021@sjtu.edu.cn (X.Y.); Tel.: +86-29-8708-2604 (S.H.); +86-21-5492-4111 (Y.H.); Fax: +86-21-5492-4111 (Y.H.)
† These authors contribute equally to this work.

Abstract: Previous research has shown that miR398 contributed to plant thermotolerance by silencing its target gene *COPPER/ZINC SUPEROXIDE DISMUTASE1* (*CSD1*) in *Arabidopsis thaliana*. However, the phylogenesis of miR398 and *CSD1* in Brassica crop and their role in regulating leaf cell death under heat stress remains unexplored. Here, we characterized the homologous genes of miR398a and *CSD1* in *Brassica rapa* ssp. *pekinensis* (Chinese cabbage) and found miR398a abundance was accumulated under heat stress (38 °C and 46 °C for 1 h) in Chinese cabbage, while the expression level of its targets *BraCSD1-1* and *BraCSD2-1* were downregulated. To further explore their role in heat response, we constructed the transgenic plants overexpressing artificial miR398a (aBra-miR398a), Bra-miR398a target mimic (Bra-MIM398a), and *BraCSD1-1* in Chinese cabbage for genetic study. Under high temperatures, *p35S::aBra-miR398a* lines reduced the areas of leaf cell death and delayed the leaf cell death. By contrast, *p35S::Bra-MIM398a* and *p35S::BraCSD1-1* plants enlarged the areas of leaf cell death and displayed the earliness of leaf cell death. Finally, we found that the expression level of stress-responsive genes *BraLEA76*, *BraCaM1*, *BraPLC*, *BraDREB2A*, and *BraP5CS* increased in transgenic plants overexpressing aBra-miR398a, which may contribute to their resistance to heat-induced leaf cell death. Taken together, these results revealed the function of Bra-miR398a in attenuating leaf cell death to ensure plant thermotolerance, indicating that the miR398-CSD1 module could be potential candidates for heat-resistant crop breeding.

Keywords: *Brassica rapa*; cell death; *CSD1*; heat resistance; miR398

Citation: Cao, B.; Jiang, J.; Bai, J.; Wang, X.; Li, Y.; Shao, W.; Hu, S.; He, Y.; Yu, X. miR398 Attenuates Heat-Induced Leaf Cell Death via Its Target CSD1 in Chinese Cabbage. *Horticulturae* **2022**, *8*, 299. https://doi.org/10.3390/horticulturae8040299

Academic Editor: Young-Doo Park

Received: 10 March 2022
Accepted: 26 March 2022
Published: 31 March 2022

Publisher's Note: MDPI stays neutral with regard to jurisdictional claims in published maps and institutional affiliations.

Copyright: © 2022 by the authors. Licensee MDPI, Basel, Switzerland. This article is an open access article distributed under the terms and conditions of the Creative Commons Attribution (CC BY) license (https://creativecommons.org/licenses/by/4.0/).

1. Introduction

As an integral part of the plant lifecycle, the death of cells, organs, and eventually the whole plant is an age-dependent process. Programmed cell death (PCD) is an essential process determining plant growth and development. It is divided into two broad categories: developmentally regulated and environmentally induced and it plays a key role in the self-destruction of cells damaged by stress factors [1–4]. The leaf is the primary photosynthetic organ for energy harvesting and nutrient production at the growth and maturation stages [5]. The visible yellowing and whitening are widely used to stage the progression of senescence and leaf cell death [6,7]. Leaf senescence occurs at the final stage of leaf development and precedes cell death. At the senescence stage, nutrients accumulated in the leaves were relocated to other organs, such as developing seeds [5,7].

Plant microRNAs are a group of endogenous, 20–24 nucleotides, small non-coding RNAs, and play crucial roles in post-transcriptional regulation by binding its targeted mRNAs for cleavage or repressing translation [8–10]. miR398 is a conserved miRNA that was identified in *Arabidopsis thaliana* by sequence analysis of stress-treated *Arabidopsis thaliana* small-RNA libraries [11,12], which is encoded by three gene loci: MIR398a, MIR398b and MIR398c, and miR398 has been mainly characterized based on the role of its target genes *CSD1* (*SOD1* or *Cu/Zn SUPEROXIDE DISMUTASE1*), *CSD2* (*SOD2*) and *CCS* (*copper chaperone of CSDs*) [13,14]. miR398 play key roles in developmental processes and multiple stress responses [9,11,12,15]. *CSD1* and *CSD2* are the genes regulating the synthesis of cytosolic Cu/Zn-SOD and chloroplast Cu/Zn-SOD, respectively [16]. The two Cu/Zn superoxide dismutase enzymes are responsible for the dismutation of the toxic superoxide to molecular oxygen and hydrogen peroxide in the cytosol by CSD1 or chloroplast by CSD2 together with CCS, which are generally involved in abiotic stress responses [17]. As a negative regulator of CSDs, miR398 is inhibited by oxidative stress during high light, high concentration of heavy metal, or herbicide, resulting in the increase in their targeted mRNAs *CSD1* and *CSD2* [18]. Under other stress, such as ozone fumigation, salt, *Pseudomonas syringae* infection, the abundance of miR398 was also downregulated, implying an important post-transcriptional regulation role of miR398 in these stress responses [9,17–21].

High temperature, one of the most detrimental stresses in nature, is known to affect almost all aspects of plants during growth, causing severe retardation in development and a dramatic decrease in yield [22–24]. The ability to survive after direct extreme heat challenge is termed basal thermotolerance, which is the foundation of all approaches carried by plants to withstand or to acclimate to damage caused by heat [22]. Followed by exposure to sub-lethal high temperatures, plants obtained the power to survive at lethal heat conditions, referred to as acquired thermotolerance [22,25,26]. *B. rapa* is one of the most important leaf vegetable crops, and Chinese cabbage is highly sensitive to heat stress [27]. On the other hand, the premature leaf cell death of Chinese cabbage usually causes the losses of leaf yield and quality [27]. We have found that enhancing miR398 processing results in stronger plant thermotolerance in *Arabidopsis thaliana* [13]. In this study, we characterized the miR398 and its target *CSD1* in *B. rapa* and explored their function in plant thermotolerance-associated leaf cell death. For this purpose, we studied the effects of expression levels of miR398 and its targeted genes on leaf cell death in *Brassica rapa* ssp. *pekinensis*.

2. Materials and Methods

2.1. Plant Materials and Growth Condition

The seeds of *B. rapa* ssp. *pekinensis* (Bre) were used in our experiments [28]. All the seeds including the Bre as wild-type and the transgenic plants were surface-sterilized and sown on Petri dishes containing Murashige and Skoog (MS) medium [28]. After the Petri dishes were sealed with Parafilm, they were stratified at 4 °C in the dark for at least three days and then moved to a growth room and incubated under 16/8 h of light/darkness per day at 22 °C. Ten days later, the seedlings were transplanted to soil (PINDSTRUP, Denmark, Germany) in plastic pots and moved from a growth room to a greenhouse in the phytotron at the Shanghai Institute of Plant Physiology and Ecology (16 h light/8 h dark). Plants were watered at intervals of three to four days [28,29]. For detecting heat-responsive gene expression, the plants were grown at 38 °C for 3 h and 6 h, respectively, or grown at 38 °C and 46 °C for 1 h, respectively. For heat-induced leaf cell death measurement, plants were treated with 45 °C for 12 h followed by 35 °C for 12 h. For the copper treatment experiment, it is worth noting that MS contains trace amounts of copper, 0 μM represents no additional copper added.

2.2. Cloning and Generation of Transgenic Plants

p35S::BraCSD1-1 construct was obtained by PCR using KOD-plus polymerase (ToYoBo, Shanghai, China) with oligonucleotide pairs (BraCSD1-1S and BraCSD1-459A) as defined in the Supplementary Table S1 and the cDNA of *B. rapa* as a template. PCR product was added to deoxyadenylic acid by rTaq (Takara, Beijing, China), and linked to PMD18T vector, and then were digested with *KpnI* and *XbaI* and cloned in the *KpnI* and *XbaI* sites of pCAMBIA2301.

For construct of *p35S::aBra-miR398a*, the artificial microRNA designer WMD delivers 4 oligonucleotide sequences (I to IV), which were used to engineer artificial microRNA into the endogenous *MIR319a* precursor by site-directed mutagenesis (Supplementary Figure S1). Plasmid pRS300 was used as a template for the PCRs, which contains the *MIR319a* precursor in pBSK [13,30]. The amiRNA containing precursor is generated by overlapping PCR. The first round of amplification fragments was obtained by PCR using KOD-plus polymerase (ToYoBo, Shanghai, China) with the oligonucleotide pairs (amiRNA-A/aBra-miR398-IVa, aBra-miR398-IIa/aBra-miR398-IIIs, and aBra-miR398-Is/amiRNA-B), respectively, as defined in the Supplementary Table S1, and the pRS300 as a template. The second round of amplification was obtained using KOD-plus polymerase (ToYoBo, Shanghai, China) with the oligonucleotide pairs (amiRNA-A/amiRNA-B) as primers and three products from the first round as a template. The products were digested with *KpnI* and *XbaI* and directly cloned in the *KpnI* and *XbaI* sites of pCAMBIA2301 [30,31].

For construct of *p35S::Bra-MIM398a*, the first round of amplification were obtained by PCR using KOD-plus polymerase (ToYoBo, Shanghai, China) with the oligonucleotide pairs (IPS-1S-BamH1/Bra-MIM398a-Ia, Bra-MIM398a-IIs/IPS-522A-Sal1), respectively, as defined in the Supplementary Table S1, and the IPS as template [32]. The second round amplifies were obtained by overlapping PCR using KOD-plus polymerase (ToYoBo, Shanghai, China) with the oligonucleotide pairs (IPS-1S-BamH1/IPS-522A-Sal1) and products (first round) as a template. The products were digested with *BamHI* and *SalI* and directly cloned in the *BamHI* and *SalI* sites of pCAMBIA1301 [13,32]. Three constructs (*p35S::BraCSD1-1, p35S::aBra-miR398a, p35S::Bra-MIMR398a*) were transformed to *E-coli* DH5α competent cells, and then were delivered into *Agrobacterium tumefaciens* strain GV3101 (pMP90RK) using the freeze-thaw method [33]. The recombinant plasmid was then inserted into *B. rapa* wild-type plants (Bre) via the vernalization–infiltration method as previously described [34].

2.3. Heat Treatment and Measurement of Leaf Cell Death

The leaf with 25% of yellow color was designated as the onset of leaf senescence, while the leaf with flaccid or dried over more than half was designated as the onset of leaf cell death. The day for the onset of leaf senescence was regarded as the first day of leaf senescence while the day for the onset of leaf cell death was regarded as the first day of leaf cell death. The day for whitening of 100% leaf area was termed as the day of leaf death while the day for whitening and browning of all leaves on the plant was regarded as the day of plant death [6]. For heat stress of whole plants, two-week-old seedlings were moved into the soil to grow to the five-leaf stage. Then, these plants were subjected to a heat stress treatment of 45 °C for 12 h followed by 35 °C for 12 h, and then cultivated at 22 °C under long-day conditions (16 h light/8 h dark). Finally, these plants were photographed and analyzed on the 0th, 4th, and 8th days after the heat stress using ImageJ [27].

2.4. miRNA Isolation and Northern Blot Analysis

Total RNA was extracted from 10-day-old seedlings of all plants. Antisense sequences of miR398 were synthesized and end-labeled as probes with biotin (TaKaRa, Beijing, China). The RNA concentration was measured by Nanodrop spectrophotometer, and 15 μg of RNA was fractionated on a 15% polyacrylamide gel containing 8 M urea and transferred to a Nitran Plus membrane (Schleicher and Schuell). Hybridization was performed at 41 °C using hybridization buffer (ULTRAhyb Ultrasensitive Hybridization buffer, Ambion). Autoradiography of the membrane was performed using the LightShift Chemiluminescent EMSA Kit (Pierce). A synthesized U6 probe end-labeled with biotin (TaKaRa, Beijing, China) was used for the quantity control of total RNA content between samples [13,35].

2.5. 5′ RACE (Rapid Amplification of cDNA Ends)

RNA was obtained from 2-week-old seedlings, and 5′ RACE was performed using the RLM-RACE Kit (Invitrogen, Carlsbad, CA, USA) according to its instructions with modification [28]. The 5′ RACE PCR products were excised from the gel and cloned into a pMD18T vector (Takara, Beijing, China) for sequencing. Gene-specific primers for 5′ RACE PCR can be found in Supplementary Table S1.

2.6. Real-Time qRT-PCR

Plant tissues were homogenized in liquid nitrogen and total RNA was extracted from the wild-type and transgenic plants using TRIzol (Invitrogen, Carlsbad, California, USA) and treated with DNaseI (TaKaRa, Beijing China) to remove DNA contamination. Approximately 5 μg of RNA was used for reverse transcription with oligo (dT) primers or stem-loop primers for real-time reverse transcriptase-polymerase chain reaction (qRT-PCR) [36,37]. Real-time qRT-PCR was performed with the Bio-Rad iCycler Thermal CycleriQ5 Multicolor Real-Time PCR machine (Bio-Rad) using iQ SYBR Green Real-Time PCR Supermix (Bio-Rad) according to the manufacturer's instructions [13,29]. The expression of 7 genes (*ACTIN1, ATCIN2, ACTIN3, ACTIN4, ACTIN8, ACTIN11, ACTIN12*) were used as an internal control using degenerate primers [29], and comparative threshold cycle method was used to determine relative transcript levels in real-time qRT-PCR [38]. Real-time PCR for detecting and quantifying miRNAs was performed based on the published protocol [38]. Three biological replicates and three technical replicates were performed for each sample. All the primers used in this study were listed in Supplementary Table S1.

2.7. Sequence Alignment and Phylogenetic Analysis

Arabidopsis thaliana and *B. rapa* pre-miR398 sequences were downloaded from the miRBase (https://www.mirbase.org/search.shtml, accessed on 1 March 2022). *Arabidopsis thaliana* AtCSD1, AtCSD2 and AtCCS protein sequences were downloaded from the Arabidopsis Information Resource (https://www.arabidopsis.org/, accessed on 1 March 2022). The homologous proteins in Chinese cabbage were identified based on HMM search from Brassicaceae Database (BRAD, http://brassicadb.cn, accessed on 1 March 2022). Multiple alignments of these protein sequences from *Arabidopsis thaliana* and *Brassica rapa* were performed using ClustalW and GeneDoc [39]. Unrooted phylogenetic trees were constructed from the aligned protein sequences using the neighbor-joining method in MEGA 6.0 with minor modifications [40], and bootstrapping was carried out with 1000 iterations.

2.8. Degradome Analysis

The degradome data (SRR2149955) from flower bud of *B. rapa* ssp. *pekinensis* [41] was downloaded from NCBI. The degradome was analyzed as previously described [42]. Briefly, the adaptor was removed from raw reads using the tool cutadapt with default parameter settings [43]. The trimmed reads were mapped to *BraCSD1-1* cDNA sequence using STAR mapper with default parameter [44]. The 5′ monophosphate (5′P) end reads were extracted and plotted around the miR398 cleavage site at *BraCSD1-1* using a customized R script.

2.9. Statistical Analysis

Statistical significance was calculated by two-tailed Student's t-test and error bars indicate SE. p value < 0.05 were considered to be statistically significant.

3. Results

3.1. Characterization of the miR398 and Its Targets Genes in B. rapa

In *Arabidopsis thaliana*, *MIR398* gene family consists of *MIR398a*, *MIR398b*, and *MIR398c*, while their mature miRNA target genes are *AtCSD1*, *AtCSD2*, and *AtCCS* [13,14]. Through alignment of *Arabidopsis thaliana* miR398 and its target genes with *B. rapa* genomic sequences, we found four *Bra-MIR398* genes and six Bra-miR398-targeted homologous genes in *B. rapa* ssp. *pekinensis*. Based on their DNA sequence similarity, *Bra-MIR398a* homologs were named as *Bra-MIR398a-1* and *Bra-MIR398a-2*, and *Bra-MIR398b* homologs were regarded as *Bra-MIR398b-1* and *Bra-MIR398b-2* (Figure 1A, Table 1 and Supplementary Table S2). It is worth noting that, the 21-nt mature miRNAs produced from *MIR398a* and *MIR398b* contained 1 nt (T/G) difference at the 3′ end in both *Arabidopsis thaliana* and *B. rapa* (Figure 1A). Based on the phylogenic analysis, the miR398-targeted homologous proteins in heading Chinese cabbage were regarded as BraCSD1-1 (Bra031642), BraCSD1-2 (Bra018596), BraCSD2-1 (Bra034394), BraCSD2-2 (Bra011971), BraCCS-1 (Bra016768), and BraCCS-2 (Bra026968), respectively, (Figure 1B, Table 1, and Supplementary Table S1). Next, to explore the tissue distribution of mature miR398a, we collected root, stem, cauline leaves, and inflorescence from the *B. rapa* accession Bre to determine Bra-miR398a abundance in different tissues using northern blotting. The result indicated that Bra-miR398a was accumulated in all these tissues and was most abundant in cauline leaves (Figure 1C).

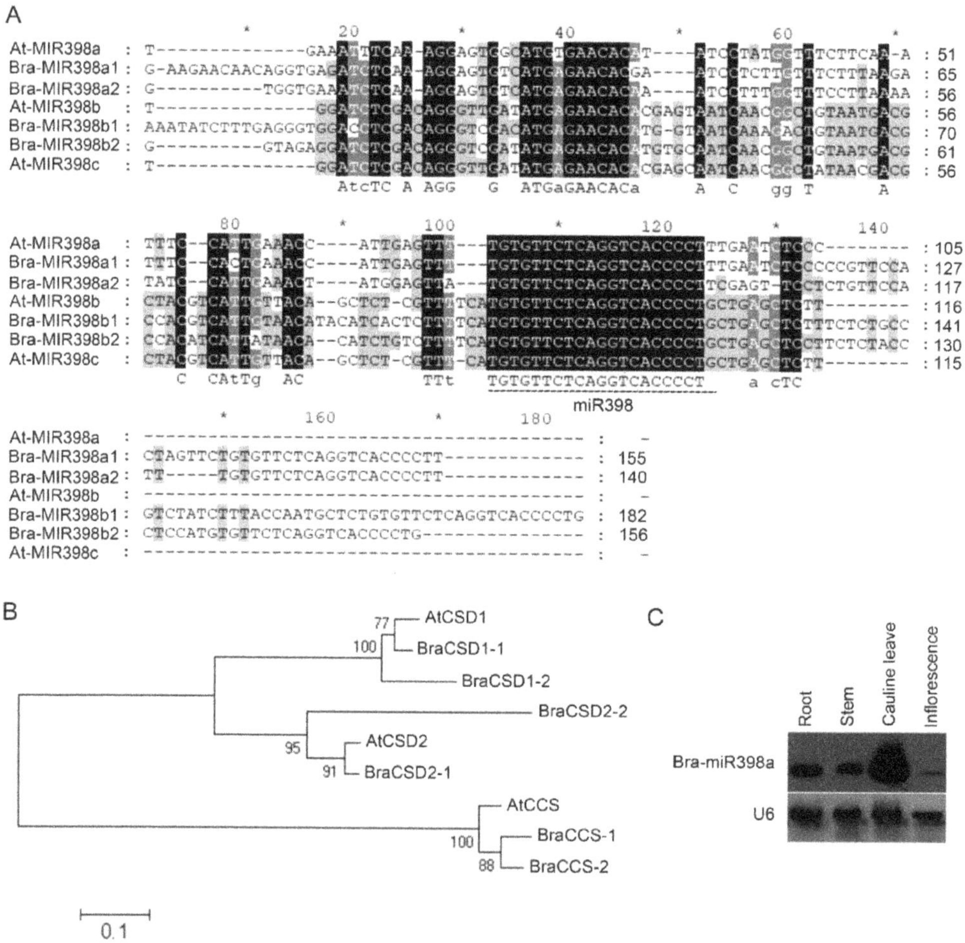

Figure 1. Characterization of miR398 and its targets in *B. rapa*. (**A**) Multiple alignment of pre-miR398 DNA sequences in *Arabidopsis thaliana* and *B. rapa*. The mature miR398 are represented by black solid lines. Asterisks represent 10 bp from the previous number. (**B**) Unrooted phylogenetic trees of miR398 targets based on their protein sequences in *Arabidopsis thaliana* and *B. rapa*. (**C**) Northern blotting showing mature miR398 abundance in different tissues of Bre plants.

Table 1. miR398 and its target genes in *B. rapa*.

miR398 and Targets in *Arabidopsis thaliana*	Homologous Genes in *B. rapa*
ath-miR398a ath-miR398b ath-miR398c	*Bra-MIR398a1/a2* *Bra-MIR398b1/b2*
AtCSD1 (At1g08830)	BraCSD1-1 (Bra031642) BraCSD1-2 (Bra018596)
AtCSD2 (At2g28190)	BraCSD2-1 (Bra034394) BraCSD2-2 (Bra011971)
AtCCS (At1g12520)	BraCCS-1 (Bra016768) BraCCS-2 (Bra026968)

3.2. Response of Bra-miR398a and Its Target Genes to Heat Stress in B. rapa

In *Arabidopsis thaliana*, mature miR398 is increased under high temperature [45], while in *B. rapa* ssp. *chinensis* (non-heading Chinese cabbage), both mature miR398a and miR398b were declined under heat shock [46]. To investigate the response of Bra-miR398a to heat stress in *B. rapa* ssp. *pekinensis* (heading Chinese cabbage), we used northern blot to detect its accumulation after being treated at 38 °C for 3 h and 6 h, respectively. We found that Bra-miR398a was induced at 38 °C for 3 h, but the accumulation decreased after treatment at 38 °C for 6 h (Figure 2A). Consistent with *Arabidopsis thaliana*, the miR398 target site at *BraCSD1-1* is located at 5′UTR with three mismatches. To confirm the miRNA cleavage at *BraCSD1-1*, 5′ RACE PCR followed by sequencing was used to detect 5′ monophosphate (5′P) end of mRNA degradation intermediates, and the 5′P end of RNAs were frequently detected at positions 10 to 11 of the target region complementary to miR398 (Figure 2B). Consistently, high-throughput degradome profiling from *B. rapa* ssp. *pekinensis* [41] also showed enrichment of 5′P end reads at miR398 cleavage stie of *BraCSD1-1* (Supplementary Figure S2). Using real-time qRT-PCR, we validated that the mature Bra-miR398a were accumulated after treatment at both 38 °C and 46 °C for 1 h (Figure 2C), while the expression of Bra-miR398-targeted *BraCSD1-1* and *BraCSD2-1* were downregulated (Figure 2C). Taken together, we found that Bra-miR398a were accumulated under heat stress in heading Chinese cabbage.

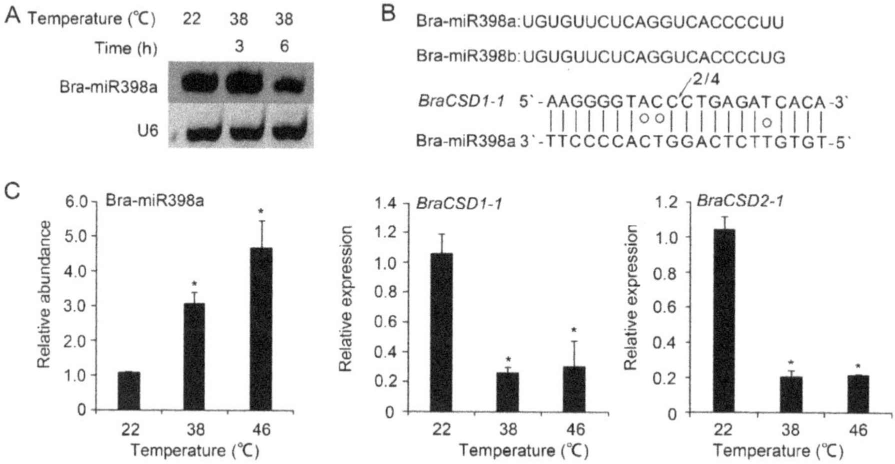

Figure 2. The response of miR398 and its target genes to heat stress. (**A**) Northern blotting showing miR398 abundance under high temperature (38 °C for 3 h and 6 h). (**B**) 5′ RACE PCR showing the cleavage sites of *BraCSD1-1*. Numbers indicate the fraction of cloned PCR products terminating at the position. (**C**) Real-time PCR showing relative abundance of Bra-miR398a and its target genes under 22 °C, 38 °C, and 46 °C for 1 h. The asterisks indicate a significant difference (* represents $p < 0.05$) (Student's *t*-test).

3.3. Bra-miR398a Aids in the Prevention of Leaf Death and Plant Death

As an integral part of plant development, senescence is directly influenced by various exogenous (environmental) factors such as high/low temperatures, drought, ozone, biotic stress, and endogenous (internal) cues including different phytohormones and reproductive development as well as development age of the leaf and plant [47]. The visible yellowing and whitening are widely used to stage the progression of senescence and leaf cell death [6,7]. However, the factors involved in aging are poorly understood. To test whether miR398 was involved in leaf cell death, we constructed transgenic plants in the background of Bre, overexpressing Bra-miR398a (*p35S::aBra-miR398a*) using the backbone of *MIR319a* gene (Supplementary Figure S2), overexpressing Bra-miR398a target mimic (*p35S::Bra-MIM398a*) by modifying the miR398 complementary sequences in *IPS1* [32] (Supplementary Figure S2) and overexpressing one of the miR398 target gene *BraCSD1-1* (*p35S::BraCSD1-1*), using the vernalization–infiltration method [34]. As shown in Figure 3A, the miR398 abundance was increased in *p35S::aBra-miR398a* lines (1#, 2#) while the expression levels of *BraCSD1-1* were up-regulated in *p35S::Bra-MIM398a* (3#, 5#) and *p35S::BraCSD1-1* plants (2#, 5#) (Figure 3A).

To accurately score and characterize the timing and extent of leaf senescence and leaf cell death in whole plants, we measured yellowing and whitening areas of the fourth leaf, counted the number of leaves with cell death in the order of occurrence, and defined leaf cell death and plant survival rates. To demonstrate the processes of leaf cell death, we treated the Bre, *p35S::aBra-miR398a* (1#, 2#), *p35S::Bra-MIM398a* (3#, 5#), and *p35S::BraCSD1-1* (2#, 5#) plants with 45 °C 12 h followed by 35 °C 12 h, and then the plants were moved into a greenhouse at 22 °C under long-day conditions (16-h light/8-h dark) for further growth, along with photographing and analysis on the 0th, 4th, and 8th days, respectively. The growth of the *p35S::Bra-MIM398a* (3#, 5#) and *p35S::BraCSD1-1* (2#, 5#) were significantly weaker than that of the Bre on the fourth day after the high-temperature treatment, while the growth of the *p35S::aBra-miR398a* (1#, 2#) were markedly stronger than that of the Bre (Figure 3B,D). The leaf senescence and leaf cell death of *p35S::Bra-MIM398a* (3#, 5#) and *p35S::BraCSD1-1* (2#, 5#) plants appeared much earlier than the wild-type while the degree of leaf cell death was much higher (Figures 3D and 4). As expected, the yellowing and whitening area of leaves of the *p35S::aBra-miR398a* (1#, 2#) were smaller than that of the Bre (Figures 3B and 4).

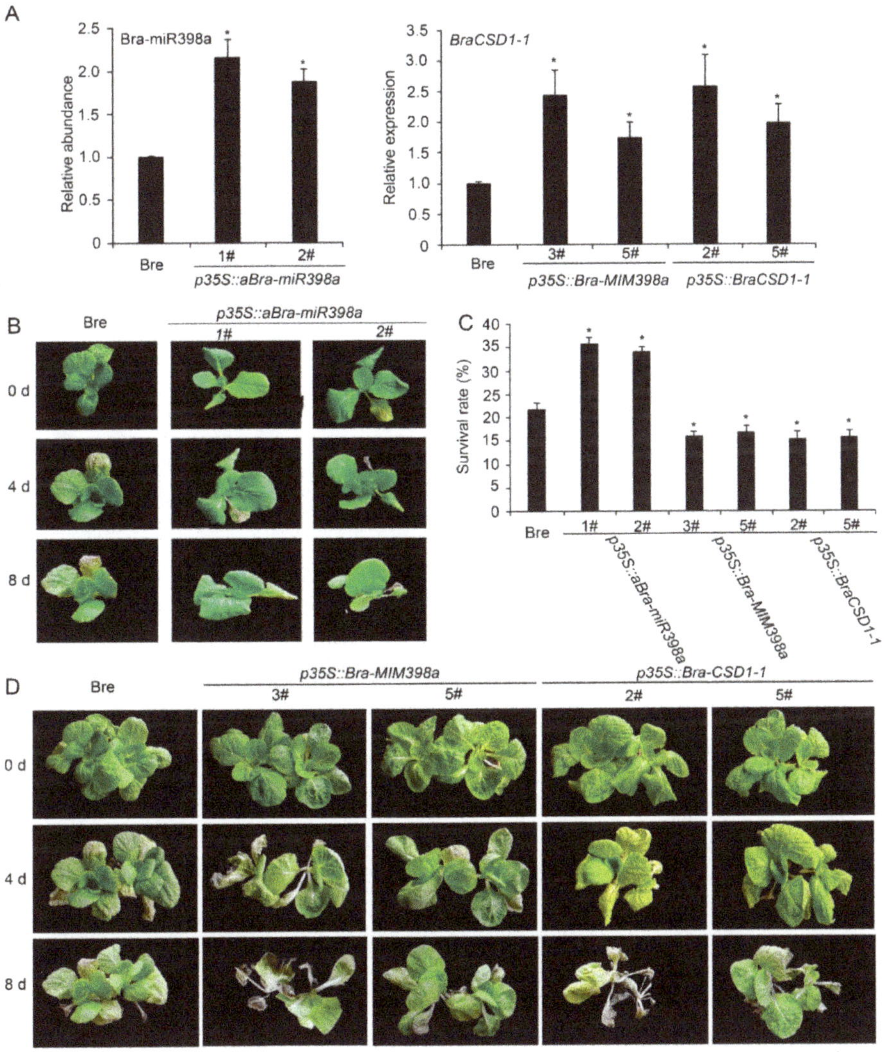

Figure 3. The expression levels of miR398 and *BraCSD1-1* and survival rates of the transgenic plants under heat stress. (**A**) Real-time PCR showing relative abundance of miR398 and relative expression *BraCSD1-1* in the transgenic plants. The asterisks indicate a significant difference (* represents $p < 0.05$) (Student's *t*-test). (**B**) Bre and *p35S::aBra-miR398a* (1#, 2#) plants were photographed and recorded on the fourth and eighth day after high-temperature treatment (45 °C 12 h followed by 35 °C 12 h). (**C**) Survival rates of Bre, *p35S::aBra-miR398a* (1#, 2#), *p35S::Bra-MIM398a* (3#, 5#), and *p35S::BraCSD1-1* (2#, 5#) plants. The asterisks indicate a significant difference (* represents $p < 0.05$) (Student's *t*-test). (**D**) Bre, *p35S::Bra-MIM398a* (3#, 5#), and *p35S::BraCSD1-1* (2#, 5#) plants were photographed and recorded on the fourth and eighth day after high-temperature treatment (45 °C 12 h followed by 35 °C 12 h).

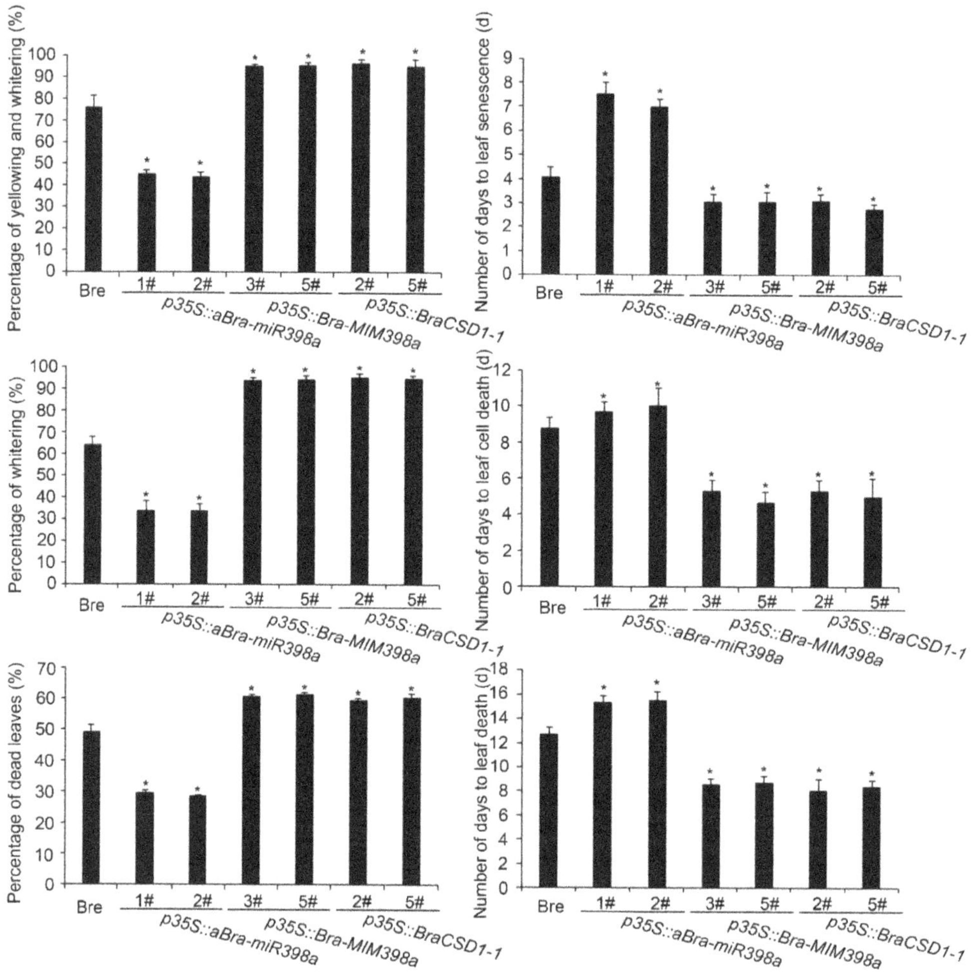

Figure 4. Leaf cell death under high-temperature treatment (45 °C 12 h followed by 35 °C 12 h). The result showed that the percentage of yellow (senescence) and white (cell death) leaf areas, percentage of white leaf areas, percentage of the dead leaves, number of days to senescence, number of days to leaf cell death, and number of days to leaf death of Bre, *p35S::aBra-miR398a* (1#, 2#), *p35S::Bra-MIM398a* (3#, 5#), and *p35S::BraCSD1-1* (2#, 5#) plants. More than 20 leaves for each treatment were harvested and measured from the fourth node of plants. The asterisks indicate a significant difference (* represents $p < 0.05$) (Student's t-test).

Leaf cell death is usually followed by leaf death and plant death. Under high-temperature stress, Bre leaves died after 14 days (Figure 4), and survival rates of Bre plants were over 20% (Figure 3C). The leaves of the *p35S::Bra-MIM398a* (3#, 5#) and *p35S::BraCSD1-1* (2#, 5#) plants died on the 10th day after being subjected to high-temperature stress (Figure 4), and the survival rate of the plants was ~15% (Figure 3C). However, the leaves of *p35S::aBra-miR398a* (1#, 2#) plants died on the 16th day after the high-temperature stress treatment (Figure 4), and the survival rate of the plants was 35% (Figure 3C). Compared to the wild-type, the *p35S::aBra-miR398a* (1#, 2#) plants showed a later leaf death, fewer dead leaves, and higher survival rate, but the *p35S::Bra-MIM398a* (3#, 5#) and *p35S::BraCSD1-1* (2#, 5#) plants were completely opposite. These results indicated that heat stress accelerated leaf senescence and leaf cell death, but the accumulation of *MIR398a* can alleviate this process.

3.4. Bra-miR398a Regulated Heat-Induced Leaf Cell Death Independent with Cu^{2+}-Mediated Pathway

miR398 is strictly regulated by Cu^{2+} levels [21]. We surveyed the germination and growth of *p35S::aBra-miR398a* (1#, 2#), *p35S::Bra-MIM398a* (3#, 5#), and *p35S::BraCSD1-1* (2#, 5#) plants on solid MS with 75 μM and 150 μM Cu^{2+} concentrations, respectively. The seedlings of the wild-type and all the transgenic lines were injured heavily under Cu^{2+} stress with 150 μM. However, the leaf color of *p35S::aBra-miR398a*, *p35S::Bra-MIM398a*, and *p35S::BraCSD1-1* transgenic plants were not different from that of the wild-type under Cu^{2+} stress (Figure 5). The *p35S::aBra-miR398a* transgenic plants were more resistant to high temperature, but showed similar sensitivity to the high level of Cu^{2+} as compared to wild-type, suggesting that Bra-miR398a potentially affected the heat-induced leaf cell death independent with Cu^{2+}-mediated pathway.

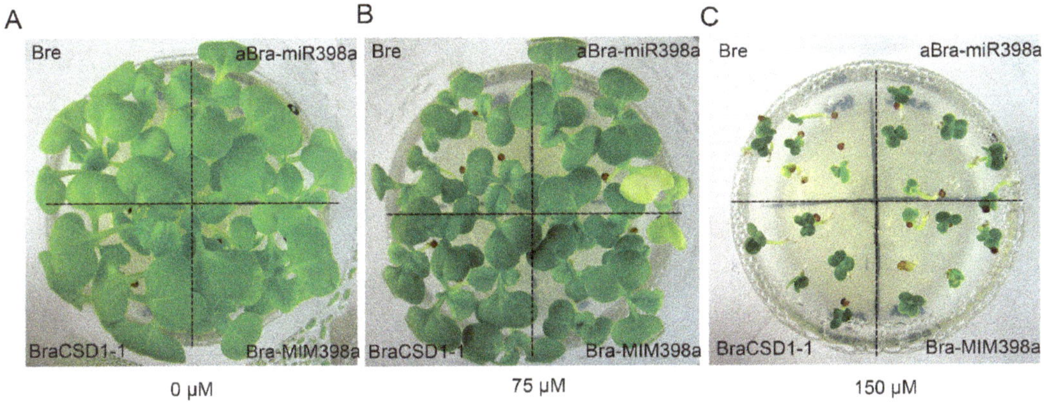

Figure 5. The phenotype of transgenic plants treated with Cu^{2+}. The seedlings of the transgenic plants growing on MS medium containing 0 μM (**A**), 75 μM (**B**), and 150 μM (**C**) Cu^{2+}, respectively.

3.5. Stress-Related Marker Genes Were Upregulated by Bra-miR398a

To further explore the role of Bra-miR398a in leaf cell death, we analyzed expression levels of some stress-related marker genes by real-time PCR. *P5CS1* gene is a rate-limiting enzyme in the biosynthesis of proline and enhances osmotic stress tolerance in transgenic plants [48]. DREB2A is a dehydration-responsive element-binding protein in plants and then activates genes that are involved in detoxification, water, and ion movement and chaperone functions [49–52]. Under high temperature, *BraLEA76*, *BraCaM1*, *BraPLC*, *BraDREB2A*, and *BraP5CS* were upregulated in *p35S::aBra-miR398a* plants (1#, 2#) (Figure 6). In addition, the expression of *BraLEA76*, *BraCaM1*, and *BraPLC* were downregulated in *35S::Bra-MIM398a* (3#, 5#) (Figure 6). These results indicated that Bra-miR398a mediated heat-induced leaf cell death possibly through these stress-related genes.

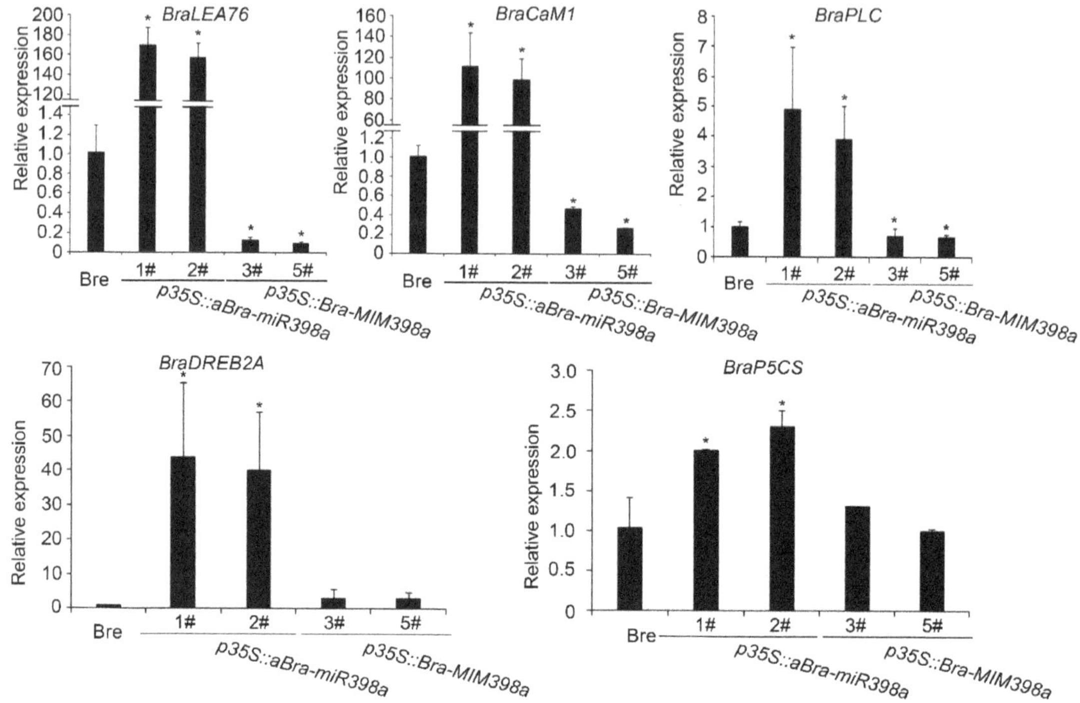

Figure 6. Expression level of cell death-related genes affected by miR398 and its targeted genes. Expression of *BraLEA76*, *BraCaM1*, *BraPLC*, *BraDREB2A*, and *BraP5CS* in Bre, *p35S::aBra-miR398a* (1#, 2#) and *p35S::Bra-MIM398a* (3#, 5#) plants at 38 °C for 1 h, respectively. The asterisks indicate a significant difference (* represents $p < 0.05$) (Student's *t*-test).

4. Discussion

miR398 is one of the miRNAs known to be involved in stress responses of the plant [53]. Enhancing of miR398 processing results in stronger plant thermotolerance [13]. In another study, the level of miR398 increased in the early senescence stage in *Arabidopsis thaliana* leaves [54]. It implies that leaf senescence and leaf cell death are related to plant thermotolerance. In this study, we provided evidence that Bra-miR398a attenuated heat-induced leaf cell death by silencing of *BraCSD1-1* gene in *B. rapa*. *B. rapa* ssp. *chinensis* and *B. rapa* ssp. *pekinensis* are two different sub-species of Brassica and the cold-resistant variety of *B. rapa* ssp. *chinensis* (Wut) is more sensitive to high temperatures as compared to *B. rapa* ssp. *pekinensis* (Bre) [46]. We found the heat response of miR398 in these two cultivars was different. The genetic variation in the *MIR398* promoter or its trans-regulatory transcription factors may contribute to the difference in the expression level of miR398 in these two sub-species. For heat stress of the whole plant, we chose 45 °C for 12 h followed by 35 °C for 12 h, which is to simulate continuous high temperature, given temperature is decreased at night. With global warming and less arable land [55], and for some high-latitude regions, it is of great significance to explore the tolerance of crops to extremely high temperatures.

Originally, we found that genetic manipulation of Bra-miR398a levels may modify the process of heat-induced leaf cell death in *B. rapa*. This conclusion is supported by the phenotype and statistical data of Bre, *p35S::Bra-MIM398a*, *p35S::BraCSD1-1*, and *p35S::aBra-miR398a* plants, as well as the expression of stress-related marker genes under heat stress. A premature senescence and leaf cell death phenotype was observed in *p35S::Bra-MIM398a* and *p35S::BraCSD1-1* plants while the senescence and leaf cell death was postponed in *p35S::aBra-miR398a* plants. Importantly, *p35S::aBra-miR398a* plants were concurrent with higher survival rates under heat stress while the plants' high expression levels of *BraCSD1-1* (*p35S::BraCSD1-1*) concomitant with lower survival rates. Our results suggested that *BraCSD1-1* in plants might be a transcription factor and might have functions beyond a simple superoxide dismutase. In addition, the plants of Bra-miR398a overexpression upregulates the expression of stress-related genes, *BraLEA76*, *BraCaM1*, *BraPLC*, *BraDREB2A*, and *BraP5CS* under high temperature. Together, these results showed that *BraCSD1-1* positively regulates senescence onset and progression under heat stress, while Bra-miR398a postpones these characteristics.

Several growth and development related genes also affect senescence to convey the developmental timing and implement the right timing of senescence. Although these genes might not be considered as specific regulators of senescence, it is very useful to understand the mechanism and downstream genes that implement these decisions, which contribute to the engineering plant senescence for diverse applications. Our findings indicated that Bra-miR398a is a strong candidate to control programmed cell death by inducing senescence. We believe that Bra-miR398a acts as a sensor for the unfavorable environmental condition to prevent senescence to keep the progeny safe by sensing the developmental age of the plant and transferring that information to this module further to help miR398/*CSD1* ensure the right timing of senescence and cell death. Senescence has a crucial impact on the final crop of agricultural products, in the sense that a longer growth period directly enhances the final yield by prolonging photosynthesis, so by detecting the genes which are important in this pathway, we can improve the final yield in many important crops.

Supplementary Materials: The following supporting information can be downloaded at: https://www.mdpi.com/article/10.3390/horticulturae8040299/s1, Figure S1: Overview of vector construction for amiR398 and MIM398.; Figure S2: The distribution of 5′P end reads around the miR398 target sites at BraCSD1-1 revealed by Degradome analysis; Table S1: Primer pairs used in this study; Table S2: The sequence conservation among Bra-MIR398a, Bra-MIR398b and Bra-CSDs to its *Arabidopsis thaliana* homologs at amino acid level.

Author Contributions: B.C., J.J., X.W., Y.L. and W.S. performed experiments and drafted the manuscript, J.B. performed the genetic transformation; X.Y., S.H. and Y.H. designed the research and wrote the manuscript. All authors have read and agreed to the published version of the manuscript.

Funding: This work was supported by National Programs for Science and Technology Development of China (Grant No. 2016YFD0101900) and CAS-TWAS (Chinese Academy of Sciences-The World Academy of Sciences) fellowship.

Institutional Review Board Statement: Not applicable.

Informed Consent Statement: Not applicable.

Data Availability Statement: Not applicable.

Conflicts of Interest: The authors declare no conflict of interest.

References

1. Cui, Y.; Peng, Y.; Zhang, Q.; Xia, S.; Ruan, B.; Xu, Q.; Yu, X.; Zhou, T.; Liu, H.; Zeng, D.; et al. Disruption of *EARLY LESION LEAF 1*, encoding a cytochrome P450 monooxygenase, induces ROS accumulation and cell death in rice. *Plant J.* **2021**, *105*, 942–956. [CrossRef] [PubMed]
2. Ramiro, D.A.; Melotto-Passarin, D.M.; Barbosa, M.; Santos, F.D.; Gomez, S.; Júnior, N.M.; Lam, E.; Carrer, H. Expression of *Arabidopsis Bax Inhibitor-1* in transgenic sugarcane confers drought tolerance. *Plant Biotechnol. J.* **2016**, *14*, 1826–1837. [CrossRef] [PubMed]
3. Lincoln, J.E.; Sanchez, J.P.; Zumstein, K.; Gilchrist, D.G. Plant and animal PR1 family members inhibit programmed cell death and suppress bacterial pathogens in plant tissues. *Mol. Plant Pathol.* **2018**, *19*, 2111–2123. [CrossRef] [PubMed]
4. Fraser, M.S.; Dauphinee, A.N.; Gunawardena, A.H.L.A.N. Determining the effect of calcium on cell death rate and perforation formation during leaf development in the novel model system, the lace plant (*Aponogeton madagascariensis*). *J. Microsc.* **2020**, *278*, 132–144. [CrossRef] [PubMed]
5. Woo, H.R.; Kim, H.J.; Lim, P.O.; Nam, H.G. Leaf Senescence: Systems and Dynamics Aspects. *Annu. Rev. Plant Biol.* **2019**, *70*, 347–376. [CrossRef] [PubMed]

6. Hensel, L.L.; Grbic, V.; Baumgarten, D.A.; Bleecker, A.B. Developmental and Age-Related Processes That Influence the Longevity and Senescence of Photosynthetic Tissues in Arabidopsis. *Plant Cell* **1993**, *5*, 553–564. [CrossRef] [PubMed]
7. Kim, J.; Kim, J.H.; Lyu, J.I.; Woo, H.R.; Lim, P.O. New insights into the regulation of leaf senescence in Arabidopsis. *J. Exp. Bot.* **2018**, *69*, 787–799. [CrossRef]
8. Bartel, D.P. MicroRNAs: Genomics, biogenesis, mechanism, and function. *Cell* **2004**, *116*, 281–297. [CrossRef]
9. Zhu, C.; Ding, Y.; Liu, H. MiR398 and plant stress responses. *Physiol. Plant* **2011**, *143*, 1–9. [CrossRef] [PubMed]
10. Song, X.W.; Li, Y.; Cao, X.F.; Qi, Y.J. MicroRNAs and their regulatory roles in plant-environment interactions. *Annu. Rev. Plant Biol.* **2019**, *70*, 489–525. [CrossRef] [PubMed]
11. Jones-Rhoades, M.W.; Bartel, D.P. Computational identification of plant microRNAs and their targets, including a stress-induced miRNA. *Mol. Cell* **2004**, *14*, 787–799. [CrossRef] [PubMed]
12. Sunkar, R.; Zhu, J.K. Novel and stress-regulated microRNAs and other small RNAs from Arabidopsis. *Plant Cell* **2004**, *16*, 2001–2019. [CrossRef]
13. Li, Y.; Li, X.; Yang, J.; He, Y. Natural antisense transcripts of *MIR398* genes suppress microR398 processing and attenuate plant thermotolerance. *Nat. Commun.* **2020**, *11*, 5351. [CrossRef] [PubMed]
14. Bouché, N. New insights into miR398 functions in Arabidopsis. *Plant Signal. Behav.* **2010**, *6*, 684–686. [CrossRef] [PubMed]
15. Bonnet, E.; Wuyts, J.; Rouze, P.; Van, D.P.Y. Detection of 91 potential conserved plant microRNAs in *Arabidopsis thaliana* and *Oryza sativa* identifies important target genes. *Proc. Natl. Acad. Sci. USA* **2004**, *101*, 11511–11516. [CrossRef] [PubMed]
16. Guan, Q.; Lu, X.; Zeng, H.; Zhang, Y.; Zhu, J. Heat stress induction of *miR398* triggers a regulatory loop that is critical for thermotolerance in Arabidopsis. *Plant J.* **2013**, *74*, 840–851. [CrossRef]
17. Jagadeeswaran, G.; Saini, A.; Sunkar, R. Biotic and abiotic stress down-regulate miR398 expression in *Arabidopsis. Planta* **2009**, *229*, 1009–1014. [CrossRef] [PubMed]
18. Sunkar, R.; Kapoor, A.; Zhu, J.K. Posttranscriptional induction of two Cu/Zn superoxide dismutase genes in *Arabidopsis* is mediated by downregulation of miR398 and important for oxidative stress tolerance. *Plant Cell* **2006**, *18*, 2051–2065. [CrossRef]
19. Abdel-Ghany, S.E.; Pilon, M. MicroRNA-mediated systemic down-regulation of copper protein expression in response to low copper availability in *Arabidopsis. J. Biol. Chem.* **2008**, *283*, 15932–15945. [CrossRef]
20. Jia, X.; Wang, W.X.; Ren, L.; Chen, Q.J.; Mendu, V.; Willcut, B.; Dinkins, R.; Tang, X.; Tang, G. Differential and dynamic regulation of miR398 in response to ABA and salt stress in *Populus tremula* and *Arabidopsis thaliana. Plant Mol. Biol.* **2009**, *71*, 51–59. [CrossRef] [PubMed]
21. Yamasaki, H.; Abdel-Ghany, S.E.; Cohu, C.M.; Kobayashi, Y.; Shikanai, T.; Pilon, M. Regulation of copper homeostasis by micro-RNA in *Arabidopsis. J. Biol. Chem.* **2007**, *282*, 16369–16378. [CrossRef]
22. Larkindale, J.; Mishkind, M.; Vierling, E. *Plant Abiotic Stress*; Jenks, M.A., Hasegawa, P.M., Eds.; Blackwell: Burlington, NC, USA, 2005; pp. 100–144.
23. Tian, X.; Wang, F.; Zhao, Y.; Lan, T.; Yu, K.; Zhang, L.; Qin, Z.; Hu, Z.; Yao, Y.; Ni, Z.; et al. Heat shock transcription factor A1b regulates heat tolerance in wheat and Arabidopsis through OPR3 and jasmonate signalling pathway. *Plant Biotechnol. J.* **2020**, *18*, 1109–1111. [CrossRef] [PubMed]
24. Battisti, D.S.; Naylor, R.L. Historical warnings of future food insecurity with unprecedented seasonal heat. *Science* **2009**, *323*, 240–244. [CrossRef] [PubMed]
25. Sunkar, R.; Li, Y.F.; Jagadeeswaran, G. Functions of microRNAs in plant stress responses. *Trends Plant Sci.* **2012**, *17*, 196–203. [CrossRef] [PubMed]
26. Zhai, Z.H.; Lin, Z.G.; Chen, H.H.; Chen, Z.H.; Center, G.C. Temporal and spatial variation of temperature suitability index for Brassica parachinesis in Guangdong. *Guangdong Agric. Sci.* **2016**, *3*, 66–71.
27. Jiang, J.; Bai, J.; Li, S.; Li, X.; Yang, L.; He, Y. HTT2 promotes plant thermotolerance in *Brassica rapa. BMC Plant Biol.* **2018**, *18*, 127. [CrossRef] [PubMed]
28. Li, X.; Zhang, S.; Bai, J.; He, Y. Tuning growth cycles of Brassica crops via natural antisense transcripts of BrFLC. *Plant Biotechnol. J.* **2016**, *14*, 905–914. [CrossRef] [PubMed]
29. Ren, W.; Wu, F.; Bai, J.; Li, X.; Yang, X.; Xue, W.; Liu, H.; He, Y. BcpLH organizes a specific subset of microRNAs to form a leafy head in Chinese cabbage (*Brassica rapa* ssp. *pekinensis*). *Hortic. Res.* **2020**, *7*, 1. [CrossRef]
30. Schwab, R.; Ossowski, S.; Riester, M.; Warthmann, N.; Weigel, D. Highly specific gene silencing by artificial microRNAs in *Arabidopsis. Plant Cell* **2006**, *18*, 1121–1133. [CrossRef]
31. Schwab, R.; Palatnik, J.F.; Riester, M.; Schommer, C.; Schmid, M.; Weigel, D. Specific effects of microRNAs on the plant transcriptome. *Dev. Cell* **2005**, *8*, 517–527. [CrossRef]
32. Franco-Zorrilla, J.M.; Valli, A.; Todesco, M.; Mateos, I.; Puga, M.I.; Rubio-Somoza, I.; Leyva, A.; Weigel, D.; Garcia, J.A.; Paz-Ares, J. Target mimicry provides a new mechanism for regulation of microRNA activity. *Nat. Genet* **2007**, *39*, 1033–1037. [CrossRef] [PubMed]
33. Nooden, L.D.; Hillsberg, J.W.; Schneider, M.J. Induction of leaf senescence in *Arabidopsis thaliana* by long days through a light-dosage effect. *Physiol. Plant.* **1996**, *96*, 491–495. [CrossRef]
34. He, Y.; Bai, J.; Wu, F.; Mao, Y. *In planta* transformation of *Brassica rapa* and *B. napus* via vernalization-infiltration methods. *Protoc. Exch.* **2013**. [CrossRef]

35. Liu, Z.; Jia, L.; Wang, H.; He, Y. HYL1 regulates the balance between adaxial and abaxial identity for leaf flattening via miRNA-mediated pathways. *J. Exp. Bot.* **2011**, *62*, 4367–4381. [CrossRef] [PubMed]
36. Chen, C.; Ridzon, D.A.; Broomer, A.J.; Zhou, Z.; Lee, D.H.; Nguyen, J.T.; Barbisin, M.; Xu, N.L.; Mahuvakar, V.R.; Andersen, M.R.; et al. Real-time quantification of microRNAs by stem-loop RT-PCR. *Nucleic Acids Res.* **2005**, *33*, e179. [CrossRef] [PubMed]
37. Varkonyi-Gasic, E.; Wu, R.; Wood, M.; Walton, E.F.; Hellens, R.P. Protocol: A highly sensitive RT-PCR method for detection and quantification of microRNAs. *Plant Methods* **2007**, *3*, 12. [CrossRef]
38. Livak, K.J.; Schmittgen, T.D. Analysis of relative gene expression data using real-time quantitative PCR and the $2^{-\Delta\Delta CT}$ Method. *Methods* **2001**, *25*, 402–408. [CrossRef] [PubMed]
39. Piast, M.; Kustrzeba-Wojcicka, I.; Matusiewicz, M.; Banas, T. Molecular evolution of enolase. *Acta Biochim. Pol.* **2005**, *52*, 507–513. [CrossRef] [PubMed]
40. Li, S.; Chen, L.; Zhang, L.; Li, X.; Liu, Y.; Wu, Z.; Dong, F.; Wan, L.; Liu, K.; Hong, D.; et al. BnaC9.SMG7b Functions as a Positive Regulator of the Number of Seeds per Silique in *Brassica napus* by Regulating the Formation of Functional Female Gametophytes. *Plant Physiol.* **2015**, *169*, 2744–2760. [CrossRef] [PubMed]
41. Wei, X.; Zhang, X.; Yao, Q.; Yuan, Y.; Li, X.; Wei, F.; Zhao, Y.; Zhang, Q.; Wang, Z.; Jiang, W.; et al. The miRNAs and their regulatory networks responsible for pollen abortion in Ogura-CMS Chinese cabbage revealed by high-throughput sequencing of miRNAs, degradomes, and transcriptomes. *Front. Plant Sci.* **2015**, *6*, 894. [CrossRef]
42. Yu, X.; Willmann, M.R.; Anderson, S.J.; Gregory, B.D. Genome-Wide Mapping of Uncapped and Cleaved Transcripts Reveals a Role for the Nuclear mRNA Cap-Binding Complex in Cotranslational RNA Decay in Arabidopsis. *Plant Cell* **2016**, *28*, 2385–2397. [CrossRef] [PubMed]
43. Martin, M. Cutadapt removes adapter sequences from high-throughput sequencing reads. *EMBnet. J.* **2011**, *17*, 10. [CrossRef]
44. Dobin, A.; Davis, C.A.; Schlesinger, F.; Drenkow, J.; Zaleski, C.; Jha, S.; Batut, P.; Chaisson, M.; Gingeras, T.R. STAR: Ultrafast universal RNA-seq aligner. *Bioinformatics* **2013**, *29*, 15–21. [CrossRef]
45. Lu, X.; Guan, Q.; Zhu, J. Downregulation of *CSD2* by a heat-inducible *miR398* is required for thermotolerance in *Arabidopsis*. *Plant Signal Behav.* **2013**, *8*, e54952. [CrossRef] [PubMed]
46. Yu, X.; Wang, H.; Lu, Y.; de Ruiter, M.; Cariaso, M.; Prins, M.; van Tunen, A.; He, Y. Identification of conserved and novel microRNAs that are responsive to heat stress in *Brassica rapa*. *J. Exp. Bot.* **2012**, *63*, 1025–1038. [CrossRef]
47. He, Y.H.; Tang, W.N.; Swain, J.D.; Green, A.L.; Jack, T.P.; Gan, S.S. Networking senescence-regulating pathways by using Arabidopsis enhancer trap lines. *Plant Physiol.* **2001**, *126*, 707–716. [CrossRef]
48. Kishor, P.; Hong, Z.; Miao, G.H.; Hu, C.; Verma, D. Overexpression of [delta]-Pyrroline-5-Carboxylate Synthetase Increases Proline Production and Confers Osmotolerance in Transgenic Plants. *Plant Physiol.* **1995**, *108*, 1387–1394. [CrossRef] [PubMed]
49. Shinozaki, K.; Yamaguchi-Shinozaki, K. Gene networks involved in drought stress response and tolerance. *J. Exp. Bot.* **2007**, *58*, 221–227. [CrossRef]
50. Wang, W.; Vinocur, B.; Altman, A. Plant responses to drought, salinity and extreme temperatures: Towards genetic engineering for stress tolerance. *Planta* **2003**, *218*, 1–14. [CrossRef]
51. Xiong, L.; Zhu, J.K. Molecular and genetic aspects of plant responses to osmotic stress. *Plant Cell Env.* **2002**, *25*, 131–139. [CrossRef]
52. Yoshida, T.; Mogami, J.; Yamaguchi-Shinozaki, K. ABA-dependent and ABA-independent signaling in response to osmotic stress in plants. *Curr. Opin. Plant Biol.* **2014**, *21*, 133–139. [CrossRef]
53. Ding, Y.F.; Wang, G.Y.; Fu, Y.P.; Zhu, C. The role of miR398 in plant stress responses. *Yi Chuan* **2010**, *32*, 129–134. [CrossRef] [PubMed]
54. Huo, X.; Wang, C.; Teng, Y.; Liu, X. Identification of miRNAs associated with dark-induced senescence in *Arabidopsis*. *BMC Plant Biol.* **2015**, *15*, 266. [CrossRef]
55. Zhu, T.; De Lima, C.F.F.; De Smet, I. The Heat is On: How Crop Growth, Development and Yield Respond to High Temperature. *J. Exp. Bot.* **2021**, *72*, 7359–7373. [CrossRef]

horticulturae

MDPI

Article

QTL Analysis of the Content of Some Bioactive Compounds in *Brassica rapa* L. Grown under Light Culture Conditions

Ksenia V. Egorova [1,*], Nadezhda G. Sinyavina [1], Anna M. Artemyeva [1,2,*], Natalia V. Kocherina [1] and Yuriy V. Chesnokov [1]

1 Agrophysical Research Institute, 195220 St. Petersburg, Russia; sinad@inbox.ru (N.G.S.); alle007@mail.ru (N.V.K.); yuv_chesnokov@agrophys.ru (Y.V.C.)
2 Federal Research Center N.I. Vavilov All-Russian Institute of Plant Genetic Resources, 190000 St. Petersburg, Russia
* Correspondence: kseniia.v.egorova@gmail.com (K.V.E.); akme11@yandex.ru (A.M.A.)

Abstract: The article presents the results of biochemical and QTL (**Q**uantitative **T**rait **L**oci) analysis of dry matter content, nutrient and biologically active compounds: sugars, ascorbic acid, chlorophylls *a* and *b*, anthocyanins and carotenoids in populations of doubled haploid lines of leaf, root crops, and oilseeds of the *Brassica rapa* L. species grown in optimal light culture conditions, but with different photoperiod durations. The purpose of this study was to evaluate the effect of the photoperiod on the transition to bolting and the accumulation of biologically active substances, as well as how the localization and identification of chromosomal loci determined the content of certain phytochemicals. The influence of the length of daylight hours on the content of components of the biochemical composition was assessed. It was shown that growing under conditions of a 16 h photoperiod increased the content of dry matter, sugars, vitamin C, and anthocyanins. On the contrary, the content of photosynthetic pigments was higher under the conditions of a 12 h photoperiod. Valuable lines that can be sources of biologically active compounds were revealed. Based on the results of the obtained data, 102 QTLs were mapped, which determine the manifestation of the studied biochemical quality traits in the *B. rapa* doubled haploid lines under conditions of short and long daylight hours. Molecular markers genetically linked to the selected QTLs were determined. It was revealed that the identified loci controlling all the studied biochemical traits were mainly in the fifth, sixth, seventh, and ninth linkage groups, which correlated with the data obtained in the field and greenhouse. Most of the identified loci controlled several studied traits simultaneously. The identified QTLs and identified molecular markers are of interest for further study of the genetic control of the economically valuable traits determined by them and for the implementation of marker-assisted selection in *B. rapa*. The data obtained can be used in genetic and breeding work, including for the obtaining of new genotypes, lines and cultivars with a valuable biochemical composition, adapted for cultivation under specific photoperiodic conditions.

Keywords: *B. rapa* doubled haploid lines; biochemical composition; photoperiod; quantitative trait loci; controlled conditions

Citation: Egorova, K.V.; Sinyavina, N.G.; Artemyeva, A.M.; Kocherina, N.V.; Chesnokov, Y.V. QTL Analysis of the Content of Some Bioactive Compounds in *Brassica rapa* L. Grown under Light Culture Conditions. *Horticulturae* **2021**, *7*, 583. https://doi.org/10.3390/horticulturae7120583

Academic Editors: Xiaowu Wang, Jian Wu and Xu Cai

Received: 26 November 2021
Accepted: 13 December 2021
Published: 16 December 2021

Publisher's Note: MDPI stays neutral with regard to jurisdictional claims in published maps and institutional affiliations.

Copyright: © 2021 by the authors. Licensee MDPI, Basel, Switzerland. This article is an open access article distributed under the terms and conditions of the Creative Commons Attribution (CC BY) license (https://creativecommons.org/licenses/by/4.0/).

1. Introduction

The multifaceted family *Brassicaceae* Burnet. includes a large number of economically valuable agricultural crops that are widespread throughout the world: varieties of *B. oleracea* L. (white cabbage, broccoli, cauliflower and others), crops of *B. rapa* L. (turnip, Chinese cabbage and pakchoi, oilseed rape), vegetable and oilseed varieties of Indian mustard (*B. juncea* Czern.), oilseed and leaf rape and swede (*B. napus* L.), spicy crops (horseradish *Armoracea rusticana* G. Gaertn., wasabi *Eutrema japonicum* Miq.), black mustard (*B. nigra* L.) and others [1–3]. Over the past decades, there has been an increasing consumer demand for foods rich in nutrients and biologically active substances that have a beneficial effect on human health. Plants of the *Brassicaceae* family are sources of vital natural biologically active substances—enzymes, pigments, vitamins, as well as specific secondary metabolites, which exhibit anticarcinogenic, cardioprotective, antioxidant and anti-inflammatory effects, and stimulate the immune system [4–6].

An important representative of the *Brassica* L. genus, the *Brassica rapa* L. species, is distinguished by a high botanical and agrobiological diversity, including a source of valuable nutrients and secondary metabolites, which makes it a unique model object for genetic and molecular studies of various economically valuable traits [1,7–9]. It should be noted that understanding the mechanisms of genetic control of economically valuable traits is necessary to increase the efficiency and accuracy of traditional breeding [10,11]. At the same time, one of the priority tasks of modern breeding is selection for the required content of biologically active compounds [12].

Most of the significant breeding traits refer to quantitative traits, the manifestation of which is largely influenced by environmental factors. To study quantitative traits in plants, including brassica plants, it is promising to use mapping populations of double haploids (DH)—a set of genetically fixed informative recombinant lines, all loci of which are in a homozygous state [13]. Currently, information on the genetic determinants that are responsible for the manifestation of economically valuable biochemical traits of quality and their inheritance in *B. rapa* is not sufficiently studied [14]. However, molecular genetic mapping represents a powerful method to use for this aim, being able to determine the relative positions of DNA markers on linkage groups [15]. The method of molecular genetic mapping allows to determine the relative positions of DNA markers on linkage groups. Mapping of chromosomal loci is carried out through the identification for relationships of molecular markers with traits and describes parallel genotypic and phenotypic variability in mapping populations. Methodologically, identification and mapping are carried out using QTL mapping using specially created two-parent segregating populations [16,17].

There are published data on previous studies devoted to the analysis of the main biochemical traits of quality in *B. rapa*. Artemyeva et al. [5] presented the results of QTL analysis of five biochemical quality traits (total protein, ascorbic acid, β-carotene, chlorophylls *a* and *b*) using two mapping populations of *Brassica rapa* double haploid lines. For each studied trait, QTLs, the effects of the identified QTLs, the proportion of phenotypic variability determined by each QTL, and molecular markers genetically linked to the identified QTLs were established. Jan et al. [18] revealed a significant variation in the total seed protein content among different *Brassica rapa* ecotypes, which is important for improvement and efficient use of crops. Ullah et al. [19] investigated genetic variation, heritability and correlation between various biochemical traits in *Brassica rapa* advanced lines and revealed significant differences were observed for glucosinolate, oil content, protein content, oleic acid, linolenic acid, etc.

Despite the apparent importance of crops of this species for humans, only limited information has been obtained to date on the genetic nature and inheritance of economically valuable traits of quality in *B. rapa*. In one study, QTL mapping of glucosinolate accumulation traits, which has an important protective function in plants and determines the nutritional quality of *B. rapa*, was carried out [20]. In another publication, the QTL search for phytate and phosphate content in seeds and leaves of *B. rapa* was performed [21]. QTL analysis of metabolites in seedlings made it possible to find the location of QTLs associated with the content of organic acids, amino acids, sugars, glucosinolates, and aromatic compounds [22]. In addition, studies of the genotype–environment interaction of morphological and physiological characteristics were carried out under conditions of normal moisture and drought [23].

The populations of the *B. rapa* doubled haploid lines DH 30 and DH 38 used in our investigation were previously studied in the field and greenhouse conditions in order to map the chromosome loci responsible for the manifestation of the main morphological and biochemical traits of quality. In total, 140 QTLs were mapped. Molecular markers genetically linked to identified QTLs have been identified [1,2,4,5]. At the same time, no one study was carried out to map the loci of quantitative traits of *B. rapa* under controlled conditions of light culture. The study of the mechanisms of manifestation of quantitative traits under controlled conditions allows one to reveal the reaction of the genotype to the action of certain environmental factors by modeling specific parameters (temperature, lighting, humidity, etc.) while maintaining the remaining factors in the optimum zone and assessing their influence on the implementation of the studied traits. Investigations of quantitative traits are especially relevant, the degree of manifestation of which significantly depends on the genotype–environment interaction [24,25]. It is known that the photoperiod is one of the most important environmental factors that significantly affect the course of physiological processes in the plant and its transition to the generative stage of development [26,27].

The present paper aim is at identifying the chromosomal loci and the molecular markers which are linked to the content of dry matter and biologically active compounds in *B. rapa*. To reach this goal, *B. rapa* mapping populations were grown at long and short light days in regulated conditions of light culture.

2. Materials and Methods

2.1. Mapping Populations of *Brassica rapa* L. Doubled Haploid Lines

We used the lines of two mapping populations of *Brassica rapa* L. doubled haploids from the collection of the Federal Research Center All-Russian Institute of Plant Genetic Resources named after N.I. Vavilov (VIR). The mapping populations of DH lines were created in the Plant Breeding laboratory of the University of Wageningen (Netherlands) on the basis of significantly different genotypically and phenotypically accessions, belonging to different botanical subspecies [1,28].

- Population DH 30—obtained by crossing Japanese root turnip (Kairyou Hakata) and oilseed yellow sarson (YS-143).
- Population DH 38—is a result of hybridization of pakchoi (Nai Bai Cai) and oilseed yellow sarson (YS-143).

For this study, 68 lines (23 lines DH 30, and 45 lines DH 38), adapted to the conditions of light culture (artificial light and low-volume growing technologies described below), were selected.

2.2. Growing Plants under Controlled Conditions

The relation to the photoperiod is one of the key physiological features of plants. To determine the dependence of the components of the biochemical composition from the photoperiod, the *B. rapa* mapping populations were grown in the biopolygon of the Agrophysical Research Institute (AFI) under light culture conditions (Figure 1) with modeling of a short and long photoperiod—under 12 h and 16 h illumination, as described earlier [29]. Seeds, previously germinated on filter paper [30], were planted in seedling pots 10 × 9 × 9 cm filled with Agrofit substrate [31] based on high-moor peat ("Pindstrup" LLC, Moscow region, Russia) with the addition of Cambrian clay and chalk; substrate pH 6.0–6.2.

Three replicates and five plants per replicate for each line were used. The plants were grown in the original vegetation light installation. High-pressure sodium lamps (DNaZ-400, "Reflax" LLC, Moscow, Russia) were used as a light source. The plant illumination was 15–20 KLX. The plants were watered daily in the morning: 2 times a week (at the beginning and end of the week) with Knop's solution, on other days with water. The temperature was maintained at 21 ± 3 °C.

Figure 1. Plants of DH lines *Brassica rapa* L. grown under light culture conditions: (**A**) seedlings; (**B**) flowering.

2.3. Biochemical Analyses

Biochemical analyses were carried out in the laboratory of biochemistry of soil–plant systems of the Agrophysical Research Institute. For biochemical analysis, plant material (mature leaves from nine plants) was taken at the stage of the start of bolting. The assessment of the content of the main biochemical substances was carried out according to generally accepted methods.

The dry matter content was determined by the thermostat-weight method by weighing the averaged sample before and after drying in a thermostat at a temperature of 105 °C for 6 h; mono- and disaccharides—according to the Bertrand method [32]; ascorbic acid (vitamin C)—by direct extraction from plants with 1% hydrochloric acid followed by titration with potassium iodate solution (State Standard of Russian Federation 24556-89) [33]. The method for determining the total content of anthocyanins in plant tissue is based on the spectrophotometric determination of the anthocyanin extract in 1% hydrochloric acid solution at a wavelength of 510 nm, in terms of cyanidin-3,5-diglycoside—453 nm [34,35].

Spectrophotometric quantitative determination of chlorophylls *a*, *b* and carotenoids was carried out by extraction in acetone and measuring the optical density of the obtained extracts at wavelengths of 662, 644, and 440.5 nm, respectively [36].

The above-described spectrophotometric studies were carried out using a spectrophotometer PE-3000UV ("Promekolab" LLC, St. Petersburg, Russia).

All data are presented in terms of raw material.

2.4. Statistical Processing of the Obtained Results and QTL Analysis

Statistical assessment of the obtained data was carried out by calculating the main descriptive characteristics: minimum, maximum, mean, standard deviation (SD), coefficient of variation (CV). The Tukey HSD (honestly significant difference) post hoc test was used to identify the differences between the means for each characteristic. A p-value <0.05 (error probability 5%) was considered an acceptable limit of statistical significance. Data analysis was performed using the software Microsoft Office Excel 2019 and Statistica v. 13.3 (StatSoft Inc., Tulsa, OK, USA).

QTL analysis, establishing the presence (candidates) of QTLs and their location in linkage groups (mapping interval 5 cM), LOD values (p = 0.05) and the degree of variation of traits (% Expl.), which are explained by the QTL data, for each trait and population, was performed using the MAPQTL 6.0 [37]. The significance of each LOD was established by a permutation test (1000 repetitions) [38]. The correlation coefficient "trait-marker" was calculated for the 95% significance level as a statistically significant linkage of the marker locus with the QTL, which determines the studied trait, based on the empirically obtained variances for each pair of trait-marker [39].

The graphic representation of molecular genetic maps was made on the basis of the obtained data on the mapping of identified QTLs using the MapChart 2.2 software [40].

3. Results and Discussion

The value of brassica crops for their use in functional nutrition is determined by the content of basic nutrients, as well as the presence and amount of biologically active compounds. It is known that species and varietal characteristics, weather conditions, place of cultivation, and phenological phase affect the concentration of phytochemical components [41]. Despite the clearly pronounced importance of *B. rapa* crops as sources of biologically active substances necessary for healthy human nutrition, to date, only limited information has been obtained on the genetic nature and inheritance of economically valuable traits of quality in this species [42,43].

This work presents the results of biochemical and QTL analysis of the content of some nutrients and biologically active substances using two mapping populations of *B. rapa* grown in light culture at short (12 h) and long (16 h) daylight hours.

3.1. The Timing of Start of Bolting

Many morphological and biochemical characteristics depend on the age of the plant to the beginning of the visible transition to bolting. A 12 h photoperiod (short daylight hours) is optimal for growing vegetative mass in long-day plants, and 16 daylight hours accelerate the transition to bolting [44]. The parental forms of mapping populations differ with respect to the photoperiod: the oilseed sarson is highly sensitive to the length of daylight hours, the length of the vegetative period in pakchoi depends on the photoperiod to a moderate extent, and the root turnip is relatively neutral [1]. When assessing the effect of day length, a significant splitting was revealed in terms of the relation to photoperiod in *B. rapa* DH lines (Figure 2, Tables 1 and 2). For DH 30 lines, the timing of the beginning of bolting varied from 31 to 63 days when grown on a 12 h daylight schedule (mean = 46.30 $\pm$ 7.63, CV = 16.48) and from 26 to 54 days (mean = 38.52 $\pm$ 8.51, CV = 22.08) when grown in 16 h of daylight. For DH 38 lines, an acceleration of the beginning of bolting with an increase in the length of daylight was also revealed: the variation of the trait when grown at a 12 h photoperiod was 31-55 days (mean = 45.44 $\pm$ 6.48, CV = 14.27), and at 16 h from 26 to 49 days (mean = 33.78 $\pm$ 5.27, CV = 15.59).

Similar data on the acceleration of the transition to bolting with increasing day length were obtained by Rachman et al. when studying the effect of the photoperiod on a spring *B. napus* mapping population carrying the genome content of *B. oleracea* [45]. For each population, late bolting (DH 30-28, DH 30-206, DH 38-13, DH 38-87), early bolting (DH 30-44, DH 38-80, DH 38-97), as well as highly sensitive to the length of daylight hours (DH 30-38, DH 30-164, DH 38-25) genotypes were identified. The data obtained can be used in the breeding of *B. rapa* crops to create lines, genotypes and cultivars with predicted growing periods, adapted to ecological and geographical conditions, differing in the length of daylight hours.

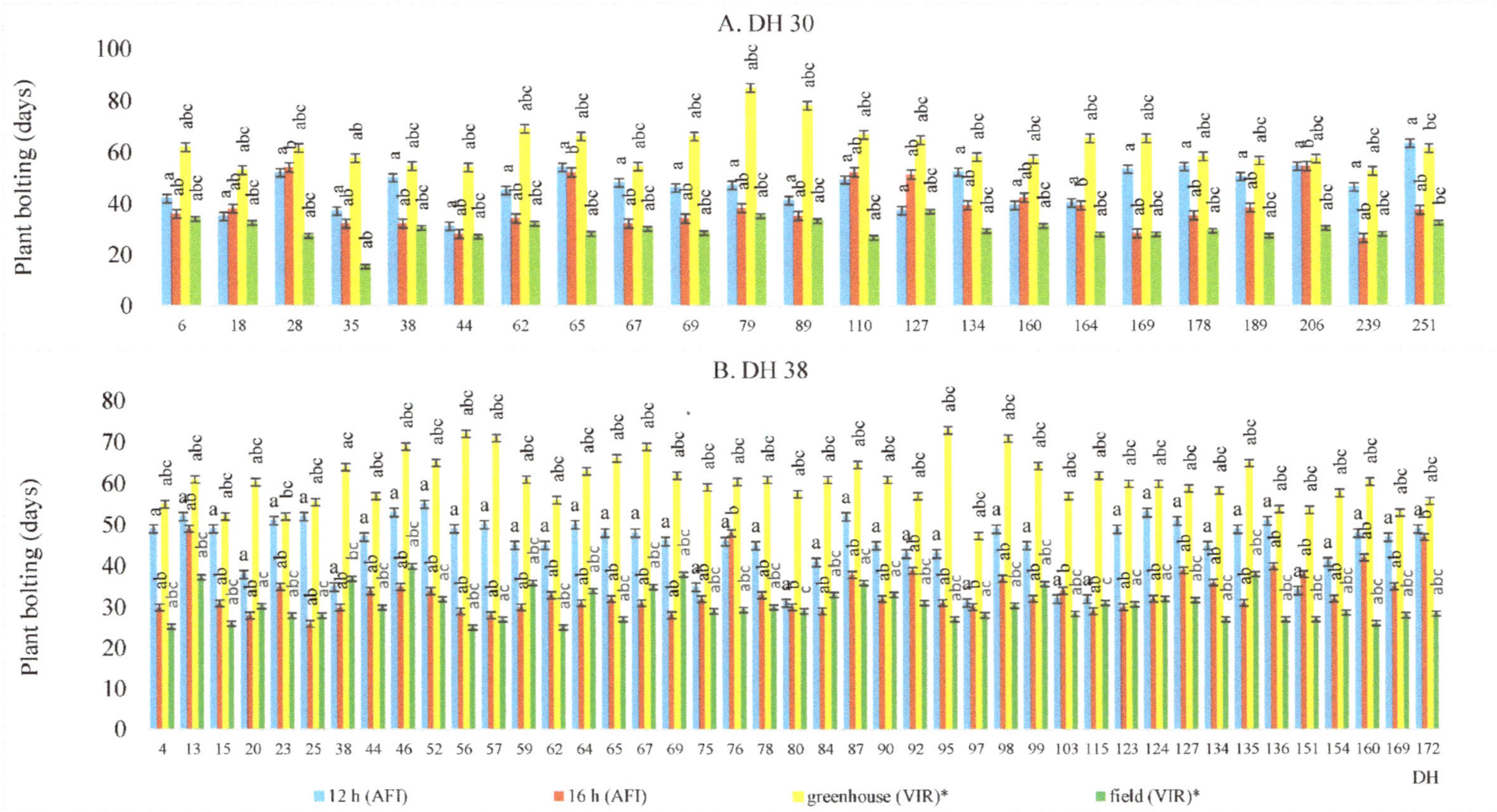

Figure 2. The timing of the start of bolting of lines of mapping populations DH 30 (**A**) and DH 38 (**B**) when grown under different conditions. Note: values with different superscript a–c in the column were significantly different ($p < 0.05$). * By Artemyeva et al., 2012 [1].

Table 1. The timing of the start of bolting and the content of nutrients and biologically active substances in the lines of the mapping population DH 30.

	Biopoligon of AFI 12 h Photoperiod			Biopoligon of AFI 16 h Photoperiod			VIR, Greenhouse *			VIR, Field *		
Trait	**Mean ± SD**	**Xmin–Xmax**	**CV, %**	**Mean ± SD**	**Xmin–Xmax**	**CV, %**	**Mean ± SD**	**Xmin–Xmax**	**CV, %**	**Mean ± SD**	**Xmin–Xmax**	**CV, %**
Days to start of bolting	46.30 ± 7.63^{a}	31.00–63.00	16.48	38.52 ± 8.51^{ab}	26.00–54.00	22.08	61.80 ± 7.99^{abc}	50.20–85.00	12.93	29.42 ± 4.16^{abc}	15.00–37.00	14.13
Dry matter (%)	11.20 ± 3.18	7.42–17.96	28.39	12.41 ± 4.22	5.61–23.97	34.00	9.72 ± 4.15	5.84–19.32	46.67	10.44 ± 1.26	7.69–12.28	12.02
Total sugar content (mg/100 g)	1.30 ± 0.69^{a}	0.11–2.63	54.08	1.88 ± 0.77^{b}	0.74–3.53	40.96	0.65 ± 0.57^{abc}	0.22–2.37	88.63	1.62 ± 0.65^{c}	0.00–2.41	40.12
Monosaccharides (mg/100 g)	0.82 ± 0.47	0.11–2.12	57.32	1.13 ± 0.62	0.20–2.33	54.87	n.d.	n.d.	n.d.	n.d.	n.d.	n.d.
Disaccharides (mg/100 g)	0.48 ± 0.47	0.00–1.38	97.92	0.74 ± 0.49	0.00–1.69	66.22	n.d.	n.d.	n.d.	n.d.	n.d.	n.d.
Ascorbic acid (mg/100 g)	63.98 ± 23.92^{a}	28.16–113.52	37.39	101.36 ± 52.76^{ab}	33.88–209.44	52.06	16.90 ± 2.24^{abc}	12.32–20.00	13.23	56.89 ± 13.30^{bc}	37.05–78.85	23.38
Chlorophyll *a* (mg/100 g)	85.71 ± 25.17^{a}	55.17–147.16	29.37	58.66 ± 5.6^{ab}	39.06–83.35	20.04	80.76 ± 22.65^{b}	35.84–117.87	28.04	64.51 ± 11.33^{a}	47.46–87.73	17.57
Chlorophyll *b* (mg/100 g)	25.04 ± 8.57^{a}	10.73–45.91	34.22	16.29 ± 2.98^{ab}	11.69–23.64	18.28	41.38 ± 12.80^{abc}	20.02–61.99	30.93	27.72 ± 5.50^{bc}	20.16–39.42	19.84
Carotenoids (mg/100 g)	30.32 ± 8.35^{a}	16.59–50.36	27.52	21.02 ± 4.24^{a}	12.88–28.77	20.17	23.60 ± 5.96^{a}	14.04–31.42	25.27	18.91 ± 2.36^{a}	16.42–23.56	12.46
Anthocyanins (mg/100 g)	1.60 ± 0.95	0.49–3.62	59.50	1.67 ± 1.43	0.58–6.34	85.33	n.d.	n.d.	n.d.	n.d.	n.d.	n.d.

Note: values with different superscript a–c in the column were significantly different ($p < 0.05$). n.d.—not detected. * By Artemyeva et al., 2012 [1].

Table 2. The timing of the start of bolting and the content of nutrients and biologically active substances in the lines of the mapping population DH 38.

	Biopoligon of AFI 12 h Photoperiod			Biopoligon of AFI 16 h Photoperiod			VIR, Greenhouse *			VIR, Field *		
Trait	Mean ± SD	Xmin–Xmax	CV, %	Mean ± SD	Xmin–Xmax	CV, %	Mean ± SD	Xmin–Xmax	CV, %	Mean ± SD	Xmin–Xmax	CV, %
Days to start of bolting	45.44 ± 6.48 ^{a}	31.00–55.00	14.27	33.78 ± 5.27 ^{ab}	26.00–49.00	15.59	60.50 ± 5.69 ^{abc}	47.00–73.00	9.40	30.62 ± 3.92 ^{abc}	25.00–40.00	12.81
Dry matter (%)	9.47 ± 2.16 ^{a}	6.26–15.77	22.84	10.25 ± 3.00 ^{b}	7.10–22.51	29.24	7.66 ± 1.88 ^{bc}	4.84–11.20	24.48	11.75 ± 2.93 ^{ac}	6.76–19.04	24.96
Total sugar content (mg/100 g)	0.82 ± 0.56 ^{a}	0.06–2.76	67.68	1.29 ± 0.57 ^{b}	0.29–2.15	48.40	0.55 ± 0.47 ^{bc}	0.18–2.68	86.62	1.32 ± 0.81 ^{ac}	0.00–3.03	63.00
Monosaccharides (mg/100 g)	0.59 ± 0.30 ^{a}	0.06–1.11	51.09	0.87 ± 0.46 ^{a}	0.10–1.67	56.81	n.d.	n.d.	n.d.	n.d.	n.d.	n.d.
Disaccharides (mg/100 g)	0.23 ± 0.40	0.00–1.83	169.88	0.42 ± 0.47	0.00–1.79	112.69	n.d.	n.d.	n.d.	n.d.	n.d.	n.d.
Ascorbic acid (mg/100 g)	43.22 ± 19.92 ^{a}	15.40–97.02	46.09	55.40 ± 27.13 ^{b}	14.08–102.08	48.97	16.15 ± 3.24 ^{abc}	8.00–20.00	20.08	55.37 ± 23.76 ^{c}	20.90–138.70	42.91
Chlorophyll *a* (mg/100 g)	73.19 ± 20.95 ^{a}	37.09–112.08	28.63	56.39 ± 14.42 ^{ab}	31.01–105.65	25.57	83.47 ± 31.75 ^{b}	41.54–150.44	38.03	73.42 ± 17.37 ^{b}	50.99–140.21	23.57
Chlorophyll *b* (mg/100 g)	22.25 ± 6.38 ^{a}	10.08–34.91	28.70	16.09 ± 4.42 ^{b}	7.86–32.49	27.50	44.18 ± 18.54 ^{abc}	23.41–86.77	41.97	32.63 ± 9.57 ^{abc}	15.24–54.49	29.32
Carotenoids (mg/100 g)	27.37 ± 7.31 ^{a}	13.22–39.97	26.72	19.94 ± 4.84 ^{a}	13.14–35.76	24.27	22.33 ± 5.51 ^{a}	13.28–31.90	24.70	17.99 ± 3.92 ^{a}	9.73–26.61	21.78
Anthocyanins (mg/100 g)	1.58 ± 0.65	0.58–3.07	41.34	1.71 ± 1.00	0.49–4.32	58.23	n.d.	n.d.	n.d.	n.d.	n.d.	n.d.

Note: values with different superscript a–c in the column were significantly different ($p < 0.05$). n.d.—not detected. * By Artemyeva et al., 2012 [1].

3.2. Analysis of the Content of Some Phytochemicals

A biochemical analysis of the content of dry matter, carbohydrates, ascorbic acid, plant pigments chlorophylls *a* and *b*, carotenoids, and anthocyanins in plants of two populations of *B. rapa* DH lines (Tables 1 and 2), grown under controlled light culture conditions, was carried out. The influence of the duration of daylight hours on the content of biochemical substances in *B. rapa* crops was assessed.

Dry matter is an important indicator characterizing the nutritional value of vegetable crops. The dry matter content affects the biomass yield and the nutritional value of plant products [46]. In our study, the total range of variability of dry matter content in plants was rather high (Tables 1 and 2). The dry matter content in the lines of the DH 30 population varied from 5.61% to 23.97%, depending on the growing conditions, and in the lines of the DH 38 population, from 4.84% to 19.04%. Under the conditions of a 16 h photoperiod (light culture, field), a slight increase in the average dry matter content (by 7–54%) was revealed in comparison with the 12 h photoperiod (light culture, greenhouse, respectively). The *B. rapa* lines with a high dry matter content independent of photoperiodic conditions were identified—DH 30-18 (up to 17.61%), DH 30-192 (up to 23.92%), 38-75 (up to 22.53%), which can be used in breeding programs as genotypes with high nutritional value.

Carbohydrates (mono- and disaccharides, starch, fiber, pectin, etc.) are a significant component of dry matter. Up to 80% of dry matter can be represented by mono- and disaccharides, and therefore this indicator is important for a comparative phytobiochemical assessment [47]. The high variability of this biochemical trait, confirmed in our study, is associated with both genetic characteristics and growing conditions (Tables 1 and 2). The total sugar content in the DH 30 population varied from 0.00 to 3.53 mg/100 g, in the DH 38 population from 0.00 to 3.03 mg/100 g; monosaccharides—0.11–2.33 mg/100 g (DH 30) and 0.06–1.67 mg/100 g (DH 38); disaccharides—0.00–1.69 mg/100 g (DH 30) and 0.00–1.83 mg/100 g (DH 38).

For both studied populations, a significant increase in sugar content was found when grown on a 16 h daylight hours schedule (for total sugars—by 45–149%, monosaccharides—38–47%, disaccharides—52–82%), which is probably due to a longer light phase of photosynthesis. The studied populations in general contain a high amount of sugars, most of which are represented by monosaccharides (monosaccharides in plants, on average, contain twice as much as disaccharides, in some plants carbohydrates are represented only by simple sugars—monosaccharides). Lines with a high carbohydrate content were identified: DH 30-35 (total sugar content—up to 3.53 mg/100 g); DH 30-65 (total sugar content—up to 3.00 mg/100 g); DH 38-159 (total sugar content—up to 2.76 mg/100 g).

Ascorbic acid (vitamin C) is a unique multifunctional compound. Possessing the ability to reversibly oxidize and recover, it takes part in the most important energy processes of the plant cell, photosynthesis and respiration, and is a recognized antioxidant. Its participation in the processes of growth, flowering, and vegetative and reproductive differentiation, in water exchange, regulation of enzymatic activity, and stimulation of metabolic reactions associated with the exchange of nucleic acids and protein synthesis, in the defense reactions of plants, has been shown. The antiviral and antitumor effects of ascorbic acid and its derivatives have been proven [48,49]. The investigated accessions showed a high variability in the content of vitamin C (12.32–209.44 mg/100 g for DH 30, 8.00–138.70 mg/100 g for DH 38); a significantly higher (by 28–243%) content of ascorbic acid was noted when grown on a 16 h photoperiod (Tables 1 and 2). Similarly, Kim and Lee [50], in their study of the effect of varying light quality and photoperiod on lettuce grown in a closed-type plant production system, also observed maximum vitamin C content under 16 h of light versus 12 and 20 h of light.

It was found that under light culture conditions the content of ascorbic acid in the studied lines was higher than when grown in a greenhouse and in the field (Tables 1 and 2), which is probably due to more optimal light–temperature conditions and the regime of mineral nutrition.

Lines with a consistently high content of ascorbic acid have been identified: DH 30-65 (up to 209.44 mg/100 g), DH 30-134 (up to 139.04 mg/100 g), DH 38-87 (up to 102.08 mg/100 g). Selected genotypes can be used as a source of this important antioxidant.

Anthocyanins are water-soluble plant pigments, phytochemicals belonging to the flavonoid family, and secondary plant metabolites. Accumulating in both vegetative and generative organs of plants, they cause red, pink, purple and blue coloration. It is also known that anthocyanins perform structural, protective, and signaling functions in plants, determine resistance to abiotic and biotic stresses, and participate in respiration and photosynthesis [51,52]. The study of the mechanisms of accumulation of anthocyanins in plant tissues makes it possible to determine the methods of plant adaptation to unfavorable environmental conditions and to understand the physiological and biochemical mechanisms of their resistance. In addition, due to their high antioxidant activity, anthocyanins have antitumor and cardioprotective functions, which makes it possible to use foods rich in them in functional nutrition [53]. Evaluation of DH 30 and DH 38 lines showed an increase in the content of anthocyanins with an increase in the length of daylight hours (Tables 1 and 2). In lines from the DH 30 population, the anthocyanin content varied from 0.49 to 6.34 mg/100 g, and in lines from the DH 38 population, from 0.58 to 4.32 mg/100 g.

Under the conditions of the 16 h photoperiod, the content of anthocyanins on average in the DH 30 population was higher by 8%, and in the DH 38 population by 4%, compared to the 12 h photoperiod. An increase in the biosynthesis of anthocyanins under conditions of longer daylight hours may be the result of a protective reaction of plant cells to photooxidative stress.

In general, significant variability in the content of anthocyanins between lines within each population was revealed. In addition, as a result of the study, homozygous lines of doubled haploids were isolated, which differed in low (DH 30-18—0.49 mg/100 g, DH 30-127—0.40 mg/100 g, DH 38-103—0.70 mg/100 g) and high (DH 30-28—2.96 mg/100 g, DH 38-134—4.32 mg/100 g, DH 38-172—3.07 mg/100 g) content of anthocyanins regardless of growing conditions, which can probably be explained by their genotypic characteristics.

Chlorophylls and carotenoids are the main photosynthetic pigments that provide absorption of light quanta and photosensitization in plants. Studies carried out to date have shown the stability of the qualitative composition of pigments in higher plants [54]. Each of these forms of pigments has a specific role and place in the photosynthetic system of plants. Since pigments are integrated into chloroplast membranes and are associated with proteins, their quantitative content and ratio in the leaf can reflect the adaptation features of the photosynthetic apparatus as a whole and provide its functional diagnostics. Quantitative changes in the pigment apparatus of plant leaves can occur in response to changes in environmental conditions. It is known that light is the main factor regulating the pigment content [55]. Changes in ambient temperature and humidity can also cause shifts in the pigment composition of plant leaves [56].

An increase in the content of photosynthetic pigments (chlorophyll *a*, *b* and carotenoids) was revealed for both studied populations on a short 12 h daylight compared to a long 16 h day (Tables 1 and 2). For plants of the DH 30 population, the content of chlorophyll *a* varied from 35.84 to 147.16 mg/100 g, chlorophyll *b* from 10.73 to 61.99 mg/100 g, and carotenoids from 12.88 to 50.36 mg/100 g; for the DH 38 population—31.05-150.44 mg/100 g, 7.86–86.77 mg/100 g and 9.73–39.97 mg/100 g, respectively. The increase in the content of photosynthetic pigments under the conditions of a 12 h photoperiod was: for chlorophyll *a* 25–46% and 14–30% for the lines DH 30 and DH 38, respectively, for chlorophyll *b* 49–54% and 35–38%, and for carotenoids 25–44% and 24–37%, respectively. A higher content of photosynthetic pigments at a short daylight hour indirectly indicates a higher activity of the photosynthetic system (plants need to most effectively use a limited period of daylight hours for photosynthesis). Similarly, Park et al. [57] showed that maximum chlorophyll fluorescence (index correlating with chlorophyll content) was observed when growing of leaf lettuce at a 12 h photoperiod, compared to a 16 and 18 h photoperiod in a closed-type plant production system.

The results obtained in our study confirm the importance of representatives of the species *B. rapa* as a source of valuable nutrients and biologically active substances [58,59], which indicates the prospects for further study of the influence of external factors on the biochemical composition of species crops for genetic and breeding investigations, especially under controlled conditions. Therefore, Raiola et al. presented a detailed overview of bioactive compounds in *Brassicaceae* vegetables, including ascorbic acid, phenolics, carotenoids, etc., and mechanisms of their action [60]. Authors also presented some ideas of biofortification to optimize the content of valuable phytochemicals in *Brassicaceae* and emphasize the importance of further studying of the factors that control the accumulation of biologically active substances. The high nutritional value of cole crops, as well as the high variability of the content of bioactive compounds, are also discussed in the research published by Tomas et al. [61].

The revealed high degree of variability is in a good correlation with previously published data, which shows a high degree of variation in biochemical characters in the studied *B. rapa* DH lines when grown in field and greenhouse conditions on the territory of the Pushkin Branch of VIR (St. Petersburg, Russia) [1,2,4,5]. At the same time, the data obtained under the conditions of a short 12 h photoperiod under light culture conditions correlate with the data obtained when growing plants in greenhouse in spring, and with a 16 h long photoperiod—with the results under field conditions in the northwest of Russia (Tables 1 and 2).This is due to the timing of growing: in the greenhouse, the plants were grown in March–April, when the daylight hours in St. Petersburg are 12–13 h, the field experiment took place in June–July, when the daylight hours are more than 18 h.

Thus, the differences in the content of the studied biochemical compounds, revealed in the field and greenhouse conditions, are largely explained by the influence of the length of daylight hours.

Based on the obtained biochemical data, a QTL analysis of the biochemical traits of both mapping populations was carried out (Figures 3 and 4, Tables 3 and 4).

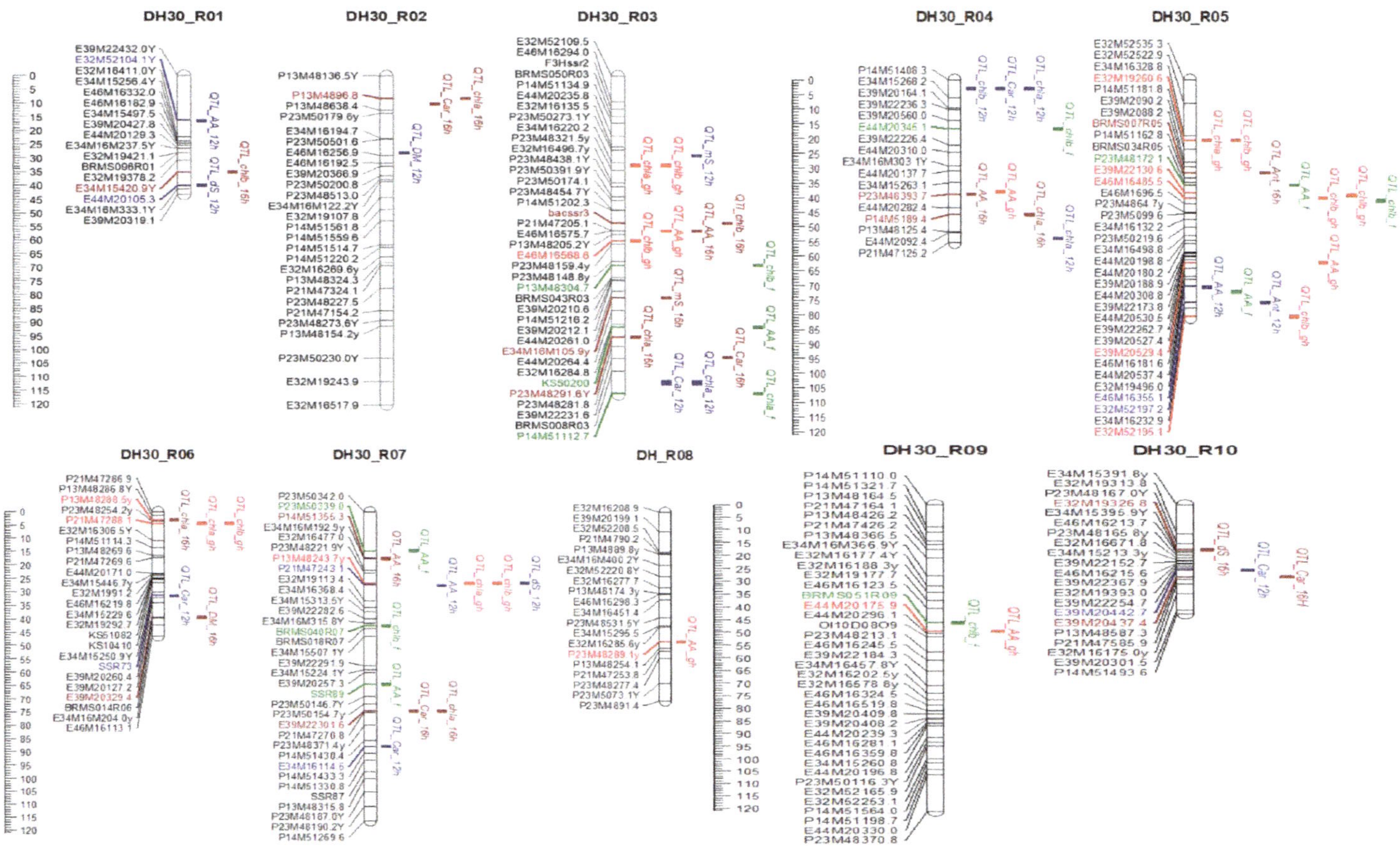

Figure 3. Mapping of QTLs determining expression of biochemical traits of in *Brassica rapa* L. (DH 30): **12 h**—AFI experiment, light culture 12 h; **16 h**—AFI experiment, light culture 16 h; **gh**—VIR experiment, greenhouse; **f**—VIR experiment, field. Trait designations: **AA**—ascorbic acid, **Car**—carotenoids, **chla**—chlorophyll *a*, **chlb**—chlorophyll *b*, **DM**—dry matter, **mS**—monosaccharides, **dS**—disaccharides.

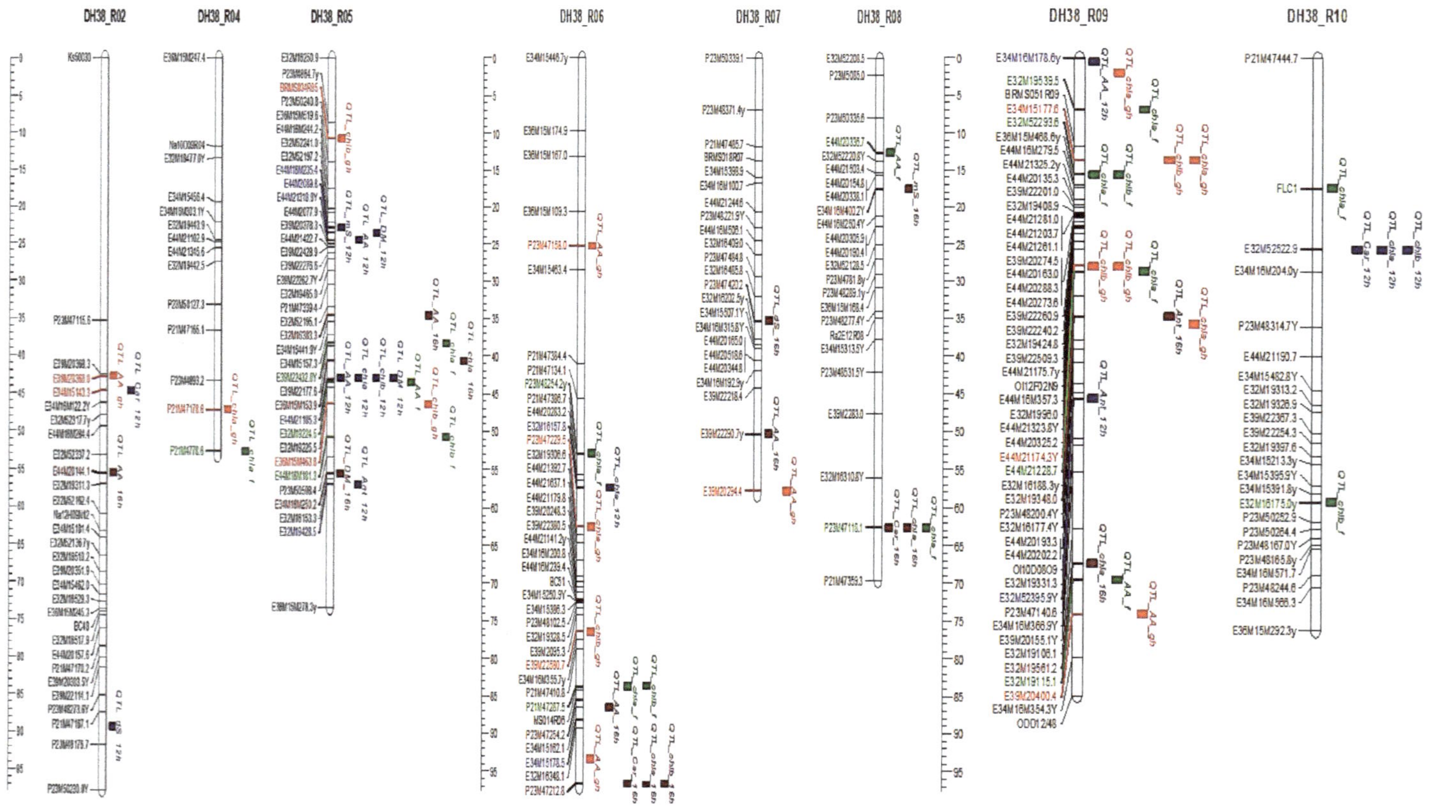

Figure 4. Mapping of QTLs determining expression of biochemical traits of in *Brassica rapa* L. (DH 38): **12 h**—AFI experiment, light culture 12 h; **16 h**—AFI experiment, light culture 16 h; **gh**—VIR experiment, greenhouse; **f**—VIR experiment, field. Trait designations: **AA**—ascorbic acid, **Car**—carotenoids, **chla**—chlorophyll *a*, **chlb**—chlorophyll *b*, **DM**—dry matter, **mS**—monosaccharides, **dS**—disaccharides.

Table 3. QTL mapping of loci that determine the content of biologically active substances and nutrients in the mapping population DH 30.

Trait	Group	Position	LOD	% Expl.	AFI		VIR *	
					12 h	16 h	Greenhouse	Field
Ascorbic acid	R 01	16.20	2.35	45.1	X			
	R 03	56.44	1.18	28.7		X		
	R 03	56.5	1.70	23.0			X	
	R 03	91.62	1.34	32.0				X
	R 04	37.89	1.46	20.0			X	
	R 04	38.91	2.02	44.1		X		
	R 05	35.74	2.20	46.9				X
	R 05	62.18	1.98	26.2			X	
	R 05	70.09	1.27	27.7	X			
	R 05	72.09	0.96	24.2				X
	R 07	14.68	0.99	24.9				X
	R 07	17.45	1.64	37.6		X		
	R 07	27.34	1.73	35.8	X			
	R 07	64.41	1.70	38.7				X
	R 08	48.18	2.13	45.8				X
	R 09	49.59	0.88	12.6			X	
Anthocyanins	R 05	31.54	1.57	33.1		X		
	R 05	75.77	1.83	34.4	X			
Carotenoids	R 02	10.26	2.92	52.6		X		
	R 03	31.63	2.68	33.7			X	
	R 03	73.54	1.17	28.7				X
	R 03	73.55	0.39	5.9			X	
	R 03	102.83	1.07	23.9		X		
	R 03	111.89	1.40	27.6	X			
	R 04	3.00	2.23	40.1	X			
	R 04	16.74	1.11	27.4				X
	R 04	22.27	0.82	11.8			X	
	R 05	20.70	0.96	24.2				X
	R 05	20.70	0.83	11.9			X	
	R 06	4.37	1.86	24.9			X	
	R 06	31.31	1.03	21.1	X			
	R07	74.49	2.82	51.4		X		
	R 07	79.92	0.94	23.6				X
	R 07	82.92	1.52	32.1	X			
	R 07	105.19	2.12	45.7				X
	R 09	31.86	1.21	29.4				X
	R 10	25.46	1.07	24.0	X			
	R 10	28.24	1.07	24.0		X		

Table 3. *Cont.*

Trait	Group	Position	LOD	% Expl.	AFI		VIR *	
					12 h	16 h	Greenhouse	Field
Chlorophyll *a*	R 02	8.26	2.44	46.5		X		
	R 03	32.63	2.34	30.2			X	
	R 03	95.20	1.77	36.4		X		
	R 03	111.89	1.39	27.3	X			
	R 03	116.01	1.23	29.9				X
	R 04	3.00	2.47	43.4	X			
	R 04	45.87	1.10	24.4		X		
	R 04	53.99	1.42	27.8	X			
	R 05	20.70	0.74	10.7			X	
	R 05	40.19	1.57	36.4				X
	R 06	3.08	1.09	24.4		X		
	R 06	4.36	1.69	22.8			X	
	R 07	26.60	0.78	11.2			X	
	R 07	74.49	2.00	40.0		X		
Chlorophyll *b*	R 01	35.22	3.17	55.6		X		
	R 03	32.63	2.65	33.4			X	
	R 03	53.66	1.64	34.2		X		
	R 03	69.15	1.29	31.1				X
	R 03	60.10	0.84	12.1			X	
	R 04	3.00	2.77	47.2	X			
	R 04	16.74	1.10	27.2				X
	R 05	20.70	0.56	8.3			X	
	R 05	39.25	0.71	10.3			X	
	R 05	41.19	1.25	30.2				X
	R 05	80.57	1.50	35.1				X
	R 06	4.36	2.10	27.5			X	
	R 07	26.60	1.15	16.2			X	
	R 07	42.40	0.62	16.3				X
	R 09	46.16	1.55	36.0				X
Dry matter	R 02	28.24	1.90	38.5	X			
	R 06	39.32	1.98	39.7		X		
Monosaccharides	R 03	29.12	1.42	29.1	X			
	R 03	80.87	1.60	38.8		X		
Disaccharides	R 01	40.02	3.01	51.8	X			
	R 03	48.68	2.26	42.2	X			
	R 07	26.60	2.81	49.4	X			
	R 10	17.44	2.79	57.5		X		

* By Artemyeva et al., 2012 [1].

Table 4. QTL mapping of loci that determine the content of biologically active substances and nutrients in the mapping population DH 38.

Trait	Group	Position	LOD	% Expl.	AFI		VIR *	
					12 h	16 h	Greenhouse	Field
Ascorbic acid	R 02	42.75	0.62	8.6			X	
		55.56	1.81	32.7		X		
	R 05	24.52	3.10	35.9	X			
		34.64	2.07	36.5		X		
		43.07	2.28	28.0	X			
		43.50	0.93	11.6				X
	R 06	25.18	2.59	31.1			X	
		86.52	1.53	28.5		X		
		93.29	1.53	19.8			X	
	R 07	50.20	1.76	32.0		X		
		57.74	0.55	7.7			X	
	R 08	12.49	1.51	18.1				X
	R 09	0.00	2.69	32.1	X			
		69.54	1.14	14.0				X
		74.03	1.66	21.2			X	
Anthocyanins	R 05	56.99	1.54	19.9	X			
	R 09	34.66	2.64	39.7		X		
		45.59	1.21	16.0	X			
Carotenoids	R 02	44.68	1.04	13.9	X			
	R 02	44.68	0.74	10.1			X	
	R04	47.16	0.88	12.0			X	
	R05	10.66	0.59	8.1			X	
	R06	62.48	1.29	17.0			X	
		96.59	1.91	30.6		X		
	R 08	62.52	1.41	23.6		X		
	R09	13.6	2.18	26.9			X	
	R 10	25.56	2.28	27.9	X			
Chlorophyll *a*	R 04	47.16	0.80	10.9			X	
		52.69	0.72	9.0				X
	R 05	38.28	1.05	12.9				X
		40.74	1.39	23.4		X		
	R 06	43.07	1.05	14.0	X			
		52.78	0.84	10.5				X
		57.27	1.26	16.6	X			
		62.48	1.44	18.7			X	
		83.67	1.22	14.8				X
		96.59	1.94	31.1		X		

Table 4. *Cont.*

Trait	Group	Position	LOD	% Expl.	AFI		VIR *	
					12 h	16 h	Greenhouse	Field
	R 08	62.51	0.80	10.0				X
		62.52	1.24	21.2		X		
	R 09	2.00	1.01	13.5			X	
		6.92	1.34	16.2				X
		13.60	1.94	24.3			X	
		15.58	1.83	21.4				X
		28.74	2.01	23.2				X
		35.65	1.31	17.2			X	
		67.28	1.00	17.4		X		
	R 10	17.43	0.98	12.1				X
		25.56	2.45	29.7	X			
Chlorophyll *b*	R 05	10.665	0.65	8.9			X	
		43.07	1.10	14.6	X			
		46.39	0.82	11.1			X	
		50.83	4.46	44.4				X
	R 06	76.37	1.37	17.9			X	
		83.67	1.2	14.6				X
		96.59	1.56	25.9		X		
	R 09	15.58	1.53	18.2				X
		13.60	2.77	32.9			X	
	R 10	25.56	2.09	25.9	X			
		27.95	2.13	26.4			X	
		59.23	1.24	15.0				X
Dry matter	R 05	23.50	2.03	25.4	X			
		43.07	1.47	19.1	X			
		55.58	1.72	28.1		X		
Monosaccharide	R 05	22.81	1.27	18.3	X			
	R 08	17.37	2.39	45.7		X		
Disaccharides	R 02	89.36	1.23	17.7	X			
	R 07	35.17	1.76	36.3		X		

* By Artemyeva et al. 2012 [1].

3.3. *Quantitative Trait Loci Mapping (QTL Analysis)*

The result of the QTL analysis was the identification of loci mapped under various growing conditions and controlling the content of dry matter, sugars, ascorbic acid, anthocyanins, chlorophylls *a* and *b*, carotenoids in populations of homozygous DH lines of *B. rapa* grown under light culture conditions with 12 h and 16 h photoperiod (Tables 3 and 4). It should be noted that there is a large total number of QTLs for the control of the studied biochemical traits (a total of 102 QTLs were identified for both populations—54 for DH 30 and 48 for DH 30) and simultaneous control of several traits by one locus (22 of the identified QTLs).

The result of the QTL analysis was the identification of loci mapped under various growing conditions and controlling the content of dry matter, sugars, ascorbic acid, anthocyanins, chlorophylls *a* and *b*, and carotenoids in populations of homozygous DH lines of *B. rapa* grown under light culture conditions with 12 h and 16 h photoperiod (Tables 3 and 4). It should be noted that there is a large total number of QTLs for the control of the studied biochemical traits (a total of 102 QTLs were identified for both populations—54 for DH 30 and 48 for DH 30) and simultaneous control of several traits by one locus (22 of the identified QTLs).

In the DH 30 population, loci at the beginning and middle of R01 were identified that control the content of ascorbic acid, chlorophyll *b*, and disaccharides, and in R02, dry matter, chlorophyll *a*. The third linkage group (R03) contained loci that control the content of ascorbic acid, chlorophylls *a* and *b*, carotenoids, and mono- and disaccharides. The loci mapped at the beginning of R04 determined the content of ascorbic acid, carotenoids, chlorophyll *a*, chlorophyll *b*, and the loci in the middle of R04, ascorbic acid, carotenoids, and chlorophyll *a*. At the beginning and middle of the fifth linkage group (R05), there were loci that control the manifestation of many of the studied biochemical characteristics: the content of ascorbic acid, anthocyanins, carotenoids, and chlorophylls *a* and *b*. The loci at the beginning of R06 determined the content of chlorophyll *a* and chlorophyll *b*, and in the middle, carotenoids. In the middle of the seventh (R07) and ninth (R09) linkage groups, there were loci that generally determine the content of ascorbic acid, chlorophyll *b*, and carotenoids, and in the middle of R07, loci were mapped that determine the content of chlorophyll *a* and disaccharides. The locus mapped to the eighth clutch group (R08) controlled the ascorbic acid content. In addition, the locus at the beginning of R10 controlled the content of carotenoids.

In the DH 38 population, the loci identified on the second linkage group (R02) controlled the content of ascorbic acid, carotenoids and disaccharides, and in the middle of R04, carotenoids. In the fifth linkage group (R05), loci were mapped that determine many biochemical parameters in the studied lines of *B. rapa*: the content of dry matter, ascorbic acid, carotenoids, chlorophylls *a* and *b*, anthocyanins, dry matter, and monosaccharides. In the sixth linkage group (R06), loci were also identified that determine the content of ascorbic acid, carotenoids, and chlorophylls *a* and *b*. In the middle of R07, the locus controlling the content of ascorbic acid was mapped. The loci mapped in the eighth linkage group determined the content of ascorbic acid, carotenoids, and chlorophyll *a*. In the ninth linkage group (R09), loci were identified that control the content of ascorbic acid, anthocyanins, carotenoids, and chlorophyll *a* and *b*. Loci at the beginning of R10 controlled the content of carotenoids and chlorophyll *a*; in the middle of R10, the locus determining the content of chlorophyll *b* was mapped.

Noteworthy is the relatively low LOD score for some of the identified QTLs. According to the literature, an LOD score below 3 is often referred to as a low confidence level due to the repetitiveness of QTL testing [62]. We have shown that at high LOD values, 1/A is close to the type I error and, conversely, at low LOD values, the error is stably less than 1/A, which indicates that the LOD estimate is quite conservative. In this case, the critical value 3 will correspond to the maximum value of the type I error (at $p < 0.001$), and if a very high particular (individual) type I error is selected, for example 5%, then a high level of cohesion will be reliably found randomly. At the same time, both major and minor QTLs are often localized in the same positions in different experiments and even in different years of experiments; therefore, an LOD score below 3 can also be taken into account [38]. It is also likely that the low LOD score for some of the identified QTLs is due to statistical reasons (limited sample size), since only lines adapted to light culture conditions were included in the estimate.

The identified loci controlling all the studied biochemical traits were predominantly in the fifth, sixth, seventh, and ninth linkage groups, which is consistent with the results of studies on mapping biochemical traits loci in *B. rapa*, which were carried out in the field and in a greenhouse [1,2,4,5].

4. Conclusions

Thus, for the first time, we identified QTLs that determine the content of nutrient and biologically active substances of *B. rapa* when grown under light culture conditions with simulation of a contrast photoperiod, and also identified molecular markers that are genetically linked to them, which in the future allows for molecular genetic screening of accessions from the collection and breeding material of the species *B. rapa* according to these economically valuable biochemical characters.

The influence of daylight hours on the content of some phytochemical compounds (dry matter content, total content of sugars, mono- and disaccharides, ascorbic acid, chlorophylls *a* and *b*, carotenoids, anthocyanins) in 68 lines of two mapping populations of *B. rapa* was evaluated. The high variability of the biochemical composition in the lines of the studied populations of brassica crops was confirmed. A significant effect of different photoperiod durations on the manifestation of the studied signs was revealed. Lines with a high content of the studied compounds were identified, regardless of the growing conditions. Based on the data obtained, QTL analysis revealed 102 QTLs that control the content of nutrients and biologically active components in the studied populations of *B. rapa*, which are located mainly in the fifth, sixth, seventh and ninth linkage groups. Some of the identified loci controlled several biochemical traits simultaneously. In general, the results obtained can be of practical importance for increasing the breeding efficiency of *B. rapa* and can be used to create new genotypes, lines and cultivars of crops with increased biochemical value, high content of biologically active substances and adapted to specific photoperiodic conditions of the growing region, as well as to growing in light culture.

Author Contributions: Conceptualization, Y.V.C., A.M.A. and K.V.E.; methodology, K.V.E., N.G.S. and A.M.A.; software, N.V.K.; validation N.V.K., K.V.E. and Y.V.C.; resources, A.M.A.; writing—original draft preparation, K.V.E., Y.V.C. and A.M.A.; writing—review and editing, K.V.E., Y.V.C., A.M.A. and N.G.S.; visualization, K.V.E., N.V.K., N.G.S. and Y.V.C.; project administration, Y.V.C. and A.M.A. All authors have read and agreed to the published version of the manuscript.

Funding: The study was financially supported by the Ministry of Education and Science of the Russian Federation (Agreement no. 075-15-2020-805 from 2 October 2020).

Data Availability Statement: The data presented in this study are available upon request from the corresponding author.

Acknowledgments: The authors are grateful for the help in carrying out biochemical analyses of the research group of the laboratory of biochemistry of soil–plant systems of the Agrophysical Research Institute.

Conflicts of Interest: The authors declare no conflict of interest. The funders had no role in the design of the study; in the collection, analyses, or interpretation of data; in the writing of the manuscript, or in the decision to publish the results.

References

1. Artemyeva, A.M.; Solovieva, A.E.; Kocherina, N.V.; Rudneva, E.N.; Volkova, E.N.; Chesnokov, Y.V. *DNA Marked Brassica rapa L. Double Haploid Lines and Identified QTLs that Control Economically Valuable Traits for Use in Breeding Cabbage Leaf Crops*; Catalogue of the VIR Global Collection, Bull. 810; VIR: St. Petersburg, Russia, 2012; 174p. (In Russian)
2. Artemyeva, A.M.; Solovieva, A.E.; Kocherina, N.V.; Berensen, F.A.; Rudneva, E.N.; Chesnokov, Y.V. Mapping of chromosome loci determined manifestation of morphological and biochemical traits of quality in *Brassica rapa* L. *crops. Russ. J. Plant Physiol.* **2016**, *63*, 259–272. [CrossRef]
3. Favela-González, K.M.; Hernández-Almanza, A.Y.; De la Fuente-Salcido, N.M. The value of bioactive compounds of cruciferous vegetables (Brassica) as antimicrobials and antioxidants: A review. *J. Food Biochem.* **2020**, *44*, e13414. [CrossRef] [PubMed]
4. Artemyeva, A.M.; Solovieva, A.Y.; Chesnokov, Y.V. Estimating morphological and biochemical characters of quality in doubled haploid lines *Brassica rapa* L. *Vestn. Russ. Agric. Sci.* **2014**, *3*, 38–41, (In Russian with English abstract)
5. Artemyeva, A.M.; Solovieva, A.E.; Kocherina, N.V.; Chesnokov, Y.V. QTL analysis of biochemical traits of quality in *Brassica rapa* L. *Veg. Crop. Russ.* **2014**, *1*, 10–13, (In Russian with English abstract) [CrossRef]
6. El-Mogy, M.M.; Mahmoud, A.W.M.; El-Sawy, M.B.Y.; Parmar, A. Pre-harvest foliar application of mineral nutrients to retard chlorophyll degradation and preserve bio-active compounds in broccoli. *Agronomy* **2019**, *9*, 711. [CrossRef]

7. Artemyeva, A.M.; Rudneva, E.N.; Kocherina, N.V.; Chesnokov, Y.V. QTL analysis of morphological traits of quality in *Brassica rapa* L. *Veg. Crop. Russ.* **2015**, *2*, 14–17, (In Russian with English abstract) [CrossRef]
8. Verma, K.; Tripathi, M.K.; Tiwari, S.; Tripathi, N. Analysis of genetic diversity among Brassica juncea genotypes using morpho-physiological and SSR markers. *Int. J. Curr. Microbiol. Appl. Sci.* **2021**, *10*, 1108–1117. [CrossRef]
9. Qi, X.; An, H.; Hall, T.E.; Di, C.; Blischak, P.D.; McKibben, M.T.W.; Hao, Y.; Conant, G.C.; Pires, J.C.; Barker, M.S. Genes derived from ancient polyploidy have higher genetic diversity and are associated with domestication in Brassica rapa. *New Phytol.* **2021**, *230*, 372–386. [CrossRef]
10. Collard, B.C.Y.; Jahufer, M.Z.Z.; Brouwer, J.B.; Pang, E.C.K. An introduction to markers, quantitative trait loci (QTL) mapping and marker-assisted selection for crop improvement: The basic concepts. *Euphytica* **2005**, *142*, 169–196. [CrossRef]
11. Chesnokov, Y.V. *Plant Genetic Resources and Modern Methods of DNA-Fingerprinting*; VIR: St. Petersburg, Russia, 2007; 80p. (In Russian)
12. Witzel, K.; Kurina, A.B.; Artemyeva, A.M. Opening the treasure chest: The current status of research on Brassica oleracea and B. *rapa vegetables from ex situ germplasm collections. Front. Plant Sci.* **2021**, *12*, 643047. [CrossRef] [PubMed]
13. Pink, D.; Bailey, L.; McClement, S.; Hand, P.; Mathas, E.; Buchanan-Wollaston, V.; Astley, D.; King, G.; Teakle, G. Double haploids, markers and QTL analysis in vegetable brassicas. *Euphytica* **2008**, *164*, 509–514. [CrossRef]
14. Lou, P.; Woody, S.; Brenham, K.; VanBuren, R.; Colle, M.; Edger, P.P.; Sartor, R.; Zheng, Y.; Levendoski, N.; Lim, J.; et al. Genetic and genomic resources to study natural variation in Brassica rapa. *Plant Direct* **2020**, *4*, e00285. [CrossRef]
15. Cockram, J.; Mackay, I. Genetic mapping populations for conducting high-resolution trait mapping in plants. In *Plant Genetics and Molecular Biology*; Varshney, R., Pandey, M., Chitikineni, A., Eds.; Advances in Biochemical Engineering/Biotechnology; Springer: Cham, Switzerland, 2018; Volume 164, pp. 109–138. [CrossRef]
16. Chesnokov, Y.V. *Mapping of Quantitative Trait Loci in Plants*; VIR: St. Petersburg, Russia, 2009; 100p. (In Russian)
17. Şahin, Ö.; Kavuncu, O. A Study on relationships between various quantitative characteristics for using availability quantitative trait loci mapping. *Selcuk J. Agric. Food Sci.* **2019**, *33*, 42–44. [CrossRef]
18. Jan, S.A.; Shinwari, Z.K.; Rabbani, M.A. Determining genetic divergence among *Brassica rapa* ecotypes through electrophoretic mobility of total seed proteins. *J. Anim. Plant Sci.* **2016**, *26*, 1758–1764.
19. Ullah, N.; Khan, J.; Khan, M.W.; Raza, H.; Alam, M.; Ullah, H.; Hussain, I.; Jan, I.U.; Rahman, Z.; Ali, F. Genetic variability for biochemical traits among advanced lines of Brassica. *Pure Appl. Biol* **2017**, *6*, 1–8. [CrossRef]
20. Lou, P.; Zhao, J.; He, H.; Hanhart, C.; Pino del Carpio, D.; Verkerk, R.; Custers, J.; Koornneef, M.; Bonnema, G. Quantitative trait loci for glucosinolates accumulation in *Brassica rapa* leaves. *New Phytol.* **2008**, *179*, 1017–1032. [CrossRef]
21. Zhao, J.; Paulo, M.J.; Jamar, D.; Lou, P.; van Eeuwijk, F.; Bonnema, G.; Vreugdenhil, D.; Koornneef, M. Association mapping of leaf traits, flowering time, and phytate content in Brassica rapa. *Genome* **2007**, *50*, 963–973. [CrossRef] [PubMed]
22. Bagheri, H.; El-Soda, M.; Kim, H.K.; Fritsche, S.; Jung, C.; Aarts, M.G. Genetic analysis of health-related secondary metabolites in *Brassica rapa* recombinant inbred lines population. *Int. J. Mol. Sci.* **2013**, *14*, 15561–15577. [CrossRef]
23. El-Soda, M.; Boer, M.P.; Bagheri, H.; Hanhart, C.J.; Koornneef, M.; Aarts, M.G. Genotype–environment interactions affecting preflowering physiological and morphological traits of Brassica rapa grown in two watering regimes. *J. Exp. Bot.* **2014**, *65*, 697–708. [CrossRef]
24. Kochetov, A.A.; Makarova, G.A.; Mirskaya, G.V.; Sinyavina, N.G. Agrophysical approach to the creation of new forms of cultivated plants. *Agrophysics* **2012**, *1*, 40–44, (In Russian with English abstract)
25. Panova, G.G.; Udalova, O.R.; Kanash, E.V.; Galushko, A.S.; Kochetov, A.A.; Priyatkin, N.S.; Arkhipov, M.V.; Chernousov, I.N. Fundamentals of Physical Modeling "ideal" agroecosystems. *Tech. Phys.* **2020**, *65*, 1563–1569. [CrossRef]
26. He, Y.; Chen, T.; Zeng, X. Genetic and epigenetic understanding of the seasonal timing of flowering. *Plant Commun.* **2020**, *1*, 100008. [CrossRef]
27. Wang, S.L.; An, H.R.; Tong, C.G.; Jang, S. Flowering and flowering genes: From model plants to orchids. *Hortic. Environ. Biotechnol.* **2021**, *62*, 135–148. [CrossRef]
28. Lou, P.; Zhao, J.; Kim, J.S.; Shen, S.; Del Carpio, D.P.; Song, X.; Jin, M.; Vreugdenhil, M.; Wand, X.; Koornneef, M.; et al. Quantitative trait loci for flowering time and morphological traits in multiple populations of Brassica rapa. *J. Exp. Bot.* **2007**, *58*, 4005–4016. [CrossRef] [PubMed]
29. Egorova, K.V.; Sinyavina, N.G.; Kochetov, A.A.; Chesnokov, Y.V. Assessment of significant for breeding morphological traits in the double haploid population of *Brassica rapa* L. *in controlled conditions of a regulated agroecosystem. Veg. Crop. Russ.* **2020**, *4*, 28–31, (In Russian with English abstract) [CrossRef]
30. Interstate Standard 12038-84. *Agricultural seeds. Methods for Determination of Germination*; Standartinform Publishing: Moscow, Russia, 2004. (In Russian)
31. Ermakov, E.I.; Zheltov, Y.I.; Milto, N.E.; Kucherov, V.I. Soil for Growing Plants "Agrophyte". Russian Federation Patent 2081555, 13 July 1997. (In Russian)
32. State Standard of Russian Federation 8756. *Fruit and Vegetable Products. Methods for Determination of Sugars*; Standartinform Publishing: Moscow, Russia, 2010. (In Russian)
33. State Standard of Russian Federation 24556-89. *Products of Fruits and Vegetables Processing. Methods for Determination of Vitamin C*; Standartinform Publishing: Moscow, Russia, 2003. (In Russian)

34. Interstate Standart NF V05-129-1984. *Fruit and Vegetable Products. Conventionnal Determination of Authocyanins. Detection and Identification*; AFNOR Editions: Paris, France, 1984.
35. Andersen, O.M.; Markham, K.R. *Flavonoids: Chemistry, Biochemistry and Applications*; CRC Press: Boca Raton, FL, USA, 2006; 1256p.
36. Trineeva, O.V.; Safonova, E.F.; Slivkin, A.I.; Voropaeva, S.V. Method of Spectrophotometric Quantitative Determination in the Leaves of Stinging Nettle in the Joint Presence of Chlorophyll, Carotenoids and Hydroxycinnamic Acids. Russian Federation Patent 2531940, 27 October 2014. (In Russian)
37. Van Ooijen, J.W. *MapQTL®6, Software for the Mapping of Quantitative Trait Loci in Experimental Populations of Diploid Species*; Kyazma BV: Wageningen, The Netherlands, 2009; Available online: https://www.kyazma.nl/index.php/mc.MapQTL/ (accessed on 22 November 2021).
38. Kocherina, N.V.; Artemyeva, A.M.; Chesnokov, Y.V. Using the technology of LOD score in mapping the quantitative trait loci in plants. *Russ. Agric. Sci.* **2011**, *3*, 14–17. [CrossRef]
39. Ronald, P.S.; Penner, G.A.; Brown, P.D.; Brule-Babel, A. Identification of RAPD markers for percent hull in oat. *Genome* **1997**, *40*, 873–878. [CrossRef] [PubMed]
40. Voorrips, R.E. MapChart: Software for the graphical presentation of linkage maps and QTLs. *J. Hered.* **2002**, *93*, 77–78. Available online: https://www.wur.nl/en/show/mapchart.htm (accessed on 22 November 2021). [CrossRef] [PubMed]
41. Barba, F.J.; Esteve, M.J.; Frígola, A. Bioactive components from leaf vegetable products. *Stud. Nat. Prod. Chem.* **2014**, *41*, 321–346. [CrossRef]
42. Artemyeva, A.M.; Sinyavina, N.G.; Panova, G.G.; Chesnokov, Y.V. Biological features of *Brassica rapa* L. vegetable leafy crops when growing in an intensive light culture. *Agric. Biol.* **2021**, *56*, 103–120. [CrossRef]
43. Kapusta-Duch, J.; Kopec, A.; Piatkowska, E.; Borczak, B.; Leszczynska, T. The beneficial effects of Brassica vegetables on human health. *Ann. Natl. Inst. Hyg.* **2012**, *63*, 389–395.
44. Xiao, D.; Shen, H.R.; Zhao, J.J.; Wei, Y.P.; Liu, D.R.; Hou, X.L.; Bonnema, G. Genetic dissection of flowering time in *Brassica rapa* responses to temperature and photoperiod. *Plant Sci.* **2019**, *280*, 110–119. [CrossRef]
45. Rahman, H.; Bennett, R.A.; Kebede, B. Molecular mapping of QTL alleles of Brassica oleracea affecting days to flowering and photosensitivity in spring Brassica napus. *PLoS ONE* **2018**, *13*, e0189723. [CrossRef] [PubMed]
46. Mandal, K.G.; Sinha, A.C. Nutrient management effects on light interception, photosynthesis, growth, dry-matter production and yield of Indian mustard (Brassica juncea). *J. Agron. Crop Sci.* **2004**, *190*, 119–129. [CrossRef]
47. Kurina, A.B.; Kornyukhin, D.L.; Solovyeva, A.E.; Artemyeva, A.M. Genetic diversity of phenotypic and biochemical traits in VIR radish (Raphanus sativus L.) germplasm collection. *Plants* **2021**, *10*, 1799. [CrossRef] [PubMed]
48. Chupakhina, G.N. *Plant Ascorbic Acid System: Monograph*; Immanuel Kant Baltic Federal University: Kaliningrad, Russia, 1997; 120p. (In Russian)
49. Arrigoni, O.M.; De Tullio, C. Ascorbic acid: Much more than just an antioxidant. *Biochim. Biophys. Acta Gen. Subj.* **2002**, *1569*, 1–9. [CrossRef]
50. Kim, Y.H.; Lee, J.S. Growth and contents of anthocyanins and ascorbic acid in lettuce as affected by supplemental UV-A LED irradiation with different light quality and photoperiod. *Korean J. Hortic. Sci. Technol.* **2016**, *34*, 596–606. [CrossRef]
51. Welch, C.R.; Wu, Q.; Simon, J.E. Recent advances in anthocyanin analysis and characterization. *Curr. Anal. Chem.* **2008**, *4*, 75–101, (In Russian with English abstract) [CrossRef]
52. Babak, O.G.; Nekrashevich, N.A.; Nikitinskaya, T.V.; Yatsevich, K.K.; Kilchevsky, A.V. Study of the Myb-factor polymorphism based on comparative genomics of vegetable Solanaceae crops (tomato, pepper, eggplant) to search for DNA markers that differentiate samples by the anthocyans accumulation. *Dokl. Natl. Acad. Sci. Belarus* **2020**, *63*, 721–729, (In Russian with English abstract) [CrossRef]
53. Šamec, D.; Urlic, B.; Salopek-Sondi, B. Kale (Brassica oleracea var. acephala) as a superfood: Review of the scientific evidence behind the statement. *Crit. Rev. Food Sci. Nutr.* **2019**, *59*, 2411–2422. [CrossRef] [PubMed]
54. Esteban, R.; Barrutia, O.; Artetxe, U.; Fernandez-Marin, B.; Hernandez, A.; Garcia-Plazaola, J.I. Internal and external factors affecting photosynthetic pigment composition in plants: A meta-analytical approach. *New Phytol.* **2015**, *206*, 268–280. [CrossRef] [PubMed]
55. Lambers, H.; Chapin, F.S.; Pons, T.L. *Plant Physiological Ecology*; Springer: New York, NY, USA, 2008; p. 623, (In Russian with English abstract) [CrossRef]
56. Ivanov, L.A.; Ronzhina, D.A.; Yudina, P.K.; Zolotareva, N.V.; Kalashnikova, I.V.; Ivanova, L.A. Seasonal dynamics of the content of chlorophylls and carotenoids in the leaves of steppe and forest plants at the species and community level. *Russ. J. Plant Physiol.* **2020**, *67*, 453–462. [CrossRef]
57. Park, J.E.; Park, Y.G.; Jeong, B.R.; Hwang, S.J. Growth and anthocyanin content of lettuce as affected by artificial light source and photoperiod in a closed-type plant production system. *Kor. J. Hort. Sci. Technol.* **2012**, *30*, 673–679. [CrossRef]
58. Fernandes, F.; Valentão, P.; Sousa, C.; Pereira, J.A.; Seabra, R.M.; Andrade, P.B. Chemical and antioxidative assessment of dietary turnip (*Brassica rapa* var. rapa L.). *Food Chem.* **2007**, *105*, 1003–1010. [CrossRef]
59. Egorova, K.V.; Dubovitskaya, V.I.; Khomyakov, Y.V.; Chesnokov, Y.V. Study of photoperiod effect on microelements content in cabbage plants of *Brassica rapa* L. *Agrophysica* **2021**, *2*, 37–44, (In Russian with English abstract) [CrossRef]
60. Raiola, A.; Errico, A.; Petruk, G.; Monti, D.M.; Barone, A.; Rigano, M.M. Bioactive compounds in Brassicaceae vegetables with a role in the prevention of chronic diseases. *Molecules* **2018**, *23*, 15. [CrossRef] [PubMed]

61. Tomas, M.; Zhang, L.; Zengin, G.; Rocchetti, G.; Capanoglu, E.; Lucini, L. Metabolomic insight into the profile, in vitro bioaccessibility and bioactive properties of polyphenols and glucosinolates from four Brassicaceae microgreens. *Food Res. Int.* **2021**, *140*, 110039. [CrossRef]
62. Lander, E.S.; Botstein, D. Mapping mendelian factors underlying quantitative traits using RFLP linkage maps. *Genetics* **1989**, *121*, 185–199. [CrossRef] [PubMed]

horticulturae

MDPI

Article

Construction of an Intragenic SSR-Based Linkage Map and QTL Mapping for Agronomic Traits in Chinese Cabbage (*Brassica rapa* L. ssp. *pekinensis*)

Hanzhong Gao [1,2,†], Xiaogang Yang [1,3,†], Hongxia Wang [1], Nianwei Qiu [3], Yanan Chen [1,4], Fengde Wang [1,4], Yihui Zhang [1,4], Huayin Li [1], Jingjuan Li [1,4,*] and Jianwei Gao [1,4,*]

1 Institute of Vegetables, Shandong Academy of Agricultural Sciences, Jinan 250100, China; gghans726@gwu.edu (H.G.); yangxiaogang2022@126.com (X.Y.); 18401154792@163.com (H.W.); chenyanan0043@163.com (Y.C.); wfengde@163.com (F.W.); zyh_0923@163.com (Y.Z.); lihuanyin1@163.com (H.L.)
2 Data Science Program, Columbian College of Arts & Sciences, The George Washington University, Phillips Hall, 801 22nd St. NW, Washington, DC 20052, USA
3 College of Life Science, Qufu Normal University, Qufu 273165, China; nianweiqiu@163.com
4 College of Life Science, Shandong Normal University, Jinan 250014, China
* Correspondence: lijj0620@163.com (J.L.); jwg_738@163.com (J.G.); Tel.: +86-531-66659060 (J.L.); +86-531-66659193 (J.G.)
† These authors contributed equally to this work.

Abstract: Chinese cabbage (*Brassica rapa* L. ssp. *pekinensis*) is one of the most widely cultivated and economically important vegetables in China. Constructing an effective genetic linkage map and mapping quantitative trait loci (QTLs) related to yield and leafy head morphology is of great importance for molecular breeding of Chinese cabbage. Using two diverse Chinese cabbage inbred lines, ZHB and G291, as parents, an F_2 segregating population consisting of 240 individuals was prepared for genetic map construction and phenotype investigation in this study. The two parents are significantly different in both shape and size. Sixteen important agronomic traits of F_2 individuals were investigated. A genetic map of 105 intragenic simple sequence repeat (SSR) markers distributed across 10 linkage groups (LGs) was constructed, which was 2034.1 cM in length and had an average inter-locus distance of 21.75 cM. We identified 48 QTLs for the tested important agronomic traits on the studied LGs, with LOD scores of 2.51–12.49, which explained the phenotypic variance of 3.41–26.66%. The QTLs identified in this study will facilitate further genetic analysis and marker-assisted genetic improvement of Chinese cabbage.

Keywords: Chinese cabbage; intragenic SSR; quantitative trait loci mapping; agronomic traits

Citation: Gao, H.; Yang, X.; Wang, H.; Qiu, N.; Chen, Y.; Wang, F.; Zhang, Y.; Li, H.; Li, J.; Gao, J. Construction of an Intragenic SSR-Based Linkage Map and QTL Mapping for Agronomic Traits in Chinese Cabbage (*Brassica rapa* L. ssp. *pekinensis*). *Horticulturae* **2022**, *8*, 165. https://doi.org/10.3390/horticulturae8020165

Academic Editor: Dilip R. Panthee

Received: 19 January 2022
Accepted: 9 February 2022
Published: 15 February 2022

Publisher's Note: MDPI stays neutral with regard to jurisdictional claims in published maps and institutional affiliations.

Copyright: © 2022 by the authors. Licensee MDPI, Basel, Switzerland. This article is an open access article distributed under the terms and conditions of the Creative Commons Attribution (CC BY) license (https://creativecommons.org/licenses/by/4.0/).

1. Introduction

Chinese cabbage (*Brassica rapa* L. ssp. *pekinensis*, AA, 2n = 20), which originated in China, is one of the most widely cultivated and economically important vegetables in eastern Asia. After hundreds of years of evolution and breeding, the important agronomic traits related to yield vary greatly among different Chinese cabbage varieties. The inheritance of these yield-related agronomic traits is of great importance for genetic improvement of Chinese cabbage. However, most of these traits are complex quantitative traits, and the expression of the controlling genes is influenced by the internal and/or external environment [1,2]. The genetic base and molecular mechanisms involved in regulating these agronomic traits are yet to be understood.

Genetic linkage maps are effective tools for studying and locating the genetic loci of interesting traits in the genomes of plants. In previous studies, some genetic maps of Chinese cabbage have been constructed using different molecular markers, such as AFLP [3],

RFLP [4], STS [5], simple sequence repeat (SSR) [6–9], InDel [7,8], and SNP [2,10,11], in different genetic populations such as F_2 [2,3,6,7,9], F_3 [5], RIL [4,8,10], and DH lines [1,11]. A number of quantitative trait loci (QTLs) have been identified in Chinese cabbage recently, including trichome number [12–14], flowering time [11,15], flower color [16], anthocyanin accumulation [17], plant morphological traits [7,15,18–21], orange inner leaves [6,22], seed coat color [3,9,23,24], bolting trait [8], floral stalk length [25], disease resistance [5,26–29], reproductive fitness traits [4], and yield-related traits [1,30]. For some genetically simple traits such as trichome number [12–14], seed coat color [3,9,25], and orange inner leaves [6,22,31], many candidate genes have been identified according to map-based cloning methods, and efficient molecular markers have been developed for marker-assisted selection (MAS). However, for plant morphological and yield-related traits, only a handful of candidate genes and efficient molecular markers have been used for MAS [1].

Intragenic SSRs are more conserved and transferable than extragenic SSRs [32–34], especially the expressed sequence tags SSRs (EST-SSRs) found in transcribed sequences. These SSRs are potentially more efficient for QTL mapping, gene targeting, and marker-assisted breeding than genomic-SSRs [35], which have been widely used in genetic linkage map construction in plants [36–39]. In Chinese cabbage, some EST-SSRs have been used for genetic linkage map construction with other molecular markers [30].

In previous studies, genomic SSRs and EST-SSRs have been identified and analyzed at the whole genome and transcriptome levels in Chinese cabbage [40,41]. In the present study, we aimed to construct a genetic linkage map using intragenic SSRs and map QTLs for important agronomic traits in Chinese cabbage. This study will provide useful information for better understanding of the molecular bases of these complex quantitative traits and molecular breeding in Chinese cabbage.

2. Materials and Methods

2.1. Plant Materials and Trait Measurements

A F_2 segregating population was developed by crossing two Chinese cabbage inbred lines, ZHB and G291. The two parents are significantly different in both size and shape. The parents and the F_2 generations were planted in the field at the normal sowing time (15 August 2016) in Changqing, Jinan, China, with a row spacing of 50 cm and plant spacing of 50 cm. The plants were harvested in mid-November 2016. A total of 240 F_2 individuals were selected randomly for trait measurements and genetic linkage map construction. Sixteen agronomic traits, including plant height (PH), plant width (PW), gross weight (GW), number of non-wrapper leaves (NNL), head weight (HW), head height (HH), head diameter (HD), number of head leaves (NHL), number of all leaves (NAL), maximum leaf length (MLL), maximum leaf width (MLW), petiole length (PEL), petiole width (PEW), petiole thickness (PET), stem length (SL), and stem width (SW), were measured following the descriptions for *Brassica* by the International Bureau of Plant Genetic Resources (IBPGR, 1990) (Table 1). The mean values and standard deviations for the agronomic parameters and the correlations between agronomic traits were analyzed by SPSS v13.0 for Windows (SPSS Inc., Chicago, IL, USA).

Table 1. Summary of the agronomic traits and their measurements.

Traits	Measurement
Plant height	Height of the highest point of the plant from the ground at the time of harvest (cm)
Plant width	The maximum distance of the outer leaves of the plant at the time of harvest (cm)
Gross weight	Gross weight of plant at the time of harvest (kg)
Number of non-wrapper leaves	Extant number of external leaves of leaf head at the time of harvest
Head weight	Weight of head at the time of harvest period (kg)
Head height	Height of the head measured at the highest point at the time of harvest (cm)
Head diameter	Width of the head measured at the widest point (cm)
Number of head leaves	Extant number of leaves of leaf head at the time of harvest period (>2 cm)

Table 1. *Cont.*

Traits	Measurement
Number of all leaves	The sum of number of non-wrapper leaves and number of head-forming leaves
Maximum leaf length	Length of the largest leaf at the longest point including petiole (cm)
Maximum leaf width	Width of the largest leaf at the widest point (cm)
Petiole length	Length of petiole of largest leaf at longest point (cm)
Petiole width	Width of petiole of largest leaf at widest point (cm)
Petiole thickness	Thickness of petiole of largest leaf at thickest point (cm)
Stem length	Stem length in head measured (cm)
Stem width	Stem diameter at head base (cm)

2.2. *DNA Isolation and Marker Genotyping*

The total DNA of the parental lines and F_2 individuals was extracted from young leaves (two weeks old) using the modified CTAB method [42]. DNA quantity and quality were assessed using a NanoDrop ND-1000 Spectrophotometer (Nano-Drop, Wilmington, DE, USA) and electrophoresis in 1.0% agarose gel using 0.5× TBE electrophoresis buffer, respectively. The DNA was diluted to 10 ng/μL and amplified by polymerase chain reaction (PCR). Five hundred SSRs distributed across 10 chromosomes of the *Brassica rapa* A genome were selected for polymorphic survey between the parental lines according to Shi et al. (2014) and Ding et al. (2015) [40,41]. PCR reactions were performed in a 96-well plate at 95 °C for 5 min, followed by 35 cycles of reaction (95 °C for 30 s, followed by 55–60 °C for 30 s and 72 °C for 30 s), and a final step of 72 °C for 5 min. The appropriate annealing temperatures depend on each primer pair. The PCR products were resolved by 6% denaturing polyacrylamide gel electrophoresis. Codominant polymorphic SSR markers with a single PCR brand selected from the parental polymorphism were used for marker genotyping in the F_2 lines.

2.3. *Genetic Map Construction and QTL Analysis*

We used the QTL Icimapping software V4.1 to construct the genetic map [43]. Redundant markers and markers with a missing rate greater than 20% were deleted using the "BIN" functionality of the software. The chi-square ($\chi 2$) test was used to check all the polymorphic SSRs for the goodness of fit against a 1:2:1 segregation ratio ($p < 0.01$). For genetic map construction, the genotype data of the homozygous alleles from the parent lines ZHB and G291 were recorded as "2" and "0", respectively. The heterozygous genotypes were recorded as "1", and all missing data were recorded as "−1". All markers were grouped at LOD = 2.5 for genetic map construction. The ICIM-ADD mapping method at a LOD threshold of 2.5 was used for QTL mapping with the Icimapping software V4.1 [43].

3. Results

3.1. *Construction of the Brassica rapa Linkage Map*

A total of 500 intragenic SSR markers, distributed across 10 chromosomes of the *Brassica rapa* A genome (50 markers per chromosome), were randomly selected for polymorphic survey between the parental lines from the SSRs developed by Shi et al. (2014) and Ding et al. (2015) [40,41]. Among the 500 intragenic SSRs, 133 were polymorphic between ZHB and G291 with a polymorphism rate of 26.6%. Only 105 clearly visible co-dominant polymorphic SSRs were recognized as usable markers for map construction (Table S1). Of the 105 polymorphic SSRs, 62 were located in exons and 43 were located in introns (Table S1).

The polymorphic SSRs were screened on these 240 F_2 individuals, and the results showed that 98 markers (93.33%) had the expected 1:2:1 segregation for the parental alleles ($p < 0.01$), while seven markers (6.67%) were distorted from the expected segregation ratio. Of the seven distorted markers, one was on A01, A06, and A07, respectively, and two were distributed on A03 and A09, respectively. Four distorted markers had a segregation bias in favor of G291, and one in favor of ZHB, while the remaining two distorted markers were in favor of the heterozygous genotype.

A genetic map comprising these 105 polymorphic SSRs was constructed, which was 2034.1 cM in length (Table 2). The SSRs were assigned to 10 linkage groups (LGs), putatively corresponding to the haploid chromosome number of *Brassica. rapa*. The number of the SSR markers in each of the 10 linkage groups varied from 5 (A05) to 15 (A06) (Table 2). The length of LGs varied from 85.53 cM (A05) to 390.5 cM (A06), and the average linkage group size was 203.41 cM. The average inter-locus distance was 21.75 cM. The smallest marker interval of 2.33 cM was found between A10S24 and A10S23 on A10, while the largest marker interval of 94.1 cM was found on A06 between A06S6 and A06S19 (Table 2).

Table 2. Features of the EST-SSR-based genetic linkage map of *Brassica rapa*.

Linkage Groups	No. of Markers	Max Gap (cM)	Min Gap (cM)	Marker Density (Markers/cM)	Average Marker Interval (cM)	Genetic Distance (cM)
A01	11	48.78	2.4	0.06	17.7	194.2
A02	14	44.8	4.2	0.07	15.4	200.4
A03	13	26.1	7.5	0.07	15.7	187.6
A04	10	47.1	3.3	0.07	15.5	139.3
A05	5	35.0	9.3	0.06	21.4	85.5
A06	15	94.1	2.6	0.04	26.0	390.5
A07	10	34.7	6.2	0.07	15.6	140.4
A08	8	96.1	10.5	0.03	38.9	272.0
A09	9	88.0	4.7	0.03	38.7	311.1
A10	10	25.1	2.3	0.09	12.6	113.1
Total/Average	105	54.0	5.3	0.06	21.8	2034.1

3.2. Investigation and Statistical Analysis of Agronomic Parameters

A total of 240 F_2 individuals were randomly selected for trait measurements and genetic linkage map construction. Sixteen important agronomic traits were investigated and analyzed thoroughly (Table 1). The results showed that the parents were significantly different in all the sixteen agricultural traits (Table 3). The plant and leafy head of the paternal parent 'G291' were larger and heavier than those of the maternal parent 'ZHB', while 'ZHB' had longer CL than 'G291' (Table 3). The F_1 line exhibited strong heterosis, as almost all the tested traits were larger than the parents (Table 3). In the F_2 population, all the 16 traits investigated in the study showed a continuous distribution and a wide genetic variation (Table 3, Figure S1). Coefficient of Variation (CV) was used to evaluate the genetic variation of the traits in these 240 F_2 lines, and the results showed that head weight, stem length, and gross weight had wider variations than the other traits, with a CV of 33.29%, 30.73%, and 29.52%, respectively, followed by number of non-wrapper leaves with a CV of 23.06% (Table 3). The CV of the other 13 agricultural traits varied from 10.82% to 16.63% (Table 3).

Most of the traits showed significant positive correlations with other traits (Table 4). Head weight, as the most important trait representing the yield of Chinese cabbage, showed significant positive correlations with all other traits tested in the study except number of non-wrapper leaves ($p < 0.01$). Head weight had the largest correlation coefficient with gross weight (with the correlation coefficient of 0.824), followed by that with head diameter and MLW, with correlation coefficients of 0.645 and 0.627, respectively (Table 4).

Table 3. Overview of the phenotypic traits in the parental, F_1, and F_2 lines used for mapping construction and QTL mapping in *Brassica rapa*.

Trait	P1	P2	F_1	F_2 Lines		
	ZHB	G291		Mean	Range	CV %
PH (cm)	14.8	33.7	36.3	28.8	18.0–40.0	14.84
PW (cm)	39.6	56.7	66.0	55.6	33.0–71.0	11.51
GW (kg)	0.42	1.43	3.33	1.72	0.55–3.20	29.52
NNL	9.0	5.0	8.7	8.7	4.0–18.0	23.06
HW (kg)	0.16	0.85	2.36	1.11	0.21–2.35	33.29
HH (cm)	10.4	27.8	29.4	23.1	13.9–32.0	12.71
HD (cm)	8.5	13.6	18.7	13.4	6.0–20.9	16.63
NHL	17.2	33.7	32.3	30.3	12.0–42.0	13.36
NAL	26.2	38.7	41.0	39.0	19.0–52.0	12.20
MLL (cm)	26.4	42.8	44.3	38.1	26.0–49.5	10.82
MLW (cm)	21.0	34.9	34.2	27.2	18.5–37.0	12.56
PEL (cm)	15.2	22.8	23.6	18.6	12.0–29.0	15.97
PEW (cm)	3.8	8.0	7.4	6.7	4.0–9.5	13.92
PET (cm)	0.62	1.33	1.10	0.89	0.50–1.30	15.84
SL (cm)	3.90	2.50	7.83	5.45	1.80–11.50	30.73
SW (cm)	2.42	3.17	3.83	2.87	1.50–3.90	15.77

PH: plant height; PW: plant width; GW: gross weight; NNL: number of non-wrapper leaves; HW: head weight; HH: head height; HD: head diameter; NHL: number of head leaves; NAL: number of all leaves; MLL: maximum leaf length; MLW: maximum leaf width; PEL: petiole length; PEW: petiole width; PET: petiole thickness; SL: stem length; and SW: stem width.

3.3. *QTL Analysis*

A total of 48 QTLs on ten chromosomes were detected for the 16 traits of Chinese cabbage. Two (on A04 and A09) to twelve QTLs (on A03) were detected in these ten LGs. The number of the detected QTLs ranged from 0 for stem length and head diameter to 7 for number of non-wrapper leaves, and the confidence interval covered by individual QTLs ranged from 2.33 cM (*qMLL-6*) to 96.06 cM (*qSW-2*). The percentage of phenotypic variation (R^2) explained by individual QTLs ranged from 3.41% (*qPEL-3*) to 26.66% (*qPEW-2*), and the LOD scores of individual QTLs varied from 2.51 (*qHH-4*) to 12.49 (*qPEW-2*) (Table 5).

Two and four QTLs for plant height (*qPH-1* and *qPH-2*) and plant width (*qPW-1*, *qPW-2*, *qPW-3* and *qPW-4*) were detected, respectively. Of these QTLs, *qPH-1* and *qPW-1* showed a relatively higher LOD score and R^2, suggesting that these QTLs may be a major QTL for plant height and plant width, respectively. One QTL for gross weight (*qGW-1*) was detected on A10 between the SSR A10S4-A10S24 with a LOD score of 3.43 and a confidence interval of 4.45 cM, explaining 5.26% of the phenotypic variation (Table 5).

Eight QTLs were identified for leafy head-related traits (head weight, head height and head diameter). Two of these QTLs, *qHW-1* on A05 and *qHW-2* on A10, were identified for head weight, explaining 6.17% and 5.22% of the phenotypic variation, respectively. Six QTLs for head height (*qHH-1* on A01, *qHH-2* on A02, *qHH-3* on A03, *qHH-4* on A05, *qHH-5* on A06 and *qHH-6* on A10) were detected and the individual QTL explained 4.29–13.39% of the phenotypic variation. *qHH-2* on A02 was the major QTL with a comparatively higher LOD of 6.96, explaining 13.39% of the phenotypic variation. No QTL was identified for head diameter in the study (Table 5).

Table 4. Correlation co-efficient analysis of the 16 agricultural traits tested in the study.

Trait	PH	PW	GW	NNL	HW	HH	HWD	NHL	NAL	MLL	MLW	PEL	PEW	PET	CL
PW	0.359 **														
GW	0.450 **	0.432 **													
NNL	0.288 **	0.336 **	0.200 **												
HW	0.348 **	0.365 **	0.824 **	0.008											
HH	0.493 **	0.363 **	0.640 **	0.165 *	0.583 **										
HD	0.280 **	0.386 **	0.673 **	0.079	0.645 **	0.476 **									
NHL	0.192 **	0.246 **	0.377 **	0.136 *	0.336 **	0.275 **	0.251 **								
NAL	0.285 **	0.351 **	0.405 **	0.538 **	0.289 **	0.304 **	0.247 **	0.908 **							
MLL	0.484 **	0.500 **	0.545 **	0.388 **	0.422 **	0.598 **	0.409 **	0.211 **	0.342 **						
MLW	0.437 **	0.307 **	0.680 **	0.031	0.627 **	0.507 **	0.538 **	0.207 **	0.190 **	0.557 **					
PEL	0.344 **	0.432 **	0.367 **	0.458 **	0.192 **	0.525 **	0.199 **	0.166 *	0.332 **	0.760 **	0.163 *				
PEW	0.273 **	0.200 **	0.632 **	−0.016	0.551 **	0.440 **	0.473 **	0.300 **	0.249 **	0.299 **	0.624 **	0.077			
PET	0.265 **	0.185 **	0.522 **	−0.019	0.472 **	0.287 **	0.369 **	0.188 **	0.153 *	0.332 **	0.605 **	0.043	0.564 **		
SL	0.181 **	0.261 **	0.401 **	0.247 **	0.397 **	0.432 **	0.195 **	0.092	0.182 **	0.371 **	0.232 **	0.370 **	0.189 **	0.152 *	
SW	0.088	0.232 **	0.484 **	0.071	0.455 **	0.248 **	0.365 **	0.286 **	0.274 **	0.232 **	0.384 **	0.076	0.436 **	0.391 **	0.389 **

PH: plant height; PW: plant width; GW: gross weight; NNL: number of non-wrapper leaves; HW: head weight; HH: head height; HD: head diameter; NHL: number of head leaves; NAL: number of all leaves; MLL: maximum leaf length; MLW: maximum leaf width; PEL: petiole length; PEW: petiole width; PET: petiole thickness; SL: stem length; and SW: stem width. *: Correlation is significant at the 0.05 level (2-tailed); **: Correlation is significant at the 0.01 level (2-tailed).

For leaf number-related traits (number of non-wrapper leaves, number of head leaves and number of all leaves), 11 QTLs were detected on 9 LGs (A01, A02, A03, A04, A05, A06, A07, A09 and A10). Seven QTLs were identified for number of non-wrapper leaves, of which three were located on A07, and one was located on A02, A03, A04 and A09, respectively. *qNNL-2*, *qNNL-3* and *qNNL-7* were the major QTLs for number of non-wrapper leaves with a relatively higher LOD (5.19, 5.97 and 3.55, respectively) and R^2 (10.19%, 9.98% and 13.33%, respectively). Two QTLs were detected for number of head leaves, of which qNHL-1 on A01 between A1S30 and A1S5 was the major QTL, with a LOD score of 3.23 and R^2 of 15.13% at the peak position of 58.80 cM. For number of all leaves, two QTLs were detected on A02 and A05 with an R^2 of 4.66% and 5.97%, respectively (*qNAL-1* and *qNAL-2*).

A total of 19 QTLs were identified for the five leaf-related traits (maximum leaf length, maximum leaf width, petiole length, petiole width and petiole thickness) on 8 LGs (A02, A03, A04, A06, A07, A08, A09 and A10). For maximum leaf length, 6 QTLs were detected on A02, A03, A04 and A10, of which *qMLL-3* and *qMLL-1* were the major QTLs with a relatively higher LOD (7.61 and 5.60, respectively) and R^2 (12.42% and 10.07, respectively). Two QTLs were identified for a maximum leaf width on A03 and A07, explaining 7.74% and 9.23% of the phenotypic variation, respectively. Three, three, and five QTLs were identified for petiole length, petiole width and petiole thickness, respectively. One major QTL for petiole width was detected on A09 (*qPEW-2*) with an LOD of 12.49 and R^2 of 26.66%. No major QTLs were detected for petiole length and petiole thickness as the R^2 of these QTLs were lower than 10% (Table 5).

For the stem-related traits (stem length and stem width), three QTL loci for stem width were identified on A06, A08 and A09. No QTL was detected for stem length in the study.

3.4. Clustering of QTLs

Of the 48 QTLs detected in 10 LGs, many QTLs were found to map in the same QTL region. The QTL region of A02, A03 and A10 showed multiple QTLs for three or more traits (Figure 1). Six QTLs for plant height (*qPH-1*), head height (*qHH-2*), number of head leaves (*qNHL-2*), number of all leaves (*qNAL-1*), maximum leaf length (*qMLL-1*) and petiole length (*qPEL-1*) were mapped in the middle portion of A02 (107.24–121.10 cM) (Figures 1 and S2). Four of these six QTLs, *qPH-1*, *qHH-2*, *qMLL-1* and *qPEL-1*, obtained increasing alleles from the parent line "G291", while *qNHL-2* and *qNAL-1* obtained increasing alleles from the parent line "ZHB" (Table 5). Four QTLs for plant width (*qPW-2*), head height (*qHH-3*), maximum leaf length (*qMLL-3*) and petiole length (*qPEL-2*) were mapped nearby A3S20 (132.08 cM) of A03, which derived increasing alleles from the parent line "G291" (Figure 1 and Table 5). The lower part of A03 (172.68–187.61 cM) showed mapping of QTLs for plant width (*qPW-3*), maximum leaf length (*qMLL-3*), petiole length (*qPEL-3*) and petiole thickness (*qPET-2*). Three (*qMLL-3*, *qPEL-3*, and *qPET-2*) of the four QTLs obtained increasing alleles from the parent line "G291", except *qPW-2*. We also observed three QTLs for gross weight (*qGW-1*), head height (*qHH-6*), and maximum leaf length (*qMLL-6*) nearby A10S24 (74.85 cM) of A10, all of which derived increasing alleles from the parent line "G291" (Figure 1 and Table 5).

Table 5. Details of the QTLs identified for the 16 traits of Chinese cabbage.

	QTLs Detected in the Study						QTLs Detected in Previous Studies		
Trait	QTL Name	Chro.	Flanking Markers	Marker Interval (cM)	Peak Position	Physical Interval (bp)	Flanking Markers	Marker Interval (cM) or Peak Position	Ref
PH	qPH-2	A09	A9S41–A9S42	0.00–15.31	0	209907–376210	bin7~8	10.87	[2]
							bin14	15.90	[2]
							me07em17-01-me08em11-05	61.0	[44]
PW	qPW-1	A01	A1S15–A1S30	0.00–45.62	12.2	9201283–24507688			
	qPW-2	A03	A3S48–A3S20	117.55–132.08	126	20610030–25478614			
	qPW-3	A03	A3S45–A3S13	172.68–187.61	187.6	15806729–16861135			
	qPW-4	A09	A9S42–A9S53	15.13–65.61	43.8	376195–4089557	bin48-bin70	89.50	[2]
							me12em05-07-cnu_280a	96.8	[44]
							BrID90319-BrID101055	66.65	[45]
GW	qGW-1	A10	A10S4–A10S24	70.40–74.85	74.8	2720349–3481119	me03em15-04–me06em17-02	40.6	[1]
							me07em17-01–me08em11-05	61.0	[1]
							me08em11-03–nia_m003a	74.5	[1]
NNL	qNNL-1	A02	A2S38–A2S58	137.79–147.42	147.4	19786087–23330794	nia_m105a-nia_m121a	9.2	[30]
							nia_m125a-sau_um434	28.1	[30]
							cnu_aP21M47_157-cnu_aP70M59_740	29.8-57.7	[21]
	qNNL-2	A03	A3S63–A3S68	40.11–66.22	42.9	7308495–9749038			
	qNNL-3	A04	A4S14–A4S31	47.08–50.39	50.3	14746829–15029077			
						17663481–20491491			
	qNNL-4	A07	A7S17–A7S61	0.00–12.45	2.6	5793158–12788473	me06em08-02–nia_m043a	43.6	[1]
	qNNL-5	A07	A7S36–A7S30	99.22–112.76	100.1	64164–2691678			
	qNNL-6	A07	A7S47–A7S46	129.09–140.40	129.1				
	qNNL-7	A09	A9S15–A9S19	244.85–311.10	295.01	30797306–30797446	sau_um174-cnu_m114a	91.0	[30]
							me11em11-08–me02em09-02	36.1	[1]
							aaf_mSR6755a-cnu_m148a	89.3–102.5	[21]
HW	qHW-1	A05	A5S72–A5S62	15.60–41.19	27.4	22060863–24429768			
	qHW-2	A10	A10S7–A10S40	99.02–113.06	100.2	7102069–8966807	me03em15-04–me06em17-02	40.6	[1]
HH	qHH-1	A01	A1S30–A1S5	45.62–64.50	63.1	9201283–24507688	me11em11-06–me03em15-03	109.2	[1]
							me03em09-01–cnu_m461a	80.4	[1]
	qHH-2	A02	A2S29–A2S32	107.24–121.10	118.8	9771589–13665037	BrID101167	11.84	[45]
							BrID10001	202.21	[45]
							me13em14-02–nia_m143a	77.2	[1]
	qHH-3	A03	A3S20–A3S49	132.08–141.56	132.1	22308150–25478614	Ra3-D04-BRMS042-2	27.3	[30]
	qHH-4	A05	A5S74–A5S72	0.00–15.60	0	24429748–25141232			
	qHH-5	A06	A6S83–A6S6	192.43–204.21	204.2	7203340–17994646	me13em09-02–me11em11-07	27.8	[1]
	qHH-6	A10	A10S24–A10S23	74.85–77.18	75.7	3458392–3481119	me13em09-01–me09em09-02	23.3	[1]

Table 5. *Cont.*

	QTLs Detected in the Study						QTLs Detected in Previous Studies		
Trait	QTL Name	Chro.	Flanking Markers	Marker Interval (cM)	Peak Position	Physical Interval (bp)	Flanking Markers	Marker Interval (cM) or Peak Position	Ref
NHL	qNHL-1	A01	A1S30–A1S5	45.62–64.50	58.8	4276434–9201283	cnu_m461a-BRMS-031	32.6	[30]
	qNHL-2	A02	A2S29–A2S32	107.24–121.10	108.4	9771589–13665037	cnu_m046a-cnu_m130a	46.3	[30]
							cnu_aE36M60_630-cnu_aP70M59_740	49.2–55.4	[21]
NAL	qNAL-1	A02	A2S29–A2S32	107.24–121.10	107.3	9771589–13665037	cnu_aP21M47_157-nia_m105a	12.8–50.4	[21]
	qNAL-2	A05	A5S62–A5S13	41.19–50.52	45.2	20345543–22060888			
MLL	qMLL-1	A02	A2S29–A2S32	107.24–121.10	114.9	9771589–13665037	BrFLC2	21.1–36.9	[20]
							BrGA20OX3	14.28–30.82	[20]
							BrLNG1 35	24.2–48.2	[20]
							pbc_mENA9a	116.7–118.5	[21]
	qMLL-2	A03	A3S32–A3S63	20.47–40.11	23.1	4464344–7308518	bin28~30	26.51	[2]
	qMLL-3	A03	A3S48–A3S20	117.55–132.08	132	20610030–25478614	bin32~46	34.14	[2]
	qMLL-4	A03	A3S45–A3S13	172.68–187.61	172.7	15806729–16861135	BrFLC5 9.2	2–14.3	[20]
							BrAS1 55.2	49.6–67.5	[20]
							pbc_mENA9a	116.7–118.5	[21]
							cnu_m384a	23.4–28.2	[21]
	qMLL-5	A04	A4S49–A4S27	91.16–97.85	97.8	11343471–12261503	me01em14-03-me12em05-03	3	[44]
	qMLL-6	A10	A10S24–A10S23	74.85–77.18	74.9	3458392–3481119	me06em18-01-me13em07-01	0.0	[44]
							bniclind1	5.41	[45]
							BrLNG1	65.9–82.1	[20]
MLW	qMLW-1	A03	A3S49–A3S17	141.56–151.92	150.3	21438487–22308173	BrFLC5 BrAE3	2–14.3	[20]
							BRASSICA RAPA LEAFY PETIOLE	14.3–29.2	[20]
							BrAS1	49.6–67.5	[20]
							BrKRP2	110.3–137.3	[20]
							cnu_m416a-nia_m092a	96.2–109.1	[21]
							cnu_m384a	26–27.7	[21]
							cnu_m477a	23.8–52.7	[21]
	qMLW-2	A07	A7S61–A7S63	12.45–28.20	19.1	17663481–17964228	BrID10503-BrID10351	77.20	[45]
							cnu_aP63M55_900	91.6–98.4	[21]
							nia_m030a-cnu_m179a	57.4–78.1	[21]
PEL	qPEL-1	A02	A2S29–A2S32	107.24–121.10	121.1	9771589–13665037			
	qPEL-2	A03	A3S20–A3S49	132.08–141.56	133.4	22308150–25478614			
	qPEL-3	A03	A3S45–A3S13	172.68–187.61	172.7	15806729–16861135			

Table 5. *Cont.*

	QTLs Detected in the Study						QTLs Detected in Previous Studies		
Trait	QTL Name	Chro.	Flanking Markers	Marker Interval (cM)	Peak Position	Physical Interval (bp)	Flanking Markers	Marker Interval (cM) or Peak Position	Ref
PEW	qPEW-1	A08	A8S18–A8S35	132.62–187.39	174.7	15883226–19587829			
	qPEW-2	A09	A9S15–A9S19	244.85–311.10	308.61	30797306–30797446	me07em17-01-me08em11-05	61	[44]
	qPEW-3	A10	A10S14–A10S54	26.06–30.80	28.8	11806404–12390431	me03em15-04-me06em17-02	40.6	[44]
							me02em05-01-me03em15-04	40.2	[44]
PET	qPET-1	A03	A3S68–A3S11	66.22–79.15	71.9	9749021–13690141			
	qPET-2	A03	A3S45–A3S13	172.68–187.61	187.6	15806729–16861135			
	qPET-3	A06	A6S33–A6S61	97.60–103.19	102.9	24350023–24703831	me03em16-01-nia_m049a	8.1	[44]
							cnu_m110-cnu_m111a	12.7	[44]
	qPET-4	A10	A10S54–A10S12	30.80–45.27	32.4	11003410–11806423			
	qPET-5	A10	A10S7–A10S40	99.02–113.06	104.2	7102069–8966807			
SW	qSW-1	A06	A6S61–A6S18	103.19–127.46	117.8	22730023–24350034	cnu_mBBSRC058-nia_m037a	57–67.4	[21]
	qSW-2	A08	A8S42–A8S20	0.00–96.06	96	1369822–20768736			
	qSW-3	A09	A9S15–A9S19	244.85–311.10	311.01	30797306–30797446			

PH: plant height; PW: plant width; GW: gross weight; NNL: number of non-wrapper leaves; HW: head weight; HH: head height; NHL: number of head leaves; NAL: number of all leaves; MLL: maximum leaf length; MLW: maximum leaf width; PEL: petiole length; PEW: petiole width; PET: petiole thickness; and SW: stem width.

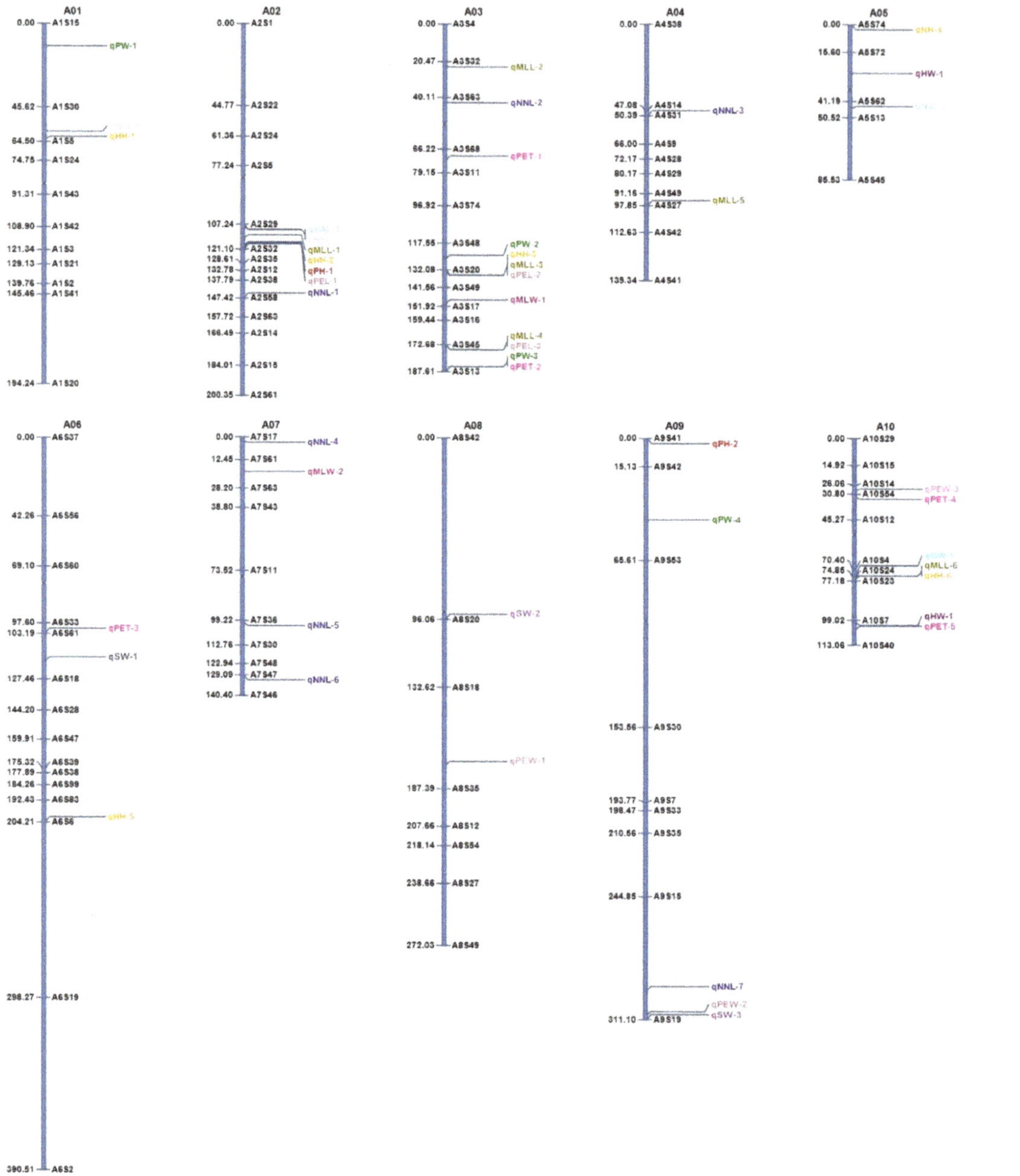

Figure 1. *Brassica rapa* genetic linkage map and QTLs for agronomic traits discovered in F_2 lines. Genetic distances (cM) are shown on the left side of the linkage group, and the names of the SSRs are shown on the right side of the linkage group.

4. Discussion

4.1. Intragenic SSR-Based Linkage Map Construction in Chinese Cabbage

With the development of high-throughput sequencing technology, the whole genome and transcriptome sequencing of Chinese cabbage have been performed in recent years [46,47], which provides great convenience for developing molecular markers and identifying genetic loci regulating qualitative and quantitative traits in Chinese cabbage. In previous studies, over 140 thousand genomic SSRs and 10 thousand EST-SSRs with a clear physical position have been developed at the whole genome and transcriptome level, respectively [40,41]. In the present study, 500 intragenic SSR markers distributed across 10 chromosomes of the *Brassica rapa* A genome (50 markers per chromosome) were randomly selected for polymorphic survey between the parental lines. Finally, 105 clearly visible co-dominant polymorphic SSRs were used for genetic map construction and QTLs mapping. This is the first genetic map construction in Chinese cabbage exclusively using distributed intragenic SSR markers. As the physical positions were clear for all the SSRs, it will provide great convenience for further map-based cloning and candidate gene screening for QTLs using this genetic map.

4.2. QTLs for Important Agronomic Traits of Chinese Cabbage

As an economically important leafy vegetable, yield and morphology-related agronomic traits, such as plant height, plant width, leafy head-related traits, leaf number-related traits and central axis-related traits, are important in Chinese cabbage breeding to achieve more attractive plants with higher yield and better architecture according to customers' demands. In previous studies, QTLs for some yield and morphology-related agronomic traits have been identified using DH, F_2, and RIL lines [1,7,15,18–21,30]. The occurrence of detectable QTL depends on polymorphisms present in the studied population, so different QTLs can be identified in different studies of the same species. Furthermore, as most of these traits were inherited, complex quantitative traits, the QTLs detected under different backgrounds and environments or by different software are usually not consistent. Here, we list the information of QTLs for important agronomic traits identified in previous studies to help us screen efficient candidate QTLs (Table 5).

Plant height and width are important traits, which are associated closely with yield and morphology of Chinese cabbage. Plant width also influences plant space in the field culture. In this study, two QTLs for plant height were identified on A02 and A09. QTLs for plant height have also been detected on A02 [30] and A09 [2,44] in previous studies. QTLs for plant height have been detected on A01, A04, A07, A08 and A10 in other studies [30,44,45]. Three QTLs for plant width are located on the linkage group A03 and A09, on which the QTL for plant width also has been identified in previous studies [2,44,45]. QTLs for plant width have been found on A05, A07 and A10 in previous studies [2,44,45]. One QTL for plant width was also identified on A01 in the study, which may be a new candidate locus for plant width (Table 5).

The leafy head is the main edible part of Chinese cabbage. Leafy head-related traits are the most important traits for breeding and production of Chinese cabbage. QTLs for head weight have been detected on A02, A03, A04, A05, A06, A07, A08, A09, and A10 in previous studies [1,10,30,45]. In this study, two QTLs for head weight were located on A05 and A10, which is consistent with the linkage group for head weight found by Yu et al. (2013) [10] and Liu et al. (2013) [1], respectively. Six QTLs for head height were identified on A01, A02, A03, A05, A06 and A10 in this study. QTLs for head height have been found on A01, A02, A06 and A10 according to Liu et al. (2013) [1], and on A03 according to Ge et al. (2011) [40] and Liu et al. (2015) [45]. No QTL for head height has been identified on A05 in previous studies, which might be a potentially new locus regulating head height (Table 5).

As an important leafy vegetable, leaf number and morphology are important in Chinese cabbage breeding. In this study, three leaf number-related traits and five leaf morphology-related traits were investigated, and 19 QTLs were identified distributed on 8 LGs (A02, A03, A04, A06, A07, A08, A09 and A10). The results indicated that the leaf traits were complex quantitative traits controlled by many genes spreading nearly all the LGs of Chinese cabbage, which is consistent with those reported in previous studies [15,20,21,30,44] (Table 5).

Stem-related traits, especially stem length, are usually used to evaluate the tolerance to bolting of Chinese cabbage. In this study, no QTL was identified for stem length, probably because the parents of the F_2 line in this study were both non-resistant bolting varieties.

As the agronomic traits were significantly correlated (Table 4), many co-localized QTLs were identified on A02, A03 and A10 in this study (Figures 1 and S2). The QTL clusters for some agronomic traits were also identified on these LGs in previous studies in Chinese cabbage [1,9,20,21,30]. It indicated that these chromosomes might carry important genes regulating more than one agronomic trait, and could be very useful for the improvement of more than one trait in breeding of Chinese cabbage.

Many QTLs detected in the study have a large region size (>10 cM); there are hundreds of gene in the region, so it is difficult to pick out the candidate genes. Further fine mapping for these QTLs should be conducted by using a high-density genetic map. For the QTLs with an interval size $\leq$ 10 cM, the gene information was taken from the Brassicaceae Database (http://www.brassicadb.cn/, accessed on 6 February 2022) (Table S2). *BraTCPs* [48] and *BraGRFs* [49] genes were reported to be involved in controlling organ size in Chinese cabbage. In the study, 3 *BraTCP (Bra032970, Bra012600* and *Bra027284)* and 1 *BraGRF* (*Bra033281*) genes were found in the QTLs regions. Genes involved in the Auxin signaling pathway play important roles in regulating leafy head formation of Chinese cabbage [50]. Here we found that 10 auxin-related genes (*Bra026598, Bra026597, Bra026596, Bra032954, Bra019369, Bra019255, Bra027232, Bra034725, Bra008615* and *Bra008722*) were located in the QTLs regions. For *qMLL-6* and *qHH-6*, only two genes, *Bra033221 (SPL8)* and *Bra033222 (NOT1)*, were found in the QTL region. These genes may be important candidates for regulating the agronomic traits of Chinese cabbage.

5. Conclusions

In summary, a genetic map comprising 105 EST-SSR markers distributing across 10 LGs were constructed, and a total of 48 QTLs regulating 16 agronomic traits were identified in Chinese cabbage in this study. QTLs consistent with previous studies could be potential candidate QTLs for further genetic analysis and marker-assisted genetic improvement of Chinese cabbage.

Supplementary Materials: The following supporting information can be downloaded at: https://www.mdpi.com/article/10.3390/horticulturae8020165/s1, Table S1: Details of the 500 SSRs used in this study. Table S2: Gene information for QTLs with the interval size $\leq$10 cM. Figure S1: Frequency distribution of agricultural traits for the F_2 population derived from a cross between G291 and ZHB. PH: plant height, PW: plant width, GW: gross weight, NNL: number of non-wrapper leaves, HW: head weight, HH: head height, HD: head diameter, NHL: number of head leaves, NAL: number of all leaves, MLL: maximum leaf length, MLW: maximum leaf width, PEL: petiole length, PEW: petiole width, PET: petiole thickness, SL: stem length, and SW: stem width. Figure S2: Brassica rapa genetic linkage map and QTLs for agronomic traits discovered in F_2 lines with LOD score.

Author Contributions: H.G., J.L., F.W. and J.G. designed the experiments; X.Y., H.W., Y.C. and J.L. performed the experiments; H.G., J.L. and N.Q. analyzed the data; F.W., H.L. and Y.Z. contributed reagents/materials/analysis tools; J.L. and J.G. wrote the main manuscript text. All authors have read and agreed to the published version of the manuscript.

Funding: This research was funded by the Agricultural Science and Technology Innovation Project of SAAS (CXGC2021A10); the Modern Agricultural Industrial Technology System Funding of Shandong Province, China (SDAIT-05); the Key R&D Program of Shandong Province, China (2019GHZ014); Shandong Upgraded Project of "Bohai Granary" Science and Technology Demonstration Engineering (2019BHLC005); Taishan Scholars Program of Shandong Province, China (tsqn201909167); Prospect of Shandong Seed Project, China (2019LZGC0060101); China Agriculture Research System (CARS-23-G13); and the National Natural Science Foundation of China, China (31401869).

Institutional Review Board Statement: Not applicable.

Informed Consent Statement: Not applicable.

Data Availability Statement: Not applicable.

Conflicts of Interest: The authors declare that there is no conflict of interest regarding the publication of this paper.

References

1. Liu, Y.; Zhang, Y.; Xing, J.; Liu, Z.; Feng, H. Mapping quantitative trait loci for yield-related traits in Chinese cabbage (*Brassica rapa* L. ssp. *pekinensis*). *Euphytica* **2013**, *193*, 221–234. [CrossRef]
2. Huang, L.; Yang, Y.; Zhang, F.; Cao, J. A genome-wide SNP-based genetic map and QTL mapping for agronomic traits in Chinese cabbage. *Sci. Rep.* **2017**, *7*, 46305. [CrossRef] [PubMed]
3. Xiao, L.; Zhao, Z.; Du, Z.; Yao, Y.; Xu, L.; Tang, G. Genetic characterization and fine mapping of a yellow-seeded gene in Dahuang (a *Brassica rapa* landrace). *Theor. Appl. Genet.* **2012**, *124*, 903–909. [CrossRef] [PubMed]
4. Dechaine, J.M.; Brock, M.T.; Weinig, C. QTL architecture of reproductive fitness characters in *Brassica rapa*. *BMC Plant Biol.* **2014**, *14*, 66. [CrossRef]
5. Saito, M.; Kubo, N.; Matsumoto, S.; Suwabe, K.; Tsukada, M.; Hirai, M. Fine mapping of the clubroot resistance gene, Crr3, in *Brassica rapa*. *Theor. Appl. Genet.* **2006**, *114*, 81. [CrossRef]
6. Feng, H.; Li, Y.; Liu, Z.; Liu, J. Mapping of or, a gene conferring orange color on the inner leaf of the Chinese cabbage (*Brassica rapa* L. ssp. *pekinensis*). *Mol. Breed.* **2012**, *29*, 235–244. [CrossRef]
7. Wang, Y.; Liu, X.; Ji, X.; Zhang, L.; Liu, Y.; Lv, X.; Feng, H. Identification and validation of a major QTL controlling the presence/absence of leaf lobes in *Brassica rapa* L. *Euphytica* **2015**, *205*, 761–771. [CrossRef]
8. Liu, Y.; Li, C.; Shi, X.; Feng, H.; Wang, Y. Identification of QTLs with additive, epistatic, and QTL × environment interaction effects for the bolting trait in *Brassica rapa* L. *Euphytica* **2016**, *210*, 427–439. [CrossRef]
9. Ren, Y.; Wu, J.; Zhao, J.; Hao, L.; Zhang, L. Identification of SSR markers closely linked to the yellow seed coat color gene in heading Chinese cabbage (*Brassica rapa* L. ssp. *pekinensis*). *Biol. Open* **2017**, *6*, 278–282. [CrossRef]
10. Yu, X.; Wang, H.; Zhong, W.; Bai, J.; Liu, P.; He, Y. QTL mapping of leafy heads by genome resequencing in the RIL population of *Brassica rapa*. *PLoS ONE* **2013**, *8*, e76059. [CrossRef]
11. Liu, J.; Liu, B.; Cheng, F.; Liang, J.; Wang, X.; Wu, J. A high density linkage map facilitates QTL mapping of flowering time in *Brassica rapa*. *Hortic. Plant J.* **2016**, *2*, 217–223. [CrossRef]
12. Zhang, J.; Lu, Y.; Yuan, Y.; Zhang, X.; Geng, J.; Chen, Y.; Cloutier, S.; McVetty, P.B.E.; Li, G. Map-based cloning and characterization of a gene controlling hairiness and seed coat color traits in *Brassica rapa*. *Plant Mol. Biol.* **2009**, *69*, 553–563. [CrossRef]
13. Ye, L.X.; Hu, F.Y.; Ren, J.; Huang, S.N.; Liu, W.J.; Feng, H.; Liu, Z.Y. Fine mapping and candidate gene analysis of Brtri1, a gene controlling trichome development in Chinese cabbage (*Brassica rapa* L. ssp. *pekinensis*). *Genet. Mol. Res.* **2016**, *15*, gmr15048924. [CrossRef]
14. Kawakatsu, Y.; Nakayama, H.; Kaminoyama, K.; Igarashi, K.; Yasugi, M.; Kudoh, H.; Nagano, A.J.; Yano, K.; Kubo, N.; Kimura, S. A GLABRA1 ortholog on LG A9 controls trichome number in the Japanese leafy vegetables Mizuna and Mibuna (*Brassica rapa* L. subsp. *nipposinica* L. H. Bailey): Evidence from QTL analysis. *J. Plant Res.* **2017**, *130*, 539–550. [CrossRef]
15. Lou, P.; Zhao, J.; Kim, J.S.; Shen, S.; Del Carpio, D.P.; Song, X.; Jin, M.; Vreugdenhil, D.; Wang, X.; Koornneef, M.; et al. Quantitative trait loci for flowering time and morphological traits in multiple populations of *Brassica rapa*. *J. Exp. Bot.* **2017**, *58*, 4005–4016. [CrossRef]
16. Zhang, N.; Chen, L.; Ma, S.; Wang, R.; He, Q.; Tian, M.; Zhang, L. Fine mapping and candidate gene analysis of the white flower gene Brwf in Chinese cabbage (*Brassica rapa* L.). *Sci. Rep.* **2020**, *10*, 6080. [CrossRef]
17. Guo, N.; Wu, J.; Zheng, S.; Cheng, F.; Liu, B.; Liang, J.; Cui, Y.; Wang, X. Anthocyanin profile characterization and quantitative trait locus mapping in zicaitai (*Brassica rapa* L. ssp. *chinensis* var. *purpurea*). *Mol. Breed.* **2015**, *35*, 113. [CrossRef]
18. Song, K.; Slocum, M.K.; Osborn, T.C. Molecular marker analysis of genes controlling morphological variation in *Brassica rapa* (syn. *campestris*). *Theor. Appl. Genet.* **1995**, *90*, 1–10. [CrossRef]
19. Kubo, N.; Saito, M.; Tsukazaki, H.; Kondo, T.; Matsumoto, S.; Hirai, M. Detection of quantitative trait loci controlling morphological traits in *Brassica rapa* L. *Breed. Sci.* **2010**, *60*, 164–171. [CrossRef]
20. Xiao, D.; Wang, H.; Basnet, R.K.; Zhao, J.; Lin, K.; Hou, X.; Bonnema, G. Genetic dissection of leaf development in *Brassica rapa* using a genetical genomics approach. *Plant Physiol.* **2014**, *164*, 1309–1325. [CrossRef]
21. Choi, S.R.; Yu, X.; Dhandapani, V.; Li, X.; Wang, Z.; Lee, S.Y.; Heon Oh, S.; Pang, W.; Ramchiary, N.; Hong, C.; et al. Integrated analysis of leaf morphological and color traits in different populations of Chinese cabbage (*Brassica rapa* ssp. *pekinensis*). *Theor. Appl. Genet.* **2017**, *130*, 1617–1634. [CrossRef]
22. Zou, L.C.; Zheng, Y.; Wang, P.; Zhang, X.; Wang, Y.H.; Liu, Z.Y.; Feng, H. Fine mapping and characterization of the or gene in Chinese cabbage (*Brassica rapa* L. ssp. *pekinensis*). *Genet. Mol. Res.* **2016**, *15*, gmr.15028370. [CrossRef]
23. Wang, Y.; Xiao, L.; Guo, S.; An, F.; Du, D. Fine mapping and whole-genome resequencing identify the seed coat color gene in *Brassica rapa*. *PLoS ONE* **2016**, *11*, e0166464. [CrossRef]
24. Zhang, Y.; Sun, Y.; Sun, J.; Feng, H.; Wang, Y. Identification and validation of major and minor QTLs controlling seed coat color in *Brassica rapa* L. *Breed. Sci.* **2019**, *69*, 47–54. [CrossRef]

25. Liu, S.; Wang, R.; Zhang, Z.; Li, Q.; Wang, L.; Wang, Y.; Zhao, Z. High-resolution mapping of quantitative trait loci con-trolling main floral stalk length in Chinese cabbage (*Brassica rapa* L. ssp. *pekinensis*). *BMC Genom.* **2019**, *20*, 437. [CrossRef]
26. Kato, T.; Hatakeyama, K.; Fukino, N.; Matsumoto, S. Fine mapping of the clubroot resistance gene CRb and development of a useful selectable marker in *Brassica rapa*. *Breed. Sci.* **2013**, *63*, 116–124. [CrossRef]
27. Yu, S.; Su, T.; Zhi, S.; Zhang, F.; Wang, W.; Zhang, D.; Zhao, X.; Yu, Y. Construction of a sequence-based bin map and mapping of QTLs for downy mildew resistance at four developmental stages in Chinese cabbage (*Brassica rapa* L. ssp. *pekinensis*). *Mol. Breed.* **2016**, *36*, 44. [CrossRef]
28. Li, Q.; Zhang, X.; Zeng, Q.; Zhang, Z.; Liu, S.; Pei, Y.; Wang, S.; Liu, X.; Xu, W.; Fu, W.; et al. Identification and mapping of a novel Turnip mosaic virus resistance gene TuRBCS01 in Chinese cabbage (*Brassica rapa* L.). *Plant Breed.* **2015**, *134*, 221–225. [CrossRef]
29. Laila, R.; Park, J.; Khan Robin, A.H.; Natarajan, S.; Kenta Shirasawa, V.H.; Isobe, S.; Kim, H.; Nou, I. Mapping of a novel clubroot resistance QTL using ddRAD-seq in Chinese cabbage (*Brassica rapa* L.). *BMC Plant Biol.* **2019**, *19*, 13. [CrossRef]
30. Ge, Y.; Ramchiary, Y.; Wang, T.; Liang, C.; Wang, N.; Wang, Z.; Choi, S.R.; Lim, Y.P.; Piao, Z.Y. Mapping quantitative trait loci for leaf and heading-related traits in Chinese cabbage (*Brassica rapa* L. ssp. *pekinesis*). *Hortic. Environ. Biotechnol.* **2011**, *52*, 494–501. [CrossRef]
31. Li, J.; Zhang, Y.; Ding, Q.; Li, H.; Liu, L.; Wang, F.; Gao, J. Transcriptome Analysis of Orange Head Chinese Cabbage (*Brassica rapa* L. ssp. *pekinensis*) and Molecular Marker Development. *Int. J. Genom.* **2017**, *2017*, 6835810. [CrossRef]
32. Eujayl, I.; Sledge, M.K.; Wang, L.; May, G.D.; Chekhovskiy, K.; Zwonitzer, J.C.; Mian, M.A.R. *Medicago truncatula* ESTSSRs reveal cross-species genetic markers for *Medicago* spp. *Theor. Appl. Genet.* **2004**, *108*, 414–422. [CrossRef]
33. Zhang, Y.L.; Bernard, M.; Leroy, P.; Feuillet, C.; Sourdille, P. High transferability of bread wheat EST-derived SSRs to other cereals. *Theor. Appl. Genet.* **2005**, *111*, 677–687. [CrossRef]
34. Saha, M.C.; Cooper, J.D.; Rouf Mian, M.A.; Chekhovskiy, K.; May, G.D. Tall fescue genomic SSR markers: Development and transferability across multiple grass species. *Theor. Appl. Genet.* **2006**, *113*, 1449–1458. [CrossRef]
35. Varshney, R.K.; Graner, A.; Sorrells, M.E. Genic microsatellite markers in plants: Features and applications. *Trends Biotechnol.* **2005**, *23*, 48–55. [CrossRef]
36. Shirasawa, K.; Oyama, M.; Hirakawa, H.; Sato, S.; Tabata, S.; Fujioka, T.; Kimizuka-Takagi, C.; Sasamoto, S.; Watanabe, A.; Kato, M.; et al. An EST-SSR Linkage Map of Raphanus sativus and Comparative Genomics of the Brassicaceae. *DNA Res.* **2011**, *18*, 221–232. [CrossRef]
37. Liu, C.; Yuan, D.; Lin, Z. Construction of an EST-SSR-based interspecific transcriptome linkage map of fibre development in cotton. *J. Genet.* **2014**, *93*, 689–697. [CrossRef]
38. El-Rodeny, W.; Kimura, M.; Hirakawa, H.; Sabah, A.; Shirasawa, K.; Sato, S.; Tabata, S.; Sasamoto, S.; Watanabe, A.; Kawashima, K.; et al. Development of EST-SSR markers and construction of a linkage map in faba bean (*Vicia faba*). *Breed. Sci.* **2014**, *64*, 252–263. [CrossRef]
39. Dhaka, N.; Mukhopadhyay, A.; Paritosh, K.; Gupta, V.; Pental, D.; Pradhan, A.K. Identification of genic SSRs and construction of a SSR-based linkage map in *Brassica juncea*. *Euphytica* **2017**, *213*, 15. [CrossRef]
40. Shi, J.; Huang, S.; Zhan, J.; Yu, J.; Wang, X.; Hua, W.; Liu, S.; Liu, G.; Wang, H. Genome-wide microsatellite characterization and marker development in the sequenced *Brassica* crop species. *DNA Res.* **2014**, *21*, 53–68. [CrossRef]
41. Ding, Q.; Li, J.; Wang, F.; Zhang, Y.; Li, H.; Zhang, J.; Gao, J. Characterization and Development of EST-SSRs by Deep Transcriptome Sequencing in Chinese Cabbage (*Brassica rapa* L. ssp. *pekinensis*). *Int. J. Genom.* **2015**, *2015*, 473028. [CrossRef]
42. Winnepenninckx, B.; Backeljau, T.; Wachter, R.D. Extraction of high molecular weight DNA from molluscs. *Trends Genet.* **1993**, *9*, 407. [CrossRef]
43. Meng, L.; Li, H.; Zhang, L.; Wang, J. QTL IciMapping: Integrated software for genetic linkage map construction and quantitative trait locus mapping in biparental populations. *Crop J.* **2015**, *3*, 269–283. [CrossRef]
44. Zhang, Y. Construction of a Molecular Genetic Map and QTL Mapping for Major Agronomic Traits in Chinese Cabbage (*Brassica campestris* spp. *pekinensis*). Doctoral Dissertation, Shenyang Agricultural University, Shenyang, China, 2012. (In Chinese)
45. Liu, J. Linkage Map Construction and QTL Analysis for Agronomic Traits in Chinese Cabbage. Master's Dissertation, Tianjing Normal University, Tianjin, China, 2015. (In Chinese)
46. Wang, X.; Wang, H.; Wang, J.; Sun, R.; Wu, J.; Liu, S.; Bai, Y.; Mun, J.; Bancroft, I.; Cheng, F.; et al. The genome of the mesopolyploid crop species *Brassica rapa*. *Nat. Genet.* **2011**, *43*, 1035–1039. [CrossRef]
47. Wang, F.; Li, L.; Li, H.; Liu, L.; Zhang, Y.; Gao, J.; Wang, X. Transcriptome analysis of rosette and folding leaves in Chinese cabbage using high-throughput RNA sequencing. *Genomics* **2012**, *99*, 299–307. [CrossRef]
48. Wang, F.; Tan, T.; Zhang, Y.; Li, J.; Li, H.; Li, L.; Liu, L.; Gao, J. Cloning and Functional Analysis of BrTCP24 Gene in Chinese Cabbage (*Brassica rapa* L. ssp. *pekinensis*). *J. Agric. Biotechnol.* **2013**, *21*, 911–919. [CrossRef]
49. Wang, F.; Qiu, N.; Ding, Q.; Li, J.; Zhang, Y.; Li, H.; Gao, J. Genome-wide identification and analysis of the growth-regulating factor family in Chinese cabbage (*Brassica rapa* L. ssp. *pekinensis*). *BMC Genom.* **2014**, *15*, 807. [CrossRef]
50. He, Y.; Xue, W.X.; Sun, Y.D.; Yu, X.H.; Liu, P.L. Leafy head formation of the progenies of transgenic plants of Chinese cabbage with exogenous auxin genes. *Cell Res.* **2000**, *10*, 151–160. [CrossRef]

horticulturae

MDPI

Article

Br*PARP1*, a Poly (ADP-Ribose) Polymerase Gene, Is Involved in Root Development in *Brassica rapa* under Drought Stress

Gangqiang Cao [1], Wenjing Jiang [1,2], Gongyao Shi [1], Zhaoran Tian [1], Jingjing Shang [1,2], Zhengqing Xie [1], Weiwei Chen [1], Baoming Tian [1], Xiaochun Wei [2], Fang Wei [1,2,*] and Huihui Gu [1,*]

1 Henan International Joint Laboratory of Crop Gene Resources and Improvements, School of Agricultural Sciences, Zhengzhou University, Zhengzhou 450001, China; caogq@zzu.edu.cn (G.C.); jiang_wenjing2021@163.com (W.J.); shigy@zzu.edu.cn (G.S.); tzr1369@126.com (Z.T.); jingjshang@163.com (J.S.); zqxie@zzu.edu.cn (Z.X.); weiwei_chen15134@zzu.edu.cn (W.C.); tianbm@zzu.edu.cn (B.T.)

2 Institute of Horticulture, Henan Academy of Agricultural Sciences, Graduate T & R Base of Zhengzhou University, Zhengzhou 450002, China; jweixiaochun@126.com

* Correspondence: fangwei@zzu.edu.cn (F.W.); hhgu@zzu.edu.cn (H.G.)

Abstract: PARP proteins are highly conserved homologs among the eukaryotic poly (ADP-ribose) polymerases. After activation, ADP-ribose polymers are synthesized on a series of ribozymes that use NAD+ as a substrate. PARPs participate in the regulation of various important biological processes, such as plant growth, development, and stress response. In this study, we characterized the homologue of *PARP1* in *B. rapa* using RNA interference (RNAi) to reveal the underlying mechanism responding to drought stress. Bioinformatics and expression pattern analyses demonstrated that two copy numbers of *PARP1* genes (Br*PARP1.A03* and Br*PARP1.A05*) in *B. rapa* following a whole-genome triplication (WGT) event were retained compared with *Arabidopsis*, but only Br*PARP1.A03* was predominantly transcribed in plant roots. Silencing of Br*PARP1* could markedly promote root growth and development, probably via regulating cell division, and the transgenic *Brassica* lines showed more tolerance under drought treatment, accompanied with substantial alterations including accumulated proline contents, significantly reduced malondialdehyde, and increased antioxidative enzyme activity. In addition, the findings showed that the expression of stress-responsive genes, as well as reactive oxygen species (ROS)-scavenging related genes, was largely reinforced in the transgenic lines under drought stress. In general, these results indicated that Br*PARP1* likely responds to drought stress by regulating root growth and the expression of stress-related genes to cope with adverse conditions in *B. rapa*.

Keywords: poly (ADP-ribose) polymerase; drought; reactive oxygen species; root growth; whole-genome triplication

Citation: Cao, G.; Jiang, W.; Shi, G.; Tian, Z.; Shang, J.; Xie, Z.; Chen, W.; Tian, B.; Wei, X.; Wei, F.; et al. Br*PARP1*, a Poly (ADP-Ribose) Polymerase Gene, Is Involved in Root Development in *Brassica rapa* under Drought Stress. *Horticulturae* **2022**, *8*, 78. https://doi.org/10.3390/horticulturae8010078

Academic Editor: Dilip R. Panthee

Received: 24 November 2021
Accepted: 12 January 2022
Published: 14 January 2022

Publisher's Note: MDPI stays neutral with regard to jurisdictional claims in published maps and institutional affiliations.

Copyright: © 2022 by the authors. Licensee MDPI, Basel, Switzerland. This article is an open access article distributed under the terms and conditions of the Creative Commons Attribution (CC BY) license (https://creativecommons.org/licenses/by/4.0/).

1. Introduction

Drought is a common abiotic stress that affects plant growth and limits crop yield and quality [1]. Long-term exposure to water-deficient conditions will affect the physiological responses of plants, such as the activities of hormone-metabolizing enzymes, the accumulation of reactive oxygen species (ROS), the opening and closing of stomata, and other characteristics, showing the phenotype of growth retardation [2–4]. ROS accumulation could cause oxidative damage to DNA and directly extract hydrogen from deoxyribose, leading to DNA single-strand and double-strand breaks, resulting in genome instability and plant aging [5]. Under drought stress, the accumulation of ROS in plants will enhance plasma membrane oxidation and protein degradation, causing oxidative damage and affecting plant growth [6]. Facing the pressure of drought, plants need complex cellular and molecular networks to establish new energy homeostasis and ensure normal growth and development. The root system is the most sensitive organ of plants to deal with drought. When the root system feels a lack of water, it will stimulate the plant's molecular and physiological responses, adjust the morphological response, and respond to drought stress [7]. At present, in order to make the plants adapt to the changeful environments and increase the productivity of crops, breeding and biotechnology are now used to reduce the sensitivity of plants to unfavorable environments [8].

Poly (ADP-ribose) polymerase (PARP) enzymes play a key role in many cellular processes, such as DNA damage repair, maintenance of genome steadiness, and cell death [9]. Poly (ADP-ribosyl)ation (PARylation) is a reversible post-translational modification catalyzed by PARP enzymes. It participates in a series of reactions consisting of DNA damage awareness and repair, cell division and death, chromatin modification, gene transcription regulation, and stress response [10,11]. PARP uses NAD+ as a substrate to continuously add poly (ADP-ribose) (PAR) fragments to the amino acid receptor residues of the target protein to catalyze PARylation [9].

Three PARPs have been found in the model plant *A. thaliana*: At*PARP1* (At2g31320), At*PARP2* (At4g02390), and At*PARP3* (At5g22470), which are all located in the nucleus. At*PARP1* and At*PARP2* act as sensors of DNA damage and participate in DNA repair and stress response [12]. Experiments have shown that the double mutants of *parp1* and *parp2* in *Arabidopsis* developed more root systems under external pressure, and the main and lateral roots grow faster [13]. At*PARP3* participates in double-strand break (DSB) repair and maintains seed vigor during seed storage [14]. The activity of PARP has also been found in other plants. For instance, *PARP1* and *PARP2* have been found in the different species such as wheat, peas, soybean, tobacco, and corn, responding to biotic and abiotic stresses and affecting the development of leaves and roots [15–18].

In plants, PARP is the essential energy consumption under exterior stress conditions. External pressure will induce the activity of PARP, leading to the synthesis and decomposition of NAD+, enhancing the respiration of mitochondria and providing ATP energy, resulting in high energy consumption. A large wide variety of studies have shown that immoderate activation of PARP can cause cell death due to energy consumption [19]. Therefore, it is necessary to maintain the steady state of cell energy and reduce the consumption of NAD+. Chemical inhibition or silencing of PARP activity can minimize the consumption of NAD+ and enhance the effective utilization rate of energy, thus making plants cope better with exterior pressure [20]. Related studies proved that transgenic Arabidopsis with low-level PARP are more resistant to a range of abiotic stresses such as drought, robust light, and heat [21].

In the present work, we characterized a homolog of *PARP1* in the major vegetable plant *B. rapa* designated as Br*PARP1*, and the underlying mechanism of Br*PARP1* responsive to drought stress was analyzed. Hopefully this study will be of great use for genetic improvements of drought resistance in *Brassica* breeding programs.

2. Materials and Methods

2.1. Plant Material and Growth Conditions

The selfing *B. rapa* DH lines (cxl-45-05) and *A. thaliana* (Col-0) were grown in a growth chamber under the following conditions: 16 h/22 °C in light, 8 h/16 °C in darkness, and relative humidity of about 45%.

2.2. Evolution and Gene Collinearity Analysis

The genome, coding sequence, and protein sequence of each species were downloaded from Ensembl plant (https://plants.ensembl.org/index.html, accessed on 27 September 2020). The amino acid sequence of PARP1 protein was aligned using the MUSCLE program with default parameters. The phylogenetic tree was constructed using MEGA7 software with the maximum likelihood (ML) method and 1000 bootstrap replicates [22]. The number of *PARP*s was counted with evolutionary retention, and the collinearity relationship between At*PARP1* and Br*PARP1* genes was visualized by Circos. The NG method was used to calculate the synonymous mutation rate (Ks) and non-synonymous mutation rate (Ka) of the coding sequence [23].

2.3. Gene Structure Protein Motif Identification and Protein Functional Domain Analysis

To identify conserved motifs, we used MEME (http://meme.sdsc.edu/meme/meme.html; accessed on 27 September 2020 [24]), with the motif length set at 10–100 and motif maximum number set at 10. We analyzed the gene structure using Gene Structure Display Server (GSDS; http://gsds.cbi.pku.edu.cn/, accessed on 28 September 2020 [25]). Moreover, the features of genes were analyzed using EBI-Tools (http://www.ebi.ac.uk/Tools/emboss/, accessed on 28 September 2020). The putatively conserved function domain was analyzed by CDD (https://www.ncbi.nlm.nih.gov/cdd, accessed on 27 September 2020).

2.4. Subcellular Location

In order to explore the location of Br*PARP1*, we constructed two expression vectors for a transient expression system in *N. benthamiana*, and the transient transformation was performed as previously described elsewhere [26]. The full-length coding sequence of Br*PARP1* was first cloned into pEASY-T1 vector for amplification and then subcloned into pCAMBIA1300-GFP vector using BamHI and XbaI restriction sites. Subsequently, they were transferred to *Agrobacterium tumefaciens* (EH105) and infiltrated with the leaves using 1 mL syringes. An Olympus fluorescence microscope (DX53, Japan) was used to detect the fluorescence signal of GFP fusion protein in leaves 2 days after transformation.

2.5. Histochemical Analysis of GUS Activity

The ProBr*PARP1*::GUS transgenic plants were subjected to hydroponic culture, and the roots, stems, and leaves of transgenic plants grown about 3 weeks were treated with PEG6000 at 5%, 10%, 15%, and 20%. The expression pattern under osmotic stress was observed by GUS tissue staining. The staining buffer was mainly prepared with 100 µL X-gluc stock solution and 5 mL X-gluc base solution. The samples were stained in the staining buffer at 37 °C for 12–24 h before destaining in ethanol, as previously described elsewhere [27].

2.6. Drought Stress Conditions and Phenotypic Observation

RNAi technology is a mechanism of silent gene transfer after exogenous short double-stranded RNA (dsRNA) mediated by RNAi technology to develop transgenic *B. rapa* lines [28]. Three independent transgenic *Brassica* lines—R1, R3, and R4—were selected with their expression level 20–30% of the wild-type. For drought treatment, about 4-week-old plants for transgene and WT were used by continuously withholding irrigation for 3 weeks before rewatering for recovery, and at least 10 planted pots for each treatment was used for phenotype recording with three biological replicates. In addition, 20% PEG6000 on 1/2MS medium was applied to simulate drought stress using each treatment with 10 seedling plantlets with 3 biological replicates. The root length and the number of lateral roots were counted at 6d, 8d, and 10d. Propidium iodide (PI) staining was used to stain and analyze the roots of both transgenic and WT plants [29]. The seedling roots were stained with 30 µg/mL PI (Sigma Aldrich, St. Louis, MI, USA) for 1 min and then observed with fluorescence microscope (Carl Zeiss, Jena, Germany).

2.7. Determination of Relative Water Content (RWC)

In order to determine relative water content (RWC), we immediately weighed the leaves separated from the *B. rapa* to determine the fresh weight (FW). After determining the FW, we kept the leaves in the distillation for 12 h and weighted them as turgid weight (TW). The dry weight (DW) was recorded by drying the sample at 65 °C for 12 h. RWC was calculated as $(FW - DW)/(TW - DW) \times 100\%$ [30].

2.8. Analysis of Proline, Malondialdehyde (MDA), and NAD+ Content

The determination of proline content was described by Leclercq et al. (2012) [31]. In brief, a leaf sample of 0.2 g was weighted and thoroughly ground in 3 mL sulfosalicylic acid (3%), and 10 mL sulfosalicylic acid was added. After boiling in a water bath for 1 h, the sample was centrifuged after cooling. A total of 1 mL supernatant was collected and mixed with 1 mL glacial acetic acid and 1 mL acid–ninhydrin, and then heated at 100 °C for 1 h. Then, 2 mL toluene was added after cooling on ice, and the mixture was stood for 2 h after shaking. After centrifugation, the upper solution was taken, and the absorbance was measured at 520 nm by spectrophotometer. The content of proline in the sample was calculated by using toluene as a blank control.

The content of malondialdehyde (MDA) was determined according to Kong et al. (2016) [32]. A total of 1 g of fresh leaves was collected and ground in 5 mL (10%) trichloroacetic acid; then, the mixture was centrifuged at 3500× *g* for 10 min. The supernatant (1 mL) was mixed with an equal amount of 0.6% thiobarbituric acid and then placed in a water bath (100 °C, 20 min). The supernatant was centrifuged and measured for the absorbance at 450, 532, and 600 nm.

The determination of NAD+ was referred to the Greiss Company's instructions. Briefly, about 0.1 g fresh leaves were collected and added with 1 mL extraction buffer, and then the sample was ground in an ice bath followed by incubation at 95 °C for 5 min. The sample was then immediately placed on ice bath for 5 min and centrifuged at 12,000 rpm at 4 °C for 10 min. The 500 μL supernatant was taken out and added with 500 μL V1 extract buffer and then centrifuged at 12,000 rpm at 4 °C for 5 min. A total of 100 μL supernatant was taken out and analyzed with the spectrophotometer at 450 nm.

2.9. Quantitative Real-Time PCR Analysis

Total RNA extractions, isolated from *B. rapa*, were performed with Plant Total RNA Isolation Kit Plus (Foregene, Chengtu, China), using 100 mg of fresh leaf tissue. The purified RNA was measured with an ultra-micro spectrophotometer (Nanodrop 2000). The cDNA was synthesized by HiFiScript cDNA synthesis Kit (CWBio, Beijing, China). The Lunar® Universal qPCR Master Mix (NEB, Beijing, China) and Light Cycler 480 system were used for qRT-PCR analysis. The PCR conditions were 94 °C for 30 s, 40 cycles at 94 °C for 10 s, and 58 °C for 30 s, followed by a melting curve to determine the specificity of the amplification. Relative expression levels were calculated using the $2^{-\Delta\Delta Ct}$ method [33], and β-actin was used as internal control with three biological replicates. All the primers used are listed in Table S1.

2.10. Statistical Analysis

The experimental data involved in this experiment have at least 3 independent biological and technical replicates. All data graphs were drawn and analyzed using GraphPad Prism 5, and one-way ANOVA was used to analyze the significance of differences between the experimental and the control with the *t*-test applied. ** $p < 0.01$ indicates that the difference between the data was very significant, whereas * $p < 0.05$ indicates that the difference was significant.

3. Results

3.1. Conservation of BrPARP Genes following Whole-Genome Triplication Event in B. rapa

Compared with *A. thaliana*, evolutionarily, the *Brassica* species has experienced an additional whole-genome triplication (WGT) event, and the duplicated genes were likely retained or lost due to the successive genomic rearrangements [34]. Thus, we constructed the phylogenetic relationship, gene structure, and functional domain of the PARP protein in three typical diploid *Brassicaceae* species, namely, *A. thaliana*, *B. rapa*, and *B. oleracea*, and analyzed the copy number of the *PARP* genes retained in their genome (Figure 1A). The results showed that the putatively duplicated genes *PARP2* and *PARP3* were both retained with single copy number in *B. rapa* and *B. oleracea*, indicating a loss of function probably induced by genomic rearrangements, but *PARP1* gene retained redundancy with two copies in *B. rapa* and three in *B. oleracea* after the WGT event (Figure 1A). We used the Circos to analyze the At*PARP1* and two Br*PARP1* genes for collinearity, which showed the retained Br*PARP1* genes were located on chromosome 3 and 5 (respectively, Br*PARP.A03* and Br*PARP1.A05*) in *B. rapa* (Figure 1B). Further, the non-synonymous/synonymous mutation rate (Ka/Ks) between gene pairs was calculated (Figure 1C), and the calculated Ka/Ks ratio between Br*PARP1* and At*PARP1* gene pairs was found to be about 0.29–0.43, which was generally believed to experience lower selection pressure and should be functionally conserved during evolution. In addition, we analyzed the *PARP1* gene structure and protein conserved domains between *B. rapa*, *B. oleracea*, and *A. thaliana* (Figure 1D). The results showed that the structure and conserved domain between Br*PARP1.A03* and Br*PARP1.A05* were highly similar, inferring both Br*PARP1* genes might be functionally redundant in *B. rapa*.

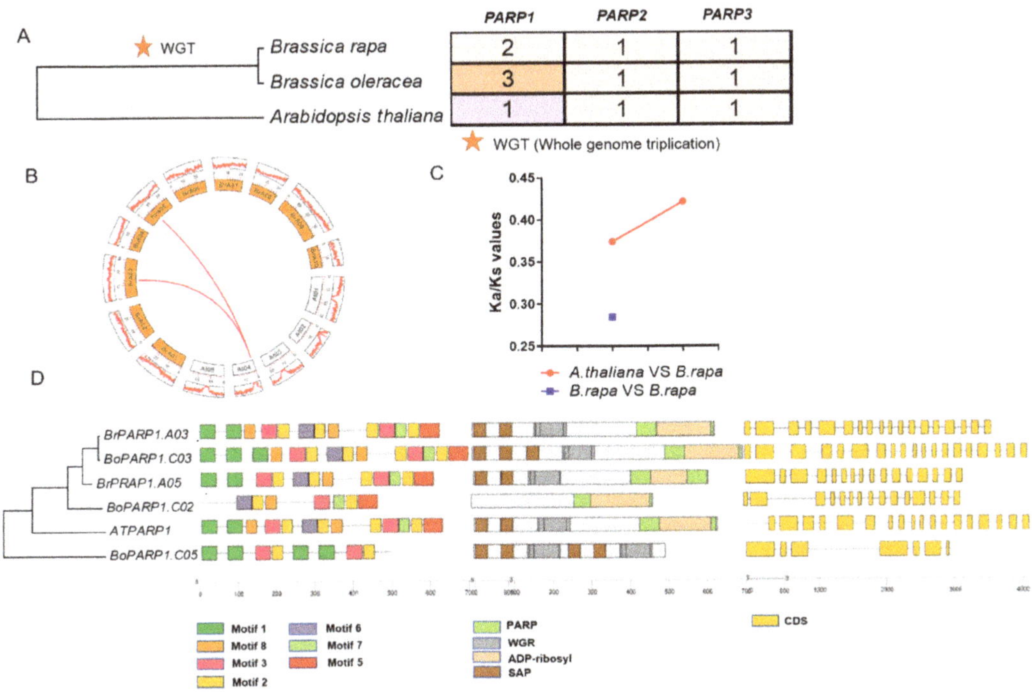

Figure 1. *PARP* genes following WGT in brassicas. (**A**) The number of *PARP* genes copies retained in *Arabidopsis* and *Brassica* species. The phylogenetic tree resulting from the evolution of species. (**B**) Collinear correlations of Br*PARP1* genes and neighboring genes in the *A. thaliana* and *B. rapa* genomes. The *B. rapa* and *A. thaliana* chromosomes are colored in accordance with the inferred ancestral chromosomes following a set up convention. The lines of Br*PARP1* genes are red and lost genes are gray. The figure was created using Circos software. (**C**) The Ka/Ks value of *B. rapa* and *A. thaliana PARP1* orthologous gene pair and paralogous gene pair. (**D**) Gene structures and conserved motifs of *PARP1*. The unrooted phylogenetic tree resulting from the full-length amino acid alignment of all of the PARP1 proteins. The tree was constructed the usage of maximum likelihood (ML) and bootstrap values calculated with 1000 replications using MEGA7.

3.2. Expression Pattern and Subcellular Localization of BrPARP1 in B. rapa

The expression level of two Br*PARP1* copies (Br*PARP1.A03* and Br*PARP1.A05*) was verified in different tissues in *B. rapa* (Figure 2A), and the results showed that transcription level of *BrPARP1.A03* (ID: Bra000883) was predominantly higher than that of Br*PARP1.A05* (ID: Bra018555) in the assayed tissues, as well as the highest expression level detected in roots, followed by stems, leaves, and flowers, which indicated that Br*PARP1.A03* played main functions in *B. rapa*. In addition, transient expression analysis of the *PARP1*.A03–GFP fusion protein in *N.benthamiana* leaf cells confirmed that the fusion protein is located in the nucleus (Figure 2B), which echoes the role of PARP in DNA repair and transcriptional regulation [12].

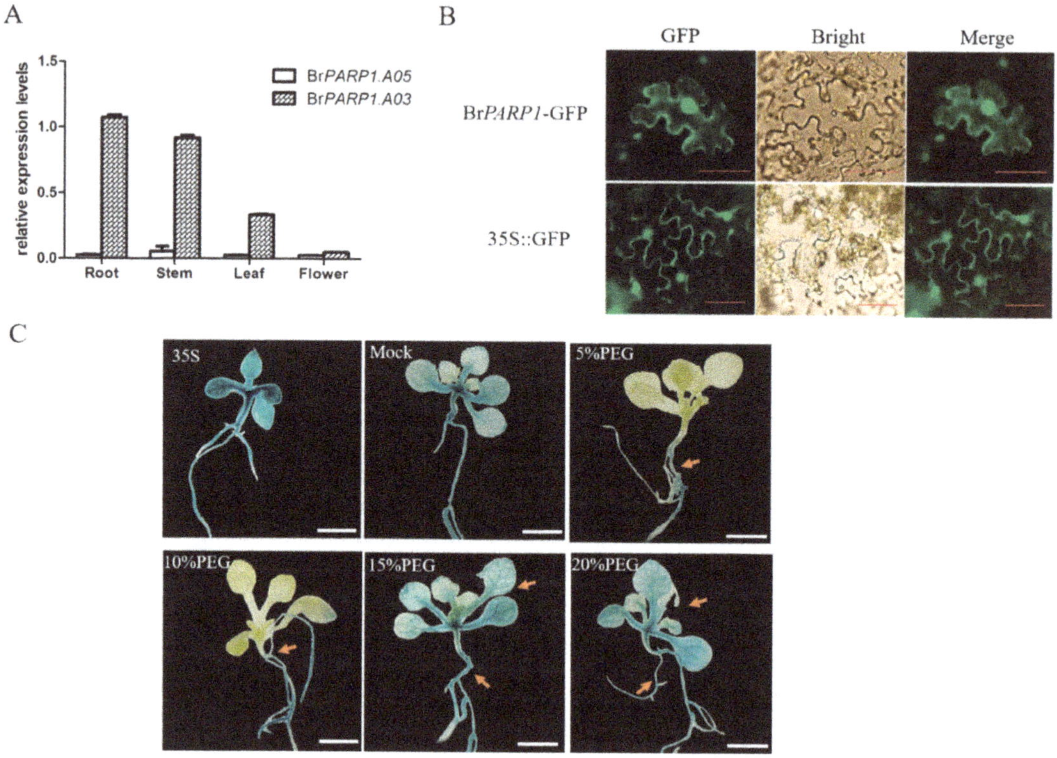

Figure 2. The expression of Br*PARP1* and subcellular location. (**A**) Analysis of Br*PARP1* for tissue-specific expression. The Actin gene was used as a reference. (**B**) Subcellular localization of the fused 35S::Br*PARP1*-GFP in *N. benthamiana* leaf cells. The 35S::GFP construct as the control. Bar = 50 μm. (**C**) Histochemical staining in transgenic *A. thaliana* plants. GUS stains analysis proCaMV35S::GUS transgenic *A. thaliana* plants. GUS stains analysis proBr*PARP1*::GUS transgenic *A. thaliana* plants after 0%, 5%, 10%, 15%, and 20% PEG6000 treatment. Bar = 5 mm.

In order to verify the expression pattern of Br*PARP1.A03* in response to drought, we used PEG6000 to simulate drought stress. The proBr*PARP1*::GUS transgenic lines were obtained, and different concentrations of PEG6000 (0%, 5%, 10%, 15%, 20%) were used for stress treatment (Figure 2C). The results showed that Br*PARP1* promoter was not easily activated under mild drought treatments (5–10% PEG6000) in the seedling leaves, but sensitively detected in roots. Especially under harsher treatments (15% to 20% PEG6000), GUS activity driven by Br*PARP1* promoter gradually increased in all tissues treated with PEG6000.

3.3. Silencing of BrPARP1 Enhanced Plant Tolerance in B. rapa under Drought Stress

To explore the response of Br*PARP1* gene under drought stress, we obtained the transgenic lines via RNA-induced gene silencing in *B. rapa*. About 4-week-old plants were used for drought treatment by withholding irrigation for 3 weeks; the data showed the transgenic *Brassica* lines exhibited a better survival ability under drought treatments, and after rewatering, the transgenic plants could better recover their growth ability (Figure 3A). The leaf relative water contents (RWC) showed less difference between transgenic lines and wild-type under normal growth conditions, but a significant difference was observed under drought treatments (Figure 3B). In addition, more proline accumulated in the leaves of transgenic lines (Figure 3C), as an osmotic protection substance to reduce potential damages to plants due to drought and water loss [35]. Reactive oxygen species (ROS) were also analyzed after exposure to drought stress, and the results showed that transgenic lines suffered less oxidative damage compared with WT (Figure 3D). Malondialdehyde (MDA) was generally considered as an indicator of oxidative damage to plant cell membranes under stresses [36,37], and the MDA content was thus determined, which was significantly lower in transgenic lines compared with WT (Figure 3E). Nicotinamide adenine dinucleotide (NAD+) is a vital pyridine nucleotide, and its depletion may occur in response to excessive DNA damage caused by free radicals. Our data showed that the transgenic lines maintained a higher NAD+ level compared with WT under drought treatments (Figure 3F), which was probably owing to reduced *PARP* activity [38,39]. Overall, the outcomes indicated that silencing of Br*PARP1* could enhance the tolerance of plants to drought stress and enlarge the survival rate of plants under adverse environmental conditions.

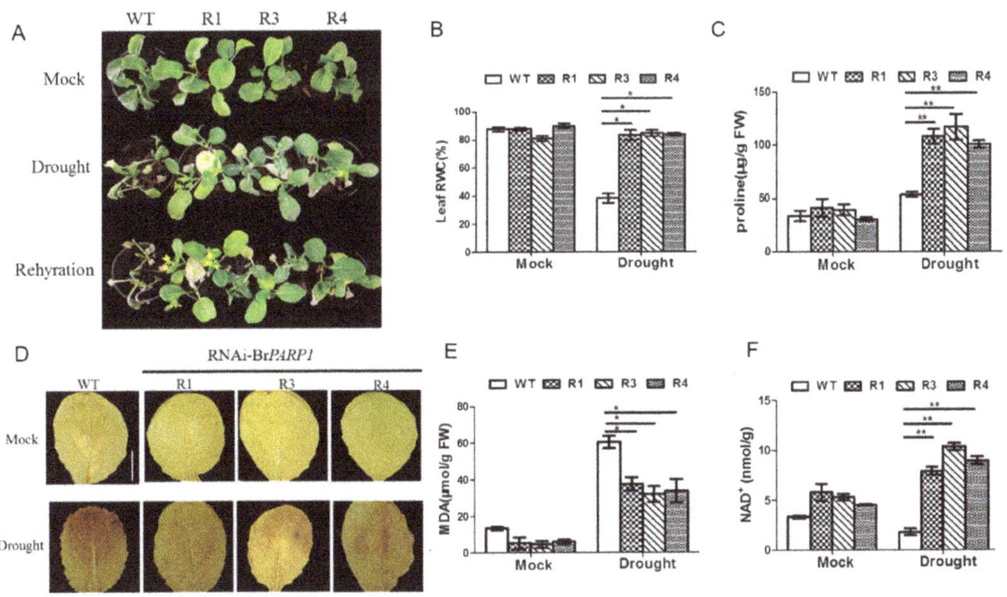

Figure 3. (**A**) Phenotype analysis of WT and RNAi-Br*PARP1* during dehydration treatment. Three RNAi-BrPARP1 transgenic lines and WT were treated with withholding irrigation at the seedling stage for about 3 weeks. (**B**) Relative water content of detached leaves of RNAi-Br*PARP1* and WT plants. (**C**) Proline content of RNAi-Br*PARP1* and WT plants under normal conditions and after water deprivation. (**D**) ROS content of RNAi-Br*PARP1* and WT plants. (**E**) MDA content of RNAi-Br*PARP1* and WT plants under normal conditions and after water shortage. (**F**) NAD+ content of RNAi-Br*PARP1* and WT plants under normal conditions and after water shortage. Values were means ± se of biological replicates (n > 3). (** $p < 0.01$, * $p < 0.05$).

3.4. BrPARP1 Regulated Root Developments in B. rapa under Drought Stress

The morphology of plant roots plays a vital role in plant growth and environmental adaptability [40]. We analyzed the root growth via silencing Br*PARP1* in the transgenic plants under drought conditions. The results showed that compared with the WT, the transgenic lines had more developed root systems under both normal conditions and drought treatment (Figure 4A), and silencing of Br*PARP1* could accelerate main root growth and lateral root developments (Figure 4B–D). In addition, we found that the genes (*LOX1, CEG,* and *TIR1*) related to lateral root development were significantly upregulated in transgenic lines, which indicated that silencing of Br*PARP1* could promote root development and enhance drought resistance in *B. rapa*.

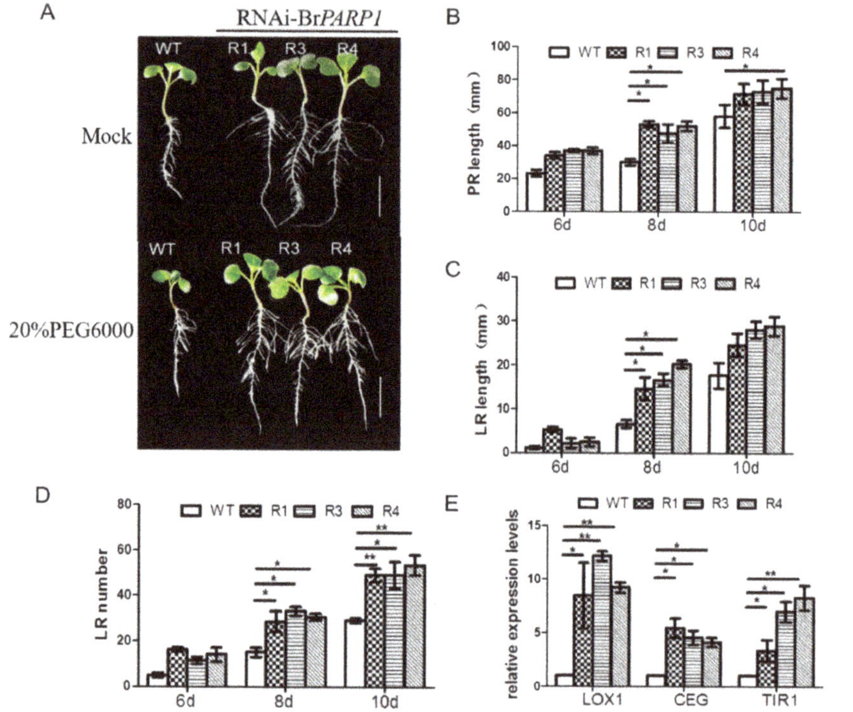

Figure 4. Analysis of the effect of Br*PARP1* on root growth. (**A**) Comparison of the root system of WT and RNAi-Br*PARP1* transgenic plants under normal conditions and after 20% PEG6000 treatment. Bar = 10 mm. (**B**) Principal root length analysis of WT and RNAi-Br*PARP1* transgenic plants at 6 days, 8 days, and 10 days. (**C**) Lateral root length analysis of WT and RNAi-Br*PARP1* transgenic plants at 6 days, 8 days, and 10 days. (**D**) Lateral root number analysis of WT and RNAi-Br*PARP1* transgenic plants at 6 days, 8 days, and 10 days. (**E**) Analysis expression of lateral root-related genes. *LOX1: Lectin-like oxidized low-density lipoprotein receptor-1; CEG: Clusters of Essential Genes; TIR1: Transport Inhibitor Response 1* (** $p < 0.01$, * $p < 0.05$).

3.5. BrPARP1 Affected Root Cell Division in B. rapa

Root growth involves cell division in the meristematic zone and cell elongation in the elongation zone [41]. We found that in comparison with the wild-type, the transgenic lines had more layers of cells in root meristematic zone (Figure 5A), and the length of the principal root meristem was longer (Figure 5B), the continuous cell division and differentiation within root tips promoted its growth and development, and the difference in the length of meristem layer revealed the rapid root growth of transgenic plants. In addition, we examined the transcription of key genes related to cell cycle control (Figure 5C), and the results showed that the expression of mitotic regulators such as type B cyclin (*CycB1;1*) increased significantly, but the transcription level of the kinase gene *CHK2* involved in apoptosis was obviously reduced. At the same time, the expression of nuclear replication-related gene *TOP6B* was significantly reduced, and the transcription level of ATM that regulates the cell cycle kinase was also downregulated in the transgenic plants. These results indicated that silencing of Br*PARP* can enhance cell division via partially inhibiting root nuclear endoreduplication in *B. rapa*.

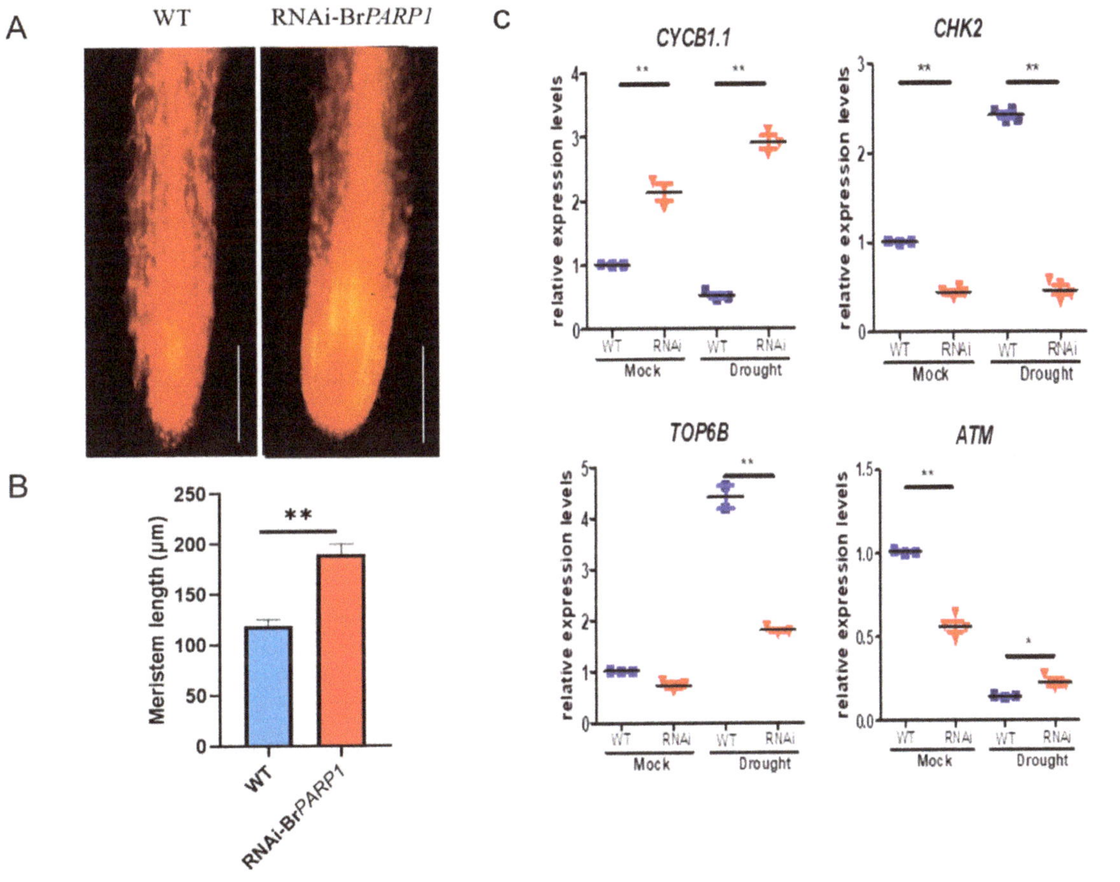

Figure 5. Analysis of Br*PARP1* affecting root growth. (**A**) Propidium iodide-stained root meristems of WT and RNAi-Br*PARP1* plants. Bar = 100 μm. (**B**) Principal root meristem length of WT and RNAi-Br*PARP1* plants. (**C**) Expression level analysis of root growth transition marker genes. *CYCB1.1: 1 Cyclin B Transcripts; CHK2: Checkpoint kinase2; TOP6B: Topoisomerase VI; ATM: Ataxia Telangiectasia Mutated.* (** $p < 0.01$, * $p < 0.05$).

3.6. BrPARP1 Regulated Expression of Stress-Related Genes in B. rapa under Drought Stress

In order to clarify the Br*PARP1*-mediated mechanisms regarding on the enhanced drought tolerance of transgenic *B. rapa* lines, we selected several key transcription factors, which were closely related to drought stress, namely, *RD26*, *SOS1*, *CBL1*, *RHL2*, and *DREB2A* (Figure 6). These results showed that these assayed genes were significantly upregulated under drought treatment in the transgenic lines compared to the wild-type.

These results indicated that Br*PARP1* may respond to drought stress by cooperating with stress-related genes, so as to better cope with external pressure.

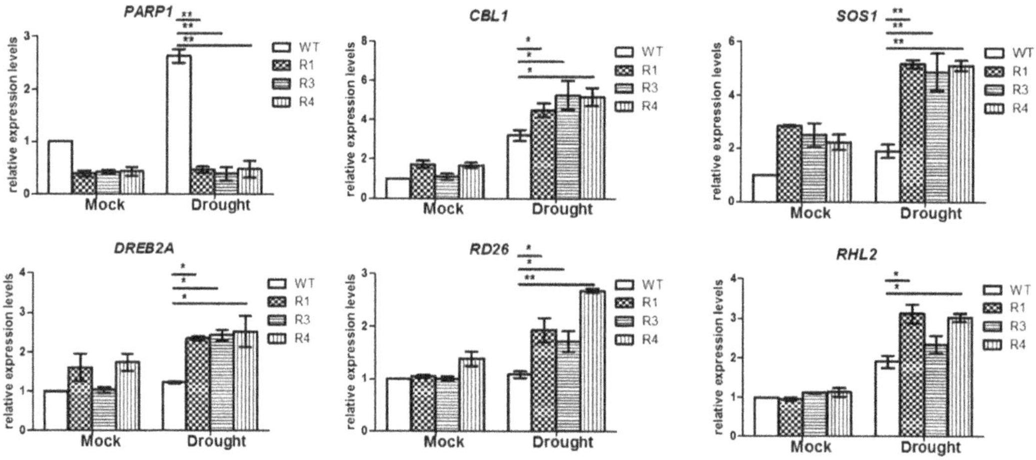

Figure 6. Expression analysis of stress-related genes in B. rapa under drought treatment. *RD26: RESPONSIVE TO DESICCATION 26; SOS1: STRESS OVERLY SENSITIVE 1; CBL1: CALCINEURIN B-LIKE PROTEIN 1; RHL2: ROOT HAIRLESS 2; DREB2A: DEHYDRATION RESPONSIVE ELEMENT-BINDING PROTEIN2A*. (** $p < 0.01$, * $p < 0.05$).

4. Discussion

4.1. BrPARP1 Probably Functions as a Single Copy Gene in B. rapa during Evolution

Poly (ADP-ribose) polymerase (PARP) is a post-translational modification enzyme in organisms. It belongs to the PARP family and has the function of catalyzing the formation of linear or branched poly ADP-ribose (PAR) polymer from ADP ribose. It is highly conservative in evolution. In the process of evolution, whole-genome duplications (WGD) do not solely cause genetic abundance but may also produce a large amount of genetic variation [42]. In *Brassica* crops, it has experienced an additional WGT event, which has an important effect on the richness and functional diversity of the species [43]. However, some studies had reported that there is a set of genes in *A. thaliana* and *O. sativa*, existing in the form of a single copy after external pressure selection [44]. The existence of single-copy genes in eukaryotes may be due to the basic functions that are often highly conserved, and the expression of single-copy genes in tissues will be higher, with higher conservation [45,46]. In order to explore the effects on the copy number of *PARP* gene after WGT event in *Brassica*, this study selected *A. thaliana*, *B. oleracea*, and *B. rapa*. Bioinformatics analysis showed that *PARP2* and *PARP3* retained one copy number in both *B. rapa* and *B. oleracea*, but *B. rapa* and *B. oleracea* retained two and three copies of the *PARP1* gene after the WGT event, respectively (Figure 1A), while *PARP*s mostly retain single copy form during the evolutionary process, which means that despite the replication event, the gene will still exist as a single-copy form. In this study, the At*PARP1* and Br*PARP1* gene pairs were analyzed for collinearity (Figure 1B). The results showed that after WGT, the Br*PARP1* gene was duplicated into three copies, and one copy was lost in the evolutionary process, leaving Br*PARP1.A03* and Br*PARP1.A05* in the subgenome. Analysis of the gene structure, conserved domains (Figure 1D), and gene transcription level (Figure 2A) of Br*PARP1.A03* and Br*PARP1.A05* showed that Br*PARP1.A03* is more conservative in evolution. Therefore, after WGT events, the duplicated Br*PARP1* gene may have undergone gene loss, and mainly functions in the form of a single copy. The communication between nuclear genes and different organelles is strictly regulated and plays an important role in maintaining the relative balance of proteins encoded between organelles [47]. The increase in gene copy number caused by WGD events may lead to the relative imbalance of protein interaction between organelles, which may have negative effects on plants [48]. It is possible that increase in copies of genes may be deleterious in some cases, and thus these genes may exist mainly in the form of a single copy. The WGT event in *B. rapa* has witnessed the power of gene evolution, which has enlightened significance for the functional analysis of multi-copy genes.

4.2. *BrPARP1 Likely Regulated Root Development Responsive to Drought Stress*

Drought stress hinders cell division and elongation, ultimately reducing crop yields and causing economic losses [49]. Plants that maintain root growth under drought conditions have a higher survival rate. Therefore, this study explored the phenotypic differences between RNAi-Br*PARP1* transgenic and wild-type plants under the drought treatment for about 3 weeks and 20% PEG6000 osmotic stress conditions. Under normal conditions, the RNAi-Br*PARP1* transgenic line and the wild-type plants showed similar growth ability. When withholding irrigation for about 3 weeks, the RNAi-Br*PARP1* transgenic lines were less wilted compared with the wild-type plants. After 3 days of rewatering, most of the RNAi-Br*PARP1* transgenic lines can survive and regain vitality, but the WT plants cannot recover themselves. Under the 20% PEG6000 osmotic stress, RNAi-Br*PARP1* transgenic plants have a more developed root system, and the main root length and the number of lateral roots increased significantly. Therefore, we used propidium iodide (PI) staining method to compare the staining of RNAi-Br*PARP1* transgenic plants and wild-type roots. The results showed that the RNAi-Br*PARP1* transgenic plants had more cells in the root meristem, which was consistent with the root phenotype observed. The response of plants to stress is often the result of multi-gene synergy. The expression level of transcripts of related stress genes is an important basis for verifying plant resistance. It has been reported that *RD26* and transcription factor *DREB2A* are significantly upregulated in *B. juncea L* under water deficit, playing an important role in improving tolerance to drought stress [50]. The high transcription levels of *SOS1* and *CBL1* can make plants more drought-tolerant and salt-tolerant, helping plants to be better able to cope with external pressures [51,52]. In this study, the stress-related marker genes *RD26*, *SOS1*, *CBL1*, *RHL2*, and *DREB2A* were thus selected for real-time quantitative PCR analysis. The results showed that these stress genes were all upregulated in RNAi-Br*PARP1* transgenic plants compared with the wild-type.

In summary, we found that after RNA interference with Br*PARP1* expression, RNAi plants had a more developed root system, which improved the water absorption capacity of plants, thereby showing a higher stress tolerance behavior. Previous studies have shown that root elongation helps plants respond to drought stress [53]. Nicotinamide adenine dinucleotide (NAD+) is an important pyridine nucleotide that plays an important cofactor and substrate role in a series of key cellular processes such as oxidative phosphorylation, ATP production, DNA repair, and epigenetic regulation of gene expression [54]. Studies have shown that increasing NAD+ levels can greatly reduce oxidative cell damage in metabolic tissues. After oxidative damage, inhibiting *PARP* activity can maintain NAD+ and ATP levels and prevent cell lysis [55]. These data support the previous results that the reduction of *PARP* activity can improve the tolerance of abiotic stress, as well as confirming the potential application of *PARP* gene in the induction of plant tolerance during breeding program [20].

Supplementary Materials: The following supporting information can be downloaded at: https://www.mdpi.com/article/10.3390/horticulturae8010078/s1, Table S1: Primer information sequence.

Author Contributions: G.C. and F.W. conceived and designed the experiments; Z.T. and J.S. performed the experiments; W.C. and Z.X. analyzed the data; W.J., X.W. and B.T. prepared the figures and tables; G.S. and H.G. drafted and revised the manuscript critically. All authors have read and agreed to the published version of the manuscript.

Funding: This work was financially supported by Henan Provincial Natural Science Foundation of China (No. 202300410366), and the Program for Science &Technology Innovation Talents in

Universities of Henan Province (No. 19HASTIT014), and Youth Innovation Project of Key Discipline of Zhengzhou University (No. XKZDQN202002), and the Fostering Project for Basic Research of Zhengzhou University (No. JC21310015).

Institutional Review Board Statement: Not applicable.

Informed Consent Statement: Not applicable.

Data Availability Statement: Not applicable.

Acknowledgments: We would like to express our thanks to the anonymous reviewers for their useful comments.

Conflicts of Interest: The authors declare no conflict of interest and absence of any commercial or financial benefits of this research.

References

1. Agurla, S.; Gahir, S.; Munemasa, S.; Murata, Y.; Raghavendra, A.S. Mechanism of stomatal closure in plants exposed to drought and cold stress. *Adv. Exp. Med. Biol.* **2018**, *1081*, 215–232. [CrossRef] [PubMed]
2. Kosar, F.; Akram, N.A.; Ashraf, M. Exogenously-applied 5-aminolevulinic acid modulates some key physiological characteristics and antioxidative defense system in spring wheat (*Triticum aestivum* L.) seedlings under water stress. *S. Afr. J. Bot.* **2015**, *96*, 71–77. [CrossRef]
3. Kamanga, R.M.; Mbega, E.; Ndakidemi, P. Drought Tolerance Mechanisms in Plants: Physiological Responses Associated with Water Deficit Stress in Solanum lycopersicum. *Adv. Crop. Sci. Technol.* **2018**, *6*, 1–8. [CrossRef]
4. Ahammed, G.J.; Li, X.; Yang, Y.; Liu, C.; Zhou, G.; Wan, H.; Cheng, Y. Tomato WRKY81 acts as a negative regulator for drought tolerance by modulating guard cell H_2O_2–mediated stomatal closure. *Environ. Exp. Bot.* **2020**, *171*, 103960. [CrossRef]
5. Cadet, J.; Richard Wagner, J. DNA base damage by reactive oxygen species, oxidizing agents, and UV radiation. *Cold Spring Harb. Perspect. Biol.* **2013**, *5*, a012559. [CrossRef] [PubMed]
6. Blokhina, O.; Virolainen, E.; Fagerstedt, K.V. Antioxidants, oxidative damage and oxygen deprivation stress: A review. *Ann. Bot.* **2003**, *91*, 179–194. [CrossRef]
7. Janiak, A.; Kwaśniewski, M.; Szarejko, I. Gene expression regulation in roots under drought. *J. Exp. Bot.* **2015**, *67*, 1003–1014. [CrossRef]
8. Zhang, H.; Xiang, Y.; He, N.; Liu, X.; Liu, H.; Fang, L.; Zhang, F.; Sun, X.; Zhang, D.; Li, X.; et al. Enhanced Vitamin C Production Mediated by an ABA-Induced PTP-like Nucleotidase Improves Plant Drought Tolerance in Arabidopsis and Maize. *Mol. Plant* **2020**, *13*, 760–776. [CrossRef]
9. Gibson, B.A.; Kraus, W.L. New insights into the molecular and cellular functions of poly(ADP-ribose) and PARPs. *Nat. Rev. Mol. Cell Biol.* **2012**, *13*, 411–424. [CrossRef]
10. Briggs, A.G.; Bent, A.F. Poly(ADP-ribosyl)ation in plants. *Trends Plant Sci.* **2011**, *16*, 372–380. [CrossRef]
11. Luo, X.; Kraus, W.L. On PAR with PARP: Cellular stress signaling through poly(ADP-ribose) and PARP-1. *Genes Dev.* **2012**, *26*, 417–432. [CrossRef] [PubMed]
12. Song, J.; Keppler, B.D.; Wise, R.R.; Bent, A.F. PARP2 Is the Predominant Poly(ADP-Ribose) Polymerase in Arabidopsis DNA Damage and Immune Responses. *PLoS Genet.* **2015**, *11*, e1005200. [CrossRef] [PubMed]
13. Liu, C.; Wu, Q.; Liu, W.; Gu, Z.; Wang, W.; Xu, P.; Ma, H.; Ge, X. Poly(ADP-ribose) polymerases regulate cell division and development in Arabidopsis roots. *J. Integr. Plant Biol.* **2017**, *59*, 459–474. [CrossRef]
14. Fenton, A.L.; Shirodkar, P.; MacRae, C.J.; Meng, L.; Anne Koch, C. The PARP3-and ATM-dependent phosphorylation of APLF facilitates DNA double-strand break repair. *Nucleic Acids Res.* **2013**, *41*, 4080–4092. [CrossRef]
15. Chen, Y.-M.; Shall, S.; O'farrell, M. Poly(ADP-Ribose) Polymerase in Plant Nuclei. *Eur. J. Biochem.* **1994**, *224*, 135–142. [CrossRef]
16. Tian, R.-H.; Zhang, G.-Y.; Yan, C.-H.; Dai, Y.-R. Involvement of poly(ADP-ribose) polymerase and activation of caspase-3-like protease in heat shock-induced apoptosis in tobacco suspension cells. *FEBS Lett.* **2000**, *474*, 11–15. [CrossRef]
17. Pham, P.A.; Wahl, V.; Tohge, T.; De Souza, L.R.; Zhang, Y.; Do, P.T.; Olas, J.J.; Stitt, M.; Araújo, W.L.; Fernie, A.R. Analysis of knockout mutants reveals non-redundant functions of poly(ADP-ribose)polymerase isoforms in Arabidopsis. *Plant Mol. Biol.* **2015**, *89*, 319–338. [CrossRef] [PubMed]
18. Adams-Phillips, L.; Briggs, A.G.; Bent, A.F. Disruption of Poly(ADP-ribosyl)ation Mechanisms Alters Responses of Arabidopsis to Biotic Stress. *Plant Physiol.* **2010**, *152*, 267–280. [CrossRef]
19. Ricci, J.E.; Waterhouse, N.; Green, D.R. Mitochondrial functions during cell death, a complex (I–V) dilemma. *Cell Death Differ.* **2003**, *10*, 488–492. [CrossRef] [PubMed]
20. De Block, M.; Verduyn, C.; De Brouwer, D.; Cornelissen, M. Poly(ADP-ribose) polymerase in plants affects energy homeotasis, cell death and stress tolerance. *Plant J.* **2005**, *41*, 95–106. [CrossRef]
21. Vanderauwera, S.; De Block, M.; Van De Steene, N.; Van De Cotte, B.; Metzlaff, M.; Van Breusegem, F. Silencing of poly(ADP-ribose) polymerase in plants alters abiotic stress signal transduction. *Pro. Nat. Aca. Sci. USA* **2007**, *104*, 15150–15155. [CrossRef]

22. Kumar, S.; Stecher, G.; Tamura, K. MEGA7: Molecular Evolutionary Genetics Analysis Version 7.0 for Bigger Datasets. *Mol. Biol. Evol.* **2016**, *33*, 1870–1874. [CrossRef]
23. Nekrutenko, A.; Makova, K.D.; Li, W.H. The KA/KS ratio test for assessing the protein-coding potential of genomic regions: An empirical and simulation study. *Genome Res.* **2002**, *12*, 198–202. [CrossRef] [PubMed]
24. Bailey, C.D.; Koch, M.A.; Mayer, M.; Mummenhoff, K.; O'Kane, S.L.; Warwick, S.I.; Windham, M.D.; Al-Shehbaz, I.A. Toward a global phylogeny of the Brassicaceae. *Mol. Biol. Evol.* **2006**, *23*, 2142–2160. [CrossRef]
25. Hu, B.; Jin, J.; Guo, A.Y.; Zhang, H.; Luo, J.; Gao, G. GSDS 2.0: An upgraded gene feature visualization server. *Bioinformatics* **2015**, *31*, 1296–1297. [CrossRef]
26. Sparkes, I.A.; Runions, J.; Kearns, A.; Hawes, C. Rapid, transient expression of fluorescent fusion proteins in tobacco plants and generation of stably transformed plants. *Nat. Protoc.* **2006**, *1*, 2019–2025. [CrossRef] [PubMed]
27. Wang, C.Q.; Guthrie, C.; Sarmast, M.K.; Dehesh, K. BBX19 interacts with CONSTANS to repress FLOWERING LOCUS T transcription, defining a flowering time checkpoint in Arabidopsis. *Plant Cell* **2014**, *26*, 3589–3602. [CrossRef] [PubMed]
28. Agrawal, N.; Dasaradhi, P.V.N.; Mohmmed, A.; Malhotra, P.; Bhatnagar, R.K.; Mukherjee, S.K. RNA Interference: Biology, Mechanism, and Applications. *Microbiol. Mol. Biol. Rev.* **2003**, *67*, 657–685. [CrossRef]
29. Riccardi, C.; Nicoletti, I. Analysis of apoptosis by propidium iodide staining and flow cytometry. *Nat. Protoc.* **2006**, *1*, 1458–1461. [CrossRef]
30. Browne, M.; Yardimci, N.T.; Scoffoni, C.; Jarrahi, M.; Sack, L. Prediction of leaf water potential and relative water content using terahertz radiation spectroscopy. *Plant Direct* **2020**, *4*, 1–16. [CrossRef]
31. Leclercq, J.; Martin, F.; Sanier, C.; Clément-Vidal, A.; Fabre, D.; Oliver, G.; Lardet, L.; Ayar, A.; Peyramard, M.; Montoro, P. Over-expression of a cytosolic isoform of the HbCuZnSOD gene in Hevea brasiliensis changes its response to a water deficit. *Plant Mol. Biol.* **2012**, *80*, 255–272. [CrossRef] [PubMed]
32. Kong, W.; Liu, F.; Zhang, C.; Zhang, J.; Feng, H. Non-destructive determination of Malondialdehyde (MDA) distribution in oilseed rape leaves by laboratory scale NIR hyperspectral imaging. *Sci. Rep.* **2016**, *6*, 35393. [CrossRef] [PubMed]
33. Livak, K.J.; Schmittgen, T.D. Analysis of relative gene expression data using real-time quantitative PCR and the $2^{-\Delta\Delta CT}$ method. *Methods* **2001**, *25*, 402–408. [CrossRef]
34. Fan, Y.; Yu, M.; Liu, M.; Zhang, R.; Sun, W.; Qian, M.; Duan, H.; Chang, W.; Ma, J.; Qu, C.; et al. Genome-wide identification, evolutionary and expression analyses of the GALACTINOL SYNTHASE gene family in rapeseed and tobacco. *Int. J. Mol. Sci.* **2017**, *18*, 2768. [CrossRef]
35. Turner, N.C.; Wright, G.C.; Siddique, K.H.M. Adaptation of grain legumes (pulses) to water-limited environments. *Adv. Agron.* **2001**, *71*, 193–231.
36. Verma, S.; Mishra, S.N. Putrescine alleviation of growth in salt stressed Brassica juncea by inducing antioxidative defense system. *J. Plant Physiol.* **2005**, *162*, 669–677. [CrossRef]
37. Hamani, A.K.M.; Wang, G.; Soothar, M.K.; Shen, X.; Gao, Y.; Qiu, R.; Mehmood, F. Responses of leaf gas exchange attributes, photosynthetic pigments and antioxidant enzymes in NaCl-stressed cotton (*Gossypium hirsutum* L.) seedlings to exogenous glycine betaine and salicylic acid. *BMC Plant Biol.* **2020**, *20*, 434. [CrossRef] [PubMed]
38. Liu, D.; Gharavi, R.; Pitta, M.; Gleichmann, M.; Mattson, M.P. Nicotinamide prevents NAD+ depletion and protects neurons against excitotoxicity and cerebral Ischemia: NAD+ consumption by sirt1 may endanger energetically compromised neurons. *Neuromol. Med.* **2009**, *11*, 28–42. [CrossRef]
39. Braidy, N.; Grant, R. Kynurenine pathway metabolism and neuroinflammatory disease. *Neural Regen. Res.* **2017**, *12*, 39–42. [CrossRef]
40. Rowe, J.H.; Topping, J.F.; Liu, J.; Lindsey, K. Abscisic acid regulates root growth under osmotic stress conditions via an interacting hormonal network with cytokinin, ethylene and auxin. *New Phytol.* **2016**, *211*, 225–239. [CrossRef]
41. Kwon, Y.R.; Lee, H.J.; Kim, K.H.; Hong, S.W.; Lee, S.J.; Lee, H. Ectopic expression of Expansin3 or Expansinβ1 causes enhanced hormone and salt stress sensitivity in Arabidopsis. *Biotechnol. Lett.* **2008**, *30*, 1281–1288. [CrossRef]
42. Huang, Z.; Tang, J.; Duan, W.; Wang, Z.; Song, X.; Hou, X. Molecular evolution, characterization, and expression analysis of SnRK2 gene family in Pak-choi (*Brassica rapa* ssp. chinensis). *Front. Plant Sci.* **2015**, *6*, 879. [CrossRef]
43. Cheng, F.; Wu, J.; Wang, X. Genome triplication drove the diversification of Brassica plants. *Hortic. Res.* **2014**, *1*, 14024. [CrossRef]
44. Paterson, A.H.; Chapman, B.A.; Kissinger, J.C.; Bowers, J.E.; Feltus, F.A.; Estill, J.C. Many gene and domain families have convergent fates following independent whole-genome duplication events in Arabidopsis, Oryza, Saccharomyces and Tetraodon. *Trends Genet.* **2006**, *22*, 597–602. [CrossRef]
45. Gout, J.-F.; Kahn, D.; Duret, L. Paramecium Post-Genomics Consortium The Relationship among Gene Expression, the Evolution of Gene Dosage, and the Rate of Protein Evolution. *PLoS Genet.* **2010**, *6*, 20. [CrossRef]
46. Yang, L.; Gaut, B.S. Factors that Contribute to Variation in Evolutionary Rate among Arabidopsis Genes. *Mol. Biol. Evol.* **2011**, *28*, 2359–2369. [CrossRef] [PubMed]
47. Pogson, B.J.; Woo, N.S.; Förster, B.; Small, I. Plastid signalling to the nucleus and beyond. *Trends Plant Sci.* **2008**, *13*, 602–609. [CrossRef] [PubMed]
48. De Smet, R.; Adams, K.L.; Vandepoele, K.; Van Montagu, M.C.E.; Maere, S.; Van de Peer, Y. Convergent gene loss following gene and genome duplications creates single-copy families in flowering plants. *Proc. Natl. Acad. Sci. USA* **2013**, *110*, 2898–2903. [CrossRef]

49. Chen, Y.; Han, Y.; Meng, Z.; Zhou, S.; Xiangzhu, K.; Wei, W. Overexpression of the wheat expansin gene TaEXPA2 improved seed production and drought tolerance in transgenic tobacco plants. *PLoS ONE* **2016**, *11*, e0153494. [CrossRef]
50. Bandeppa, S.; Paul, S.; Thakur, J.K.; Chandrashekar, N.; Umesh, D.K.; Aggarwal, C.; Asha, A. Antioxidant, physiological and biochemical responses of drought susceptible and drought tolerant mustard (*Brassica juncea* L) genotypes to rhizobacterial inoculation under water deficit stress. *Plant Physiol. Biochem.* **2019**, *143*, 19–28. [CrossRef]
51. Yue, Y.; Zhang, M.; Zhang, J.; Duan, L.; Li, Z. SOS1 gene overexpression increased salt tolerance in transgenic tobacco by maintaining a higher K^+/Na^+ ratio. *J. Plant Physiol.* **2012**, *169*, 255–261. [CrossRef] [PubMed]
52. Albrecht, V.; Weinl, S.; Blazevic, D.; D'Angelo, C.; Batistic, O.; Kolukisaoglu, Ü.; Bock, R.; Schulz, B.; Harter, K.; Kudla, J. The calcium sensor CBL1 integrates plant responses to abiotic stresses. *Plant J.* **2003**, *36*, 457–470. [CrossRef] [PubMed]
53. Muthusamy, M.; Kim, J.Y.; Yoon, E.K.; Kim, J.A.; Lee, S.I. BrEXLB1, a brassica rapa expansin-like b1 gene is associated with root development, drought stress response, and seed germination. *Genes* **2020**, *11*, 404. [CrossRef] [PubMed]
54. Williams, A.C.; Hill, L.J.; Ramsden, D.B. Nicotinamide, NAD(P)(H), and methyl-group homeostasis evolved and became a determinant of ageing diseases: Hypotheses and lessons from pellagra. *Curr. Gerontol. Geriatr. Res.* **2012**, *2012*, 302875. [CrossRef] [PubMed]
55. Altmeyer, M.; Hottiger, M.O. Poly(ADP-ribose) polymerase 1 at the crossroad of metabolic stress and inflammation in aging. *Aging* **2009**, *1*, 458–469. [CrossRef]

horticulturae

Article

Transcriptome Profiling Reveals Candidate Key Genes Involved in Sinigrin Biosynthesis in *Brassica nigra*

Yang Li [1,2,†], Youjian Yu [1,†], Liai Xu [1], Erbiao Guo [1], Yunxiang Zang [1], Yong He [1,*] and Zhujun Zhu [1,*]

[1] Key Laboratory for Quality Improvement of Agricultural Products of Zhejiang Province, Collaborative Innovation Center for Efficient and Green Production of Agriculture in Mountainous Areas of Zhejiang Province, College of Horticulture Science, Zhejiang A&F University, Wusu Street 666, Lin'an, Hangzhou 311300, China; lylily0327@163.com (Y.L.); yjyu@zafu.edu.cn (Y.Y.); 11416052@zju.edu.cn (L.X.); geb1632021@163.com (E.G.); yxzang78@163.com (Y.Z.)

[2] College of Forestry and Biotechnology, Zhejiang A&F University, Wusu Street 666, Lin'an, Hangzhou 311300, China

* Correspondence: heyong@zafu.edu.cn (Y.H.); zhuzj@zafu.edu.cn (Z.Z.); Tel.: +86-571-6374-3001 (Z.Z.)

† These authors have contributed equally to this work.

Abstract: Glucosinolates (GSLs) are important secondary metabolites in Brassicales related to insect and disease resistance, flavor formation, and human health. Here, we determined the GSL profile with sinigrin as the predominant GSL in *Brassica nigra*. A total of 184 GSL biosynthetic genes (*BniGSLs*) were identified in *B. nigra* by a genome-wide search for orthologs of 82 of the 95 known GSL genes in *Arabidopsis thaliana*. Transcriptome data demonstrated that at least one *BniGSL* was highly expressed in stems and leaves at each step of the sinigrin synthesis pathway, which ensured the synthesis of a large amount of sinigrin in *B. nigra*. Among these key candidates of *BniGSLs*, the high expression of *BniMAM1-2*, *BniCYP79F1*, and *BniAOP2-1/2*, and the absence of *MAM3* and *AOP3*, may contribute remarkably to the synthesis and accumulation of sinigrin. In addition, the low expression of some key *BniGSLs* partially explains the low content of indolic and aromatic GSLs in *B. nigra*. This study provided a genetic explanation for the formation of the unique GSL profile with sinigrin as the main GSL in *B. nigra*. The results of this study will be valuable for further functional analysis of *BniGSLs* and genetic improvement of GSLs in *B. nigra* and other *Brassica* species.

Keywords: glucosinolates; sinigrin; biosynthetic genes; gene expression; *Brassica nigra*

Citation: Li, Y.; Yu, Y.; Xu, L.; Guo, E.; Zang, Y.; He, Y.; Zhu, Z. Transcriptome Profiling Reveals Candidate Key Genes Involved in Sinigrin Biosynthesis in *Brassica nigra*. *Horticulturae* **2021**, *7*, 173. https://doi.org/10.3390/horticulturae7070173

Academic Editors: Xiaowu Wang and Jian Wu

Received: 26 May 2021
Accepted: 27 June 2021
Published: 2 July 2021

Publisher's Note: MDPI stays neutral with regard to jurisdictional claims in published maps and institutional affiliations.

Copyright: © 2021 by the authors. Licensee MDPI, Basel, Switzerland. This article is an open access article distributed under the terms and conditions of the Creative Commons Attribution (CC BY) license (https://creativecommons.org/licenses/by/4.0/).

1. Introduction

Glucosinolates (GSLs) are a group of sulfur-rich and nitrogen-containing secondary metabolites that are synthesized from amino acids and sugars in plants. Currently, over 100 different GSLs have been identified [1], most of which are exclusively found in the order Brassicales [2]. GSL is composed of three common moieties, including a β-D-thioglucose group, a sulfonated aldoxime moiety, and a variable side chain derived from a precursor amino acid [3,4]. Upon hydrolysis by the myrosinase enzyme, GSLs are degraded into different bioactive products, mainly isothiocyanates [5]. These broken down products exhibit a variety of biological activities, which not only endow *Brassica* vegetables with characteristic flavor [6] and help defend against pathogens and insect herbivores [7], but also function in preventing carcinogenesis in animals by stimulating apoptosis and regulating the cell cycle [8]. For instance, mounting studies have shown that the enzymatic hydrolysate of sinigrin has anti-cancer, anti-inflammatory, anti-oxidant, anti-bacterial, antifungal, wound healing properties, and biofumigation applications [9]. The pharmacological and therapeutic properties of GSLs that are beneficial to human health have made the *Brassica* species attract the considerable interest of many plant breeders and geneticists in the past 30 years [3,10,11].

On the basis of the chemical structure of different side chains, GSLs are classified into aliphatic, indolic, and aromatic GSLs, with methionine (or alanine, valine, leucine, and isoleucine), tryptophan, and phenylalanine (or tyrosine) as the basic amino acid precursors, respectively [3,12]. The GSL biosynthesis is a tripartite pathway involving three independent phases in at least two different locations: (i) side-chain elongation of selected precursors (only methionine and phenylalanine) in the chloroplast, (ii) formation of the core GSL structure in the cytosol interface of the endoplasmic reticulum, and (iii) side-chain modification in the cytosol [3,10,11]. To date, GSL biosynthesis has been well elucidated mainly in *Arabidopsis*, and the inventory of related genes in this process is close to completion [11]. Most GSLs in *Arabidopsis* and *Brassica* crops are synthesized from methionine, beginning with side-chain elongation and condensation, which involves branched-chain aminotransferase (BCAT), bile acid transporter 5 (BAT5), and methylthioalkylmalate synthase (MAM) [13–15]. Subsequently, core structures are formed via a five-step process that includes the conversion of 2-oxo-methylthio acid homologs into aldoximes by cytochrome P450 of the CYP79s family [16], oxidation of aldoximes by the CYP83s family [17,18], followed by *C-S* cleavage (SUR1) [19], and the formation of desulfoglucosinolate (UGT74s) [20] and the basic GSL structure (SOTs) [21,22]. Finally, the side chains of these basic GSLs are modified by oxygenation, alkenylation, hydroxylation, and benzoylation. Genes involving these modification events, including those encoding GS-ELONG, GS-OX, GS-AOP, and GS-OH, have been well studied in recent years [10,11,23,24]. Several transcription factors (TFs) have also been characterized to participate in the regulation of GSL biosynthesis [25–29].

Brassica nigra, commonly known as black mustard, is a member of the Brassicaceae family that has been cultivated for thousands of years over a wide range of climates [30]. *B. nigra* has a strong, pungent flavor and has been used as a major condiment crop in many countries. Its seed has also been used as a traditional herbal medicine [31,32]. *B. nigra* is one of the three important diploid species (*B. rapa* (AA), *B. nigra* (BB), and *B. oleracea* (CC)) in the so-called "Triangle of U" theory that explains how the three allotetraploid *Brassica* species (*B. juncea* (AABB), *B. napus* (AACC), and *B. carinata* (BBCC)) were evolved [33,34]. Although the current genomes of these three progenitor *Brassica* species were derived from a common hexaploid ancestor, the evolutionary process of gene loss and neo- or sub-functionalization of genes made the genome of *B. nigra* evolve separately from *B. rapa* and *B. oleracea* [35]. In terms of GSL synthesis, the differences in the diversity and content of GSLs in these three *Brassica* species have been confirmed. *B. oleracea* mainly synthesizes 4C and 3C GSLs [36,37], *B. rapa* accumulates 4C and 5C GSLs [38]. However, with sinigrin as the predominant GSL, the GSL profile in *B. nigra* is much different from those in *B. rapa* and *B. oleracea* [30,39–41]. Excitingly, genome-wide characterization and expression analysis of genes involved in GSL synthesis in *B. rapa* and *B. oleracea* can more or less explain the differences [42,43]. The genetic and genomic information of *B. nigra* have been published through whole-genome sequencing recently [44]. However, no systematic investigation on GSL genes have been reported in *B. nigra* to date, and the expression patterns of genes related to GSL metabolism in different organs of *B. nigra* are still limited and remain to be further explored.

In this study, we first investigated the GSL profile of black mustard and found that sinigrin was the most predominant GSL in *B. nigra*. Next, we conducted a genome-wide in silico search to determine the GSL biosynthesis genes by using the assembled genome sequence data of *B. nigra* [44], and provided detailed information of *B. nigra* GSL genes (*BniGSLs*). Combining the expression profile data and phylogenetic tree analysis of key GSL genes, we screened out the key candidate *BniGSLs* involved in the massive synthesis of sinigrin in *B. nigra*. The knowledge gained in this study will be useful for further studies on the biological functions of *BniGSLs* and the genetic improvement of GSLs in *B. nigra* and other *Brassica* species.

2. Materials and Methods

2.1. Plant Materials and Growth Conditions

The black mustard genotype used in this experiment was *B. nigra* cv. 1511-01 (an inbred line kept in our laboratory). The seeds of the mustard were planted in a controlled climate chamber at 26 °C/20 °C day/night temperature, 14/10 h light/dark photoperiod, 600 $\mu mol \cdot m^{-2} \cdot s^{-1}$ light intensity, and 60–70% relative humidity. The stems, rosette leaves, cauline leaves, inflorescences, and siliques (about 3 cm in length) used for the GSL profiling were collected during the flowering period. Cauline leaves, stems, and roots were used for RNA sequencing. All the materials were mixed and frozen in liquid nitrogen immediately and stored at −80 °C. All samples were collected from at least three plants, and both glucosinolate analysis and RNA sequencing were carried out in three biological replicates.

2.2. GSL Extraction and Analysis

GSL extraction and analysis were performed as previously described with only slight modification [45]. First, 0.25 g sample powder was boiled with 10 mL of 70% methanol after adding 200 μL of 5 mM glucotropaeolin (CAS, 5115-71-9; Code No., A5300,0050; Applichem, Darmstadt, Germany) as internal standard. Then, the supernatant was loaded onto a 1 mL mini-column to desulfate overnight with 200 μL arylsulfatase (Sigma-Aldrich Co., St. Louis, MO, USA). The mini-column containing 250 μL activated DEAE Sephadex A-25 was equilibrated at room temperature for at least 2 h prior to use. Resultant desulfoglucosinolates were eluted with ultrapure water and stored at −20 °C until analysis. Samples were analyzed by high-performance liquid chromatography (HPLC) in an Agilent 1200 HPLC system equipped with a C-18 reversed-phase column (250 × 4 μm, 5 μm, Bischoff, Leonberg, Germany). Elution was performed with ultrapure water (solvent A) and acetonitrile (solvent B) in a linear gradient from 0% to 20% B for 45 min and then constant 20% B for 6 min, followed by 100% A for 5 min prior to the injection of the next sample. The flow rate was 1 $mL \cdot min^{-1}$ (injection volume of 20 μL). The eluent was monitored by diode array detection at 229 nm. The data of GSL concentrations were analyzed using analysis of variance (ANOVA) software. Mean values were compared using the least significant difference at 0.05 significance level.

2.3. In Silico Identification of BniGSL Genes in B. nigra

Based on previous studies [11,42,46], sequences of the GSL biosynthetic genes in *A. thaliana*, *B. rapa*, and *B. oleracea* (Table S1) were acquired from The Arabidopsis Information Resource (TAIR) website (http://www.arabidopsis.org/) (accessed on 10 November 2020), The National Center for Biotechnology Information (NCBI) website (http://www.ncbi.nlm.nih.gov) (accessed on 10 November 2020), and the *Brassica* database website (http://brassicadb.cn/#/, BRAD) (accessed on 10 November 2020). The whole sequences of each *B. nigra* chromosome were downloaded from the BRAD database (http://brassicadb.cn/#/Download/) (accessed on 10 November 2020). Using the sequences of GSL genes acquired from the databases mentioned above, BLASTn was performed to search for homologous candidate genes in *B. nigra*. All candidates in the *B. nigra* genome, together with flank regions of 5000 bp upstream and downstream of each candidate, were analyzed and re-annotated using FGENESH (http://www.softberry.com/) (accessed on 10 November 2020). By comprehensively considering the candidate's re-annotation results, the collinearity relationship with the known GSL genes, the similarity of amino acid sequence, and whether it contained the corresponding key domains or sequences, it is determined whether the candidate is a GSL gene. Given that the GSL genes of *A. thaliana* have been clearly named, the nomenclature system for the *BniGSL* genes in this study was based on their homology and identities with their counterparts in *Arabidopsis*. The resulting *BniGSL* genes were further used as query sequences to determine the precise locations of each gene on chromosomes through Oligo 6.0 software.

2.4. RNA Extraction, Library Construction, Sequencing, and Gene Expression Analysis

Total RNA of leaves, stems, and roots was isolated using Trizol reagent (Invitrogen, Waltham, MA, USA) following the manufacturer's instructions. Nanodrop, Qubit 2.0, and Aglient 2100 were used separately to measure the purity, concentration, integrity, and other values of RNA to ensure the qualified samples for transcriptome sequencing. The mRNA was purified with 20 mg total RNA by using oligo (dT) magnetic beads. After purification, the mRNA was fragmented by adding a fragmentation buffer. The fragments were used to synthesize the first-strand cDNA by using random hexamer adapters and reverse transcriptase (Code No., RR047A; Takara, Japan). The second-strand cDNA was synthesized using buffer, dNTPs, RNase H, and DNA polymerase I. The cDNA fragments that went through an end-repair process were added with a single 'A' base and had the sequencing joints connected. AMPure XP beads were used to select the fragment size for the ligature of adapter sequences. The products were purified, enriched with PCR, and then used as templates for sequencing. Sequencing and assembly were performed by the Biomarker Biotechnology Corporation (Beijing, China) using the Illumina HiSeqTM 2500 platform. The Log2FoldChange and false discovery rate (FDR) value of *BniGSL* genes between two samples (Root vs Stem, Root vs Leaf, and Stem vs Leaf) were calculated, and the *BniGSL* genes with $|\text{Log2FoldChange}| > 1$ and FDR < 0.05 were regarded as differentially expressed genes.

All RNA samples used for RNA-Seq were also used for qRT-PCR analysis (Code NO., RR820A; Takara, Japan). All qRT-PCR experiments included three independent biological repetitions. *Brassica nigra* tonoplastic intrinsic protein-41 (*TIPS*) gene was used as a reference gene [47]. The $2^{-\Delta\Delta Ct}$ method was used to calculate the relative gene expression values. The gene-specific primers were listed in Table S2.

The expression data of *B. rapa* is from Tong et al. [48] (GEO accession number GSE43245). Root, stem, and leaf tissues of *B. rapa* accession Chiifu-401-42 were collected from seven-week-old plants. The expression data of *B. oleracea* is from Liu et al. [43] and Yu et al. [49] (GEO accession number GSE 42891).

3. Results

3.1. Analysis of Glucosinolate Profile in B. nigra

HPLC analyses revealed the presence of seven different types of GSLs in five organs of *B. nigra* cv. 1511-01 (Table 1, Figure S1), including two aliphatic GSLs (i.e., sinigrin and gluconapin), four indolic GSLs (i.e., 4-hydroxy glucobrassicin, glucobrassicin, 4-methoxyglucobrassicin, and 1-methoxyglucobrassicin), and one aromatic GSL (i.e., gluconasturtiin). Results showed that the stems of black mustard contained all the above-mentioned GSLs. Rosette leaves contained all GSLs except gluconapin. Cauline leaves and siliques contained all GSs except gluconasturtiin. However, only four GSLs were detected in the inflorescences of black mustard, with three indolic GSLs were absent. In terms of total GSL concentration, black mustard stems possessed the highest concentration of total GSL, followed by inflorescences, cauline leaves and siliques, and rosette leaves contained the lowest total GSL content. Notably, although the GSL profiles in different organs of black mustard were different, sinigrin was the predominant GSL in all five tested organs and accounted for 90.7–98.5% of the total GSL. Moreover, as the main GSL, the content of sinigrin in five organs ranked the same as the total GSL. These results indicated that the GSL profiles in *B. nigra* are significantly different from those in *B. rapa* and *B. oleracea*, suggesting that there may be a major GSL biosynthesis pathway in *B. nigra*, which directs the synthesis of sinigrin.

Table 1. Glucosinolate content (μmol·g^{-1} DW) in different organs of *Brassica nigra*.

Organ	Aliphatic		Indolic				Aromatic
	SIN	GNA	4OHI3M	I3M	4MOI3M	1MOI3M	2PHET
St	82.35 ± 2.12d	0.52 ± 0.3c	0.25 ± 0.02a	0.12 ± 0.02b	0.05 ± 0.02ab	0.09 ± 0.02b	0.25 ± 0.01a
RL	13.20 ± 1.08a	-	0.50 ± 0.14b	0.04 ± 0.01a	0.04 ± 0.01a	0.03 ± 0.01a	0.78 ± 0.10b
CL	53.80 ± 2.52b	0.32 ± 0.05b	0.32 ± 0.01ab	0.13 ± 0.02b	0.05 ± 0.00ab	0.08 ± 0.01b	-
In	59.14 ± 2.39c	0.32 ± 0.25b	1.50 ± 0.17c	-	-	-	0.34 ± 0.01a
Si	51.31 ± 0.25b	0.35 ± 0.11a	0.37 ± 0.07ab	0.13 ± 0.03b	0.06 ± 0.01b	0.08 ± 0.01b	-

Different letters indicate significant difference. SIN, Sinigrin; GNA, Gluconapin; 4OHI3M, 4-OH-Glucobrassicin; I3M, Glucobrassicin; 4MOI3M, 4-Methoxy-Glucobrassicin; 1MOI3M, 1-Methoxy-Glucobrassicin; 2PHET, Gluconasturtiin. St, stems; RL, rosette leaves; CL, cauline leaves; In, inflorescences; Si, siliques; DW, dry weight.

3.2. Identification and Annotation of BniGSL Genes from B. nigra Genome

We first searched for *BniGSL* genes in the whole genome of *B. nigra* before looking for clues as to why black mustard preferred to synthesize sinigrin. Preliminary BLAST searches for *BniGSL* genes in the whole-genome sequences of *B. nigra* were performed using GSL genes of *A. thaliana*, *B. rapa*, and *B. oleracea*. Using these pre-screened *BniGSL* genes, re-annotation, and BLASTP search against known GSL gene sequences of *A. thaliana*, *B. rapa*, and *B. oleracea* resulted in 184 *BniGSL* genes as orthologs of 82 of the 95 known *AtGSL* genes, with 13 *AtGSL* genes having no *B. nigra* ortholog. The number of *BniGSL* genes in *B. nigra* has expanded, with an average of 2.24 copies per gene. It is worth noting that there are thirteen copies of SOT18 in *B. nigra*, which is far more than the average.

All the identified *BniGSL* genes are listed in Tables 2 and 3, and the sequences of DNA, CDS, and amino acid are listed in Table S3. A total of 124, 48, and 12 *BniGSLs* encode enzymes involved in glucosinolate biosynthesis (Table 2), TFs with regulatory functions, and transporters involved in glucosinolate transport (Table 3), respectively. More specifically, there are 16, 55, 21, and 32 genes involved in side-chain elongation, core structure synthesis, side-chain modification, and co-substrate pathways, respectively (Table 2). Among the 48 TFs, 33, 8, and 4 TFs act as activators, repressors, and mediators, respectively. HY5, which has both activating and suppressing functions during glucosinolate biosynthesis, consists of three copies in *B. nigra* (Table 3). In addition, there are 12 *BniGSL* genes encoding five transporters in black mustard (Table 3).

Table 2. The inventory of glucosinolate biosynthetic genes (*BniGSLs*) in *Brassica nigra*.

Name1	Name2	Gene ID	Chromosome Location		AA	Identity/%	AGI ID
Side-chain elongation							
BCAT4	*BniBCAT4-1*	BniB01g044210	B1 (+)	48,590,123–48,592,361	356	82.3	AT3G19710
	BniBCAT4-2	BniB07g030730	B7 (−)	41,353,700–41,356,224	306	62.7	
BCAT6	*BniBCAT6*	BniB04g060340	B4 (+)	51,535,934–51,537,864	359	86.7	AT1G50110
MAM1	*BniMAM1-1*	BniB04g003060	B4 (+)	1,498,531–1,501,807	505	79.8	AT5G23010
	BniMAM1-2	BniB02g076300	B2 (+)	64,723,274–64,726,274	504	79.1	
	BniMAM1-3	BniB08g036280	B8 (−)	21,370,428–21,376,553	495	74.8	
	BniMAM1-4	BniB02g076310	B2 (+)	64,728,011–64,730,517	397	59.6	
	BniMAM1-5	BniB05g055400	B5 (−)	49,610,184–49,616,319	388	56.6	
MAM3	*						AT5G23020
IPMI SSU1	*BniIPMI SSU1-1*	BniB08g028010	B8 (+)	14,699,124–14,699,873	249	87.7	AT2G43090
	BniIPMI SSU1-2	BniB06g001990	B6 (+)	917,687–917,463	258	79.6	
IPMI SSU2	*BniIPMI SSU2*	BniB06g002000	B6 (+)	919,426–920,271	281	72.7	AT2G43100

Table 2. *Cont.*

Name1	Name2	Gene ID	Chromosome Location		AA	Identity/%	AGI ID
IPMI SSU3	*						AT3G58990
IIL1	*BniIIL1-1*	BniB07g011950	B7 (+)	15,245,075–15,248,497	505	94.5	AT4G13430
	BniIIL1-2	BniB07g011910	B7 (+)	15,114,671–15,118,112	505	94.3	
IMD1	*BniIMD1*	BniB05g024330	B5 (−)	11,915,893–11,917,777	408	91.7	AT5G14200
BCAT3	*BniBCAT3-1*	BniB08g063340	B8 (+)	59,567,292–59,569,478	419	87.1	AT3G49680
	BniBCAT3-2	BniB05g048020	B5 (−)	27,787,734–27,790,073	419	84.5	
Core structure formation							
CYP79F1	*BniCYP79F1*	BniB06g041700	B6 (+)	46,406,179–46,408,341	541	82.1	AT1G16410
CYP79F2	*						AT1G16400
CYP79A2	*BniCYP79A2-1*	BniB02g055340	B2 (+)	53,932,413–53,934,250	532	77.6	AT5G05260
	BniCYP79A2-2	BniB02g055360	B2 (+)	53,941,026–53,942,864	532	77.6	
	BniCYP79A2-3	BniB02g055420	B2 (+)	53,958,847–53,960,685	532	76.3	
	BniCYP79A2-4	BniB05g029220	B5 (+)	14,358,148–14,360,431	562	73.1	
	BniCYP79A2-5	BniB08g051390	B8 (+)	34,728,808–34,732,349	439	56.6	
CYP79B2	*BniCYP79B2-1*	BniB02g088940	B2 (−)	70,487,544–70,489,564	541	94.5	AT4G39950
	BniCYP79B2-2	BniB05g000130	B5 (−)	64,703–66,466	542	93.2	
	BniCYP79B2-3	BniB03g014770	B3 (−)	6,330,659–6,332,937	518	89.8	
CYP79B3	*BniCYP79B3*	BniB01g021760	B1 (−)	13,001,786–13,004,048	564	89.0	AT2G22330
CYP79C1	*BniCYP79C1-1*	BniB08g040770	B8 (−)	25,176,913–25,179,607	531	80.3	AT1G79370
	BniCYP79C1-2	BniB08g040860	B8 (−)	25,245,858–25,248,871	531	80	
	BniCYP79C1-3	BniB08g040830	B8 (−)	25,220,662–25,222,785	396	62.1	
	BniCYP79C1-4	BniB08g040890	B8 (−)	25,258,462–25,261,159	382	59.7	
CYP79C2	*BniCYP79C2-1*	BniB04g007960	B4 (−)	3,767,912–3,772,131	530	78.7	AT1G58260
	BniCYP79C2-2	BniB01g003800	B1 (−)	1,903,991–1,908,135	528	74.9	
	BniCYP79C2-3	BniB06g009140	B6 (−)	4,583,904–4,586,034	527	73.6	
CYP83A1	*BniCYP83A1-1*	BniB01g003820	B1 (+)	1,927,665–1,929,469	500	86.9	AT4G13770
	BniCYP83A1-2	BniB06g067060	B6 (−)	59,968,418–59,970,020	498	86.3	
CYP83B1	*BniCYP83B1*	BniB07g013920	B7 (−)	18,805,780–18,807,433	499	95.4	AT4G31500
CYTB5-C	*BniCYTB5-C-1*	BniB01g000830	B1 (−)	449,656–450,237	135	80.1	AT2G46650
	BniCYTB5-C-2	BniS02554g140	utg2554 +	100,663–101,227	114	57.7	
GSTF9	*BniGSTF9-1*	BniB01g013320	B1 (−)	6,927,611–6,928,497	215	97.2	AT2G30860
	BniGSTF9-2	BniB08g019200	B8 (+)	9,243,485–9,244,436	215	96.7	
GSTF10	*BniGSTF10*	BniB08g019210	B8 (+)	9,246,072–9,247,215	215	94.4	AT2G30870
GSTF11	*BniGSTF11*	BniB01g061730	B1 (+)	57,396,702–57,397,557	214	82.2	AT3G03190
GSTU13	*BniGSTU13-1*	BniB04g039630	B4 (−)	21,519,677–21,520,647	227	81.1	AT1G27130
	BniGSTU13-2	BniB03g031270	B3 (−)	14,627,558–14,628,949	227	76.2	
GSTU20	*BniGSTU20*	BniB06g050340	B6 (−)	51,398,415–51,399,277	217	88.9	AT1G78370
GGP1	*BniGGP1-1*	BniB02g084810	B2 (+)	68,616,575–68,618,095	250	88.4	AT4G30530
	BniGGP1-2	BniB05g007160	B5 (−)	3,545,600–3,546,824	250	86.4	
	BniGGP1-3	BniB03g010950	B3 (−)	4,449,737–4,451,401	228	80	
	BniGGP1-4	BniB05g007170	B5 (−)	3,548,331–3,552,009	251	62.4	
GGP3	*BniGGP3*	BniB03g010940	B3 (−)	4,446,818–4,448,424	258	84.9	AT4G30550
SUR1	*BniSUR1-1*	BniB03g070290	B3 (−)	54,745,695–54,747,733	458	88.6	AT2G20610
	BniSUR1-2	BniB02g015530	B2 (−)	9,570,907–9,578,729	445	83.3	
UGT74B1	*BniUGT74B1*	BniB04g039340	B4 (−)	21,237,953–21,239,435	465	85.2	AT1G24100
UGT74C1	*BniUGT74C1-1*	BniB06g020830	B6 (+)	11,872,313–11,874,700	456	85.1	AT2G31790
	BniUGT74C1-2	BniB01g012360	B1 (−)	6,333,513–6,335,348	456	83.6	
SOT16	*BniSOT16-1*	BniB03g052560	B3 (−)	28,106,607–28,107,620	337	92.9	AT1G74100
	BniSOT16-2	BniB05g070980	B5 (−)	64,926,130–64,927,143	337	92.9	

Table 2. *Cont.*

Name1	Name2	Gene ID	Chromosome Location		AA	Identity/%	AGI ID
SOT17	*BniSOT17*	BniB04g033130	B4 (+)	17,622,762–17,623,781	339	86.5	AT1G18590
SOT18	*BniSOT18-1*	BniB03g052540	B3 (+)	28,101,782–28,102,844	334	80.9	AT1G74090
	BniSOT18-2	BniB04g019280	B4 (−)	9,875,796–9,876,833	345	78.6	
	BniSOT18-3	BniB03g038560-1	B3 (−)	19,264,549–19,265,610	353	73.7	
	BniSOT18-4	BniB03g052550	B3 (+)	28,105,157–28,106,167	319	72.9	
	BniSOT18-5	BniB06g047660	B6 (+)	49,729,557–49,730,615	352	72.7	
	BniSOT18-6	BniB06g047670	B6 (+)	49,740,347–49,741,381	344	72.6	
	BniSOT18-7	BniB06g047650	B6 (+)	49,723,604–49,724,665	353	72.2	
	BniSOT18-8	BniB03g038560-2	B3 (−)	19,273,089–19,274,147	352	71.3	
	BniSOT18-9	BniB04g020570	B4 (−)	10,687,342–10,688,406	354	70.4	
	BniSOT18-10	BniB02g035710	B2 (+)	43,650,924–43,651,937	337	70.1	
	BniSOT18-11	BniB04g020560	B4 (+)	10,684,415–10,685,521	368	69.4	
	BniSOT18-12	BniB06g047680	B6 (+)	49,744,098–49,745,159	333	68.3	
	BniSOT18-13	BniB03g038630	B3 (−)	19,304,714–19,306,727	367	49.6	
Secondary modification							
FMO$_{GS\text{-}OX1}$	*BniFMO*$_{GS\text{-}OX1}$	BniB04g021440	B4 (+)	11,434,174–11,440,389	461	77.2	AT1G65860
FMO$_{GS\text{-}OX2}$	*						AT1G62540
FMO$_{GS\text{-}OX3}$	*						AT1G62560
FMO$_{GS\text{-}OX4}$	*						AT1G62570
FMO$_{GS\text{-}OX5}$	*BniFMO*$_{GS\text{-}OX5}$*-1*	BniB03g004970	B3 (+)	1,983,575–1,985,457	459	84.6	AT1G12140
	BniFMO$_{GS\text{-}OX5}$*-2*	BniB02g006840	B2 (+)	3,630,178–3,632,014	459	82.8	
FMO$_{GS\text{-}OX6}$	*						AT1G12130
FMO$_{GS\text{-}OX7}$	*						AT1G12160
CYP81F1	*BniCYP81F1-1*	BniB05g001500	B5 (−)	845,091–846,941	499	86.2	AT4G37430
	BniCYP81F1-2	BniB03g041020	B3 (+)	20,720,008–20,721,740	496	80.4	
	BniCYP81F1-3	BniB05g001510	B5 (−)	849,140–850,741	504	79.5	
CYP81F2	*BniCYP81F2-1*	BniB02g036600	B2 (+)	44,402,518–44,404,789	491	90.7	AT5G57220
	BniCYP81F2-2	BniB05g036790	B5 (+)	19,374,525–19,376,334	493	88.4	
	BniCYP81F2-3	BniB08g013930	B8 (−)	6,333,770–6,33,6220	493	87.2	
CYP81F3	*BniCYP81F3-1*	BniB05g001540	B5 (−)	866,140–869,499	497	89.3	AT4G37400
	BniCYP81F3-2	BniB03g014280	B3 (+)	6,049,124–6,051,489	491	86.2	
CYP81F4	*BniCYP81F4-1*	BniB05g001530	B5 (−)	859,533–861,269	501	78.7	AT4G37410
	BniCYP81F4-2	BniB03g014290	B3 (+)	6,062,864–6,065,223	519	76.9	
AOP1	*BniAOP1-1*	BniB08g045800	B8 (+)	28,827,457–28,828,861	320	76.5	AT4G03070
	BniAOP1-2	BniB08g039120	B8 (−)	23,904,179–23,911,127	423	49.2	
AOP2	*BniAOP2-1*	BniB05g055540	B5 (−)	49,961,828–49,965,927	432	59.2	AT4G03060
	BniAOP2-2	BniB08g045740	B8 (−)	28,785,908–28,787,937	475	51.3	

Table 2. *Cont.*

Name1	Name2	Gene ID	Chromosome Location		AA	Identity/%	AGI ID
AOP3	*						AT4G03050
GSL-OH	*						AT2G25450
IGMT1	*BniIGMT1-1* *BniIGMT1-2*	BniB04g035790 BniB03g008950	B4 (−) B3 (−)	19,012,288–19,013,601 3,546,629–3,547,917	372 372	90.4 89.1	AT1G21100
IGMT2	*BniIGMT2*	BniB03g008960	B3 (−)	3,550,953–3,552,247	374	92.6	AT1G21120
IGMT5	*BniIGMT5*	BniB03g055850	B3 (−)	30,178,032–3,0179,487	370	79.4	AT1G76790
			Co-substrate pathways				
TSB1	*BniTSB1-1* *BniTSB1-2*	BniB02g033500 BniB05g034690	B2 (−) B5 (−)	42,216,546–42,218,333 17,928,767–17,930,499	472 309	92.6 55.2	AT5G54810
ASA1	*BniASA1-1* *BniASA1-2*	BniB08g002350 BniB02g054940	B8 (−) B2 (+)	1,172,677–1,175,600 53,796,911–53,799,668	594 604	88.9 88.2	AT5G05730
APK1	*BniAPK1-1* *BniAPK1-2*	BniB06g037540 BniB02g011750	B6 (+) B2 (+)	43,668,583–43,669,978 68,87,476–6,888,716	273 278	89.5 88.6	AT2G14750
APK2	*BniAPK2-1* *BniAPK2-2* *BniAPK2-3*	BniB02g088950 BniB05g000140 BniB03g014780	B2 (−) B5 (−) B3 (−)	70,492,417–70,493,784 71,545–73,229 6,338,724–6,339,871	293 293 227	90.1 89.8 63.9	AT4G39940
GSH1	*BniGSH1-1* *BniGSH1-2* *BniGSH1-3* *BniGSH1-4*	BniB05g014010 BniB02g080290-2 BniB02g080290-1 BniB03g021390	B5 (+) B2 (−) B2 (−) B3 (+)	6,646,285–6,649,224 66,603,961–66,606,919 66,598,228–66,600,769 9,550,645–9,553,182	514 520 469 359	93.1 92.2 62.6 61.7	AT4G23100
GSH2	*BniGSH2-1* *BniGSH2-2*	BniB07g038410 BniB02g079220	B7 (−) B2 (+)	45,708,701–45,711,311 66,040,689–66,048,921	531 478	79.8 75.1	AT5G27380
APS1	*BniAPS1-1* *BniAPS1-2*	BniB03g024880 BniB01g039550	B3 (+) B1 (−)	11,208,027–11,211,252 44,897,233–44,899,328	465 462	92.9 92.7	AT3G22890
APS3	*BniAPS3*	BniB05g021440	B5 (+)	10,445,377–10,447,307	470	89.9	AT4G14680
APR1	*BniAPR1-1* *BniAPR1-2*	BniB08g047710 BniB05g056120	B8 (−) B5 (−)	30,542,974–30,544,615 50,700,159–50,701,310	463 223	87 35.5	AT4G04610
APR2	*						AT1G62180
APR3	*BniAPR3-1* *BniAPR3-2*	BniB02g079650 BniB05g015070	B2 (−) B5 (+)	66,283,619–66,285,188 7,144,150–7,145,729	464 464	92.5 92.2	AT4G21990
OASA1	*BniOASA1-1* *BniOASA1-2*	BniB07g020470 BniB05g021220	B7 (−) B5 (+)	33,626,673–33,628,609 10,294,312–10,296,240	322 324	97.5 94.8	AT4G14880
CHY1	*BniCHY1-1* *BniCHY1-2* *BniCHY1-3* *BniCHY1-4* *BniCHY1-5*	BniB04g000010 BniB01g014580 BniB01g014490 BniB06g018370 BniB06g018590	B4 (−) B1 (−) B1 (+) B6 (+) B6 (+)	7994–10,682 7,787,511–7,790,188 7,718,116–7,721,800 10,266,194–10,271,471 10,442,521–10,445,218	374 380 332 370 225	87 80.7 73.8 71.8 39.4	AT5G65940
AAO4	*BniAAO4*	BniB06g046290	B6 (+)	49,137,455–49,142,867	1337	86.8	AT1G04580
BZO1	*BniBZO1*	BniB03g043330	B3 (−)	22,256,842–22,258,819	566	76.4	AT1G65880
SCPL17	*BniSCPL17*	BniB07g056480	B7 (+)	55,233,985–55,236,333	436	65.3	AT3G12203

* means that the orthologous gene in *B. nigra* has been missing. In the chromosome location, the positive (+) and negative (−) signs indicate the existence of a gene on the positive and negative strand of that specific chromosome, respectively.

Table 3. The inventory of transcription factor genes and transporter genes involved in glucosinolate biosynthesis in *Brassica nigra*.

Name1	Name2	Gene ID	Chromosome Location		AA	Identity/%	AGI ID
			Activator				
MYB28	*BniMYB28-1*	BniB02g075850	B2 (+)	64,495,904–64,497,224	352	78.6	AT5G61420
	BniMYB28-2	BniB07g041200	B7 (−)	47,224,440–47,225,564	329	68.3	
	BniMYB28-3	BniB04g002670	B4 (−)	1,321,731–1,323,290	329	60.2	
MYB29	*BniMYB29-1*	BniB08g003370	B8 (+)	1,668,386–1,669,814	339	74.3	AT5G07690
	BniMYB29-2	BniB02g053290	B2 (−)	53,047,760–53,050,344	298	46.2	
MYB76	*						AT5G07700
MYB34	*BniMYB34-1*	BniB02g076220	B2 (−)	64,684,279–64,685,680	313	75.1	AT5G60890
	BniMYB34-2	BniB04g003010	B4 (−)	1,462,793–1,463,979	276	70.4	
	BniMYB34-3	BniB07g040970	B7 (−)	47,066,742–47,067,688	262	60.7	
MYB51	*BniMYB51-1*	BniB03g028100	B3 (+)	12,800,136–12,801,361	326	75.4	AT1G18570
	BniMYB51-2	BniB03g007660	B3 (+)	2,995,206–2,996,724	319	72.1	
	BniMYB51-3	BniB04g033080	B4 (+)	17,574,440–17,575,684	323	71.7	
MYB115	*						AT5G40360
MYB118	*BniMYB118-1*	BniB04g007120	B4 (+)	3,444,507–3,446,768	486	63	AT3G27785
	BniMYB118-2	BniB02g067910	B2 (−)	60,652,891–60,655,308	477	63	
MYB122	*BniMYB122-1*	BniB03g052520	B3 (+)	28,090,119–28,091,605	335	74.9	AT1G74080
	BniMYB122-2	BniB05g070970	B5 (+)	64,915,053–64,916,920	376	67.8	
MYC2	*BniMYC2-1*	BniB04g049150	B4 (−)	30,208,579–30,210,393	604	87.2	AT1G32640
	BniMYC2-2	BniB03g035750	B3 (−)	17,597,458–17,599,278	606	86.6	
	BniMYC2-3	BniB03g016310	B3 (+)	7,033,365–7,035,185	606	84.3	
MYC3	*BniMYC3-1*	BniB02g063410	B2 (+)	58,260,877–58,262,631	584	76.9	AT5G46760
	BniMYC3-2	BniB04g015560	B4 (−)	7,676,518–7,678,221	567	71.5	
MYC4	*BniMYC4*	BniB05g018490	B5 (+)	8,814,854–8,816,635	593	73.1	AT4G17880
IQD1	*BniIQD1-1*	BniB07g058100	B7 (+)	55,932,859–55,934,585	444	67.2	AT3G09710
	BniIQD1-2	BniB03g023450	B3 (−)	10,613,021–10,615,049	485	59.9	
SLIM1	*BniSLIM1-1*	BniB03g052120	B3 (−)	27,906,056–27,908,203	587	85.1	AT1G73730
	BniSLIM1-2	BniB03g038740	B3 (+)	19,356,898–19,358,816	584	82.1	
	BniSLIM1-3	BniB05g070840	B5 (−)	64,849,951–64,851,819	538	77.1	
OBP2	*BniOBP2-1*	BniB03g002340	B3 (−)	936,743–938,038	340	76.4	AT1G07640
	BniOBP2-2	BniB06g043020	B6 (+)	47,097,586–47,098,898	334	75.8	
	BniOBP2-3	BniB02g003700	B2 (−)	1,819,220–1,820,440	313	69.5	
CAMTA3	*BniCAMTA3-1*	BniB01g021790	B1 (−)	13,029,617–13,034,577	1034	86.2	AT2G22300
	BniCAMTA3-2	BniB03g062790	B3 (−)	45,548,997–45,550,544	429	34.9	
	BniCAMTA3-3	BniB02g018410	B2 (+)	11,860,712–11,862,740	358	27	
CCA1	*BniCCA1*	BniS02554g500	utg2554 (−)	328,706–331,283	572	74.4	AT2G46830
			Activator/Repressor				
HY5	*BniHY5-1*	BniB05g026170	B5 (+)	12,836,342–12,837,354	163	91.7	AT5G11260
	BniHY5-2	BniB02g050820	B2 (+)	51,949,284–51,950,286	168	91.1	
	BniHY5-3	BniB06g056490	B6 (−)	54,414,011–54,416,320	208	57.8	
			Repressor				
SD1	*BniSD1-1*	BniB02g069920	B2 (−)	61,609,407–61,611,242	306	92.5	AT5G48850
	BniSD1-2	BniB07g037550	B7 (−)	45,221,658–45,223,201	306	90.5	
	BniSD1-3	BniB04g005760	B4 (+)	2,848,170–2,850,064	299	88.9	
	BniSD1-4	BniB07g037520	B7 (−)	45,208,991–45,210,584	306	88.2	
SD2	*BniSD2-1*	BniB06g046090	B6 (+)	49,043,247–49,044,406	303	88.2	AT1G04770
	BniSD2-2	BniB03g001070	B3 (−)	446,167–447,323	299	83.8	
FRS7	*BniFRS7*	BniB01g057400	B1 (+)	55,333,501–55,335,611	534	51.1	AT3G06250
FRS12	*BniFRS12*	BniB08g009620	B8 (+)	4,493,856–4,496,198	780	86.4	AT5G18960

Table 3. *Cont.*

Name1	Name2	Gene ID	Chromosome Location		AA	Identity/%	AGI ID
			Mediator				
MED5	*BniMED5-1*	BniB01g038210	B1 (−)	43,391,779–43,397,254	1308	89.5	AT3G23590
	BniMED5-2	BniB07g048400	B7 (−)	51,150,764–51,156,016	1297	87.3	
MED25	*BniMED25-1*	BniB03g018400	B3 (−)	8,065,052–8,070,109	830	87.9	AT1G25540
	BniMED25-2	BniB04g040670	B4 (+)	22,229,344–22,231,613	396	23.7	
			Transporter				
BAT5	*BniBAT5-1*	BniB08g037040	B8 (+)	22,004,327–22,006,322	408	90	AT4G12030
	BniBAT5-2	BniB08g049400	B8 (−)	32,467,906–32,474,047	393	79.1	
SULTR1;1	*BniSULTR1;1*	BniB05g033180	B5 (−)	16,848,645–16,851,311	519	69.8	AT4G08620
SULTR1;2	*BniSULTR1;2-1*	BniB05g074030	B5 (+)	67,138,702–67,142,049	652	93.7	AT1G78000
	BniSULTR1;2-2	BniB03g057280	B3 (+)	31,113,533–31,116,846	655	92.7	
	BniSULTR1;2-3	BniB03g057270	B3 (+)	31,102,957–31,106,559	671	91.8	
GTR1	*BniGTR1-1*	BniB06g024230	B6 (+)	14,337,898–14,340,309	634	83.4	AT3G47960
	BniGTR1-2	BniB08g061180	B8 (−)	57,886,043–57,888,835	615	80.7	
	BniGTR1-3	BniB05g052360	B5 (−)	44,394,313–44,396,700	617	78	
GTR2	*BniGTR2-1*	BniB04g002130	B4 (+)	1,022,088–1,024,566	612	92	AT5G62680
	BniGTR2-2	BniB06g010940	B6 (−)	5,574,061–5,576,460	612	91.6	
	BniGTR2-3	BniB07g041760	B7 (−)	47,577,010–47,579,359	606	85.2	

* means that the orthologous gene in *B. nigra* has been missing. In the chromosome location, the positive (+) and negative (−) signs indicate the existence of a gene on the positive and negative strand of that specific chromosome, respectively.

Of the 184 *BniGSL* genes, 182 are unevenly mapped among the eight chromosomes of *B. nigra*, with 14, 28, 36, 21, 28, 20, 14, and 21 *BniGSL* genes anchoring on chromosome B1–B8, respectively (Figure 1). Two other *BniGSL* genes are distributed on a large scaffold, which have not yet been mapped onto chromosomes (Tables 2 and 3). In *A. thaliana*, there are 25 genes that constitute 11 tandem-duplicate gene modules. Here, we found that there are 20 tandem-duplicate gene modules in *B. nigra* involving 46 genes. However, the gene families involved in tandem duplication are not exactly the same between these two species (Table 2 and Figure 1).

By doing pairwise sequence alignment of GSL genes between *B. nigra* and *A. thaliana*, we found that a total of 13 homologs corresponding to *A. thaliana* GSL genes were absent in *B. nigra*, including two genes involved in side-chain elongation (i.e., *MAM3* and *IPMI-SSU3*), a gene related to core structure synthesis (i.e., *CYP79F2*), seven genes involved in side-chain modification (i.e., $FMO_{GS\text{-}OX2/3/4/6/7}$, *AOP3*, and *GSL-OH*), a gene related to co-substrate pathways (i.e., *APR2*), and two TFs (i.e., *MYB76* and *MYB115*) (Table 2). In general, 95.7% of the *BniGSL* genes identified in this study shared 51–97% amino acid sequence identity with *AtGSL* genes in *A. thaliana*, with an average of 78.4%.

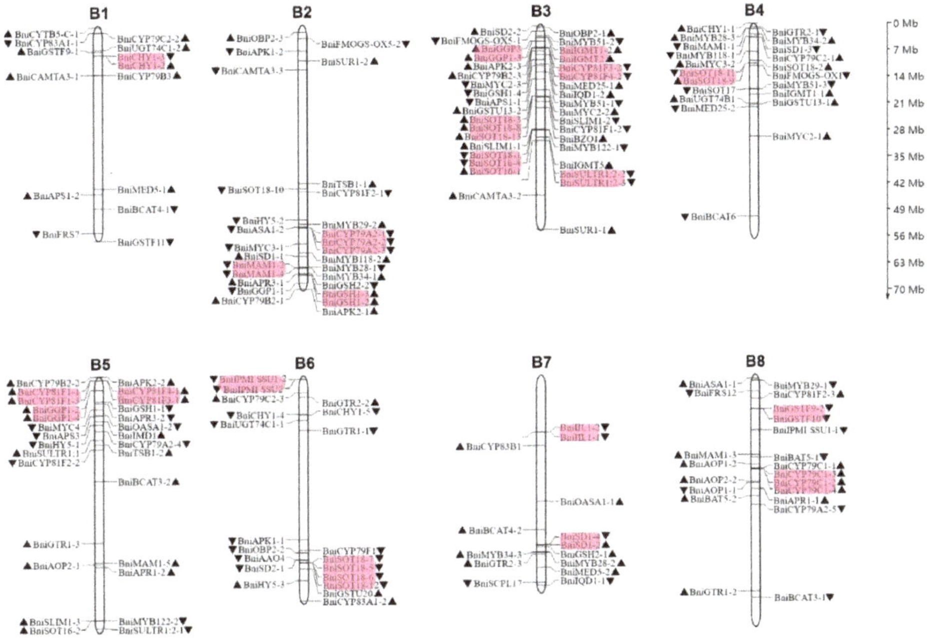

Figure 1. Genomic distribution of glucosinolate genes (*BniGSLs*) on the chromosomes of *Brassica nigra*. The arrowheads next to gene names show the direction of transcription. The chromosome numbers are demonstrated at the top of each chromosome. Tandem repeat genes are marked with a pink background.

3.3. Expression Patterns of BniGSL Genes Encoding Enzymes in Three Organs of B. nigra

Transcriptome sequencing was conducted to investigate the expression of *BniGSL* genes involved in different processes of GSL biosynthesis, and those that participated in aliphatic, indolic, and aromatic GSLs in three organs (i.e., root, stem, and leaf). First, the correlation analysis between biological replicates was performed, and the results showed that the three biological replicates of root, stem, and leaf all have a good correlation (Figure S2), indicating the high data reliability in this study.

Among 184 *BniGSL* genes determined in this study, 172 were detected in at least one organ, and 12 genes were not expressed in all three organs. Moreover, there were 34 *BniGSL* genes whose expression in roots, stems, and leaves all lower than 1 FPKM (Table S4). The low expression of these genes suggests that they may contribute less to GSL biosynthesis in these three organs.

On the basis of the GSL biosynthetic pathways involving *BniGSL* genes and their biological functions, the expression of 184 *BniGSL* genes in roots, stems, and leaves was further analyzed. The biosynthesis of aliphatic GSLs can be divided into three main phases, and dozens of enzymes have been determined to participate in the corresponding reactions (Figure 2). The side-chain elongation phase of aliphatic GSLs consists of six steps and eleven enzymes, of which MAM3 and IPMI SSU3 have been lost in *B. nigra*. Seven enzymatic steps of the core structure synthesis involved 14 enzymes, with 38 *BniGSL* genes encoding 13 enzymes in *B. nigra* (except CYP79F2). The aliphatic GSL side-chain modification phase mainly includes S-oxygenation, side-chain oxygenation, and further conversion of hydroxyalkyl GSL into benzoylated and sinapoylated GSLs, which are catalyzed by five types of enzymes. However, both *AOP3* and *GSL-OH* are absent in *B. nigra*, and there are only two of the seven Arabidopsis *FMOGS-OX* are retained in *B. nigra*. Encouragingly, although there are some aliphatic GSL synthetic *BniGSL* genes that had an extremely low expression, expression data showed that at least one *BniGSL* gene was highly expressed in every step of the synthesis of sinigrin. Moreover, the expression of these highly expressed *BniGSL* genes in leaves and stems was generally higher than that in roots (Figure 2A), and this finding was consistent with the fact that the side-chain elongation of aliphatic GSLs should be performed in green tissues containing chloroplasts.

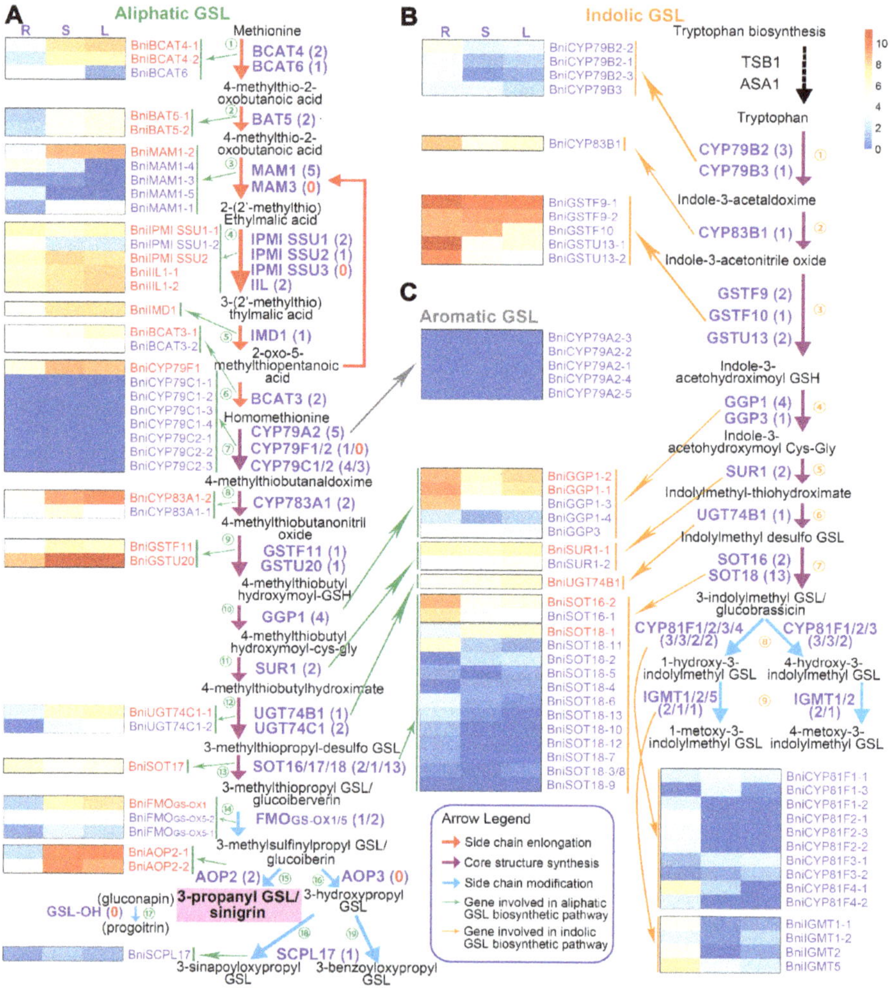

Figure 2. Biosynthetic pathways of aliphatic and indolic glucosinolates (GSLs) and the heatmap of related *BniGSLs* in *Brassica nigra*. The pathway contains 3 major phases: side-chain elongation (Steps 1–6), core structure synthesis (steps 7–13 in (**A**) and steps 1–7 in (**B**)), and side-chain modification (steps 14–19 in (**A**) and steps 8 and 9 in (**B**)). The biosynthetic pathway of aliphatic (**A**), indolic (**B**), and aromatic (**C**) GSLs and the heatmap of related *BniGSLs*. The number in parentheses represents the copy number of the gene. Abbreviations: AOP2, 2-oxoglutarate-dependent dioxygenase; BAT5, probable sodium/metabolite cotransporter BASS5; BCAT, branched-chain amino acid aminotransferase; CYP79A, phenylalanine N-monooxygenase; CYP79B, tryptophan N-monooxygenase; CYP79C/F, cytochrome P450 79C/F; CYP83B, CYP83B monooxygenase; FMO_{GS-OX}, flavincontaining monooxygenase; GGP1, γ-glutamyl peptidase 1; GSTF/U, glutathione S-transferase F/U; GTR, GSL transporter; IGMT, indole GSL O-methyltransferase; IIL1, isopropylmalate isomerase large subunit 1; IPMI-SSU, isopropylmalate isomerase small subunit; L, leaf; MAM, methylthioalkylmalate synthase; R, root; S, stem; SOT, sulfotransferase. The framework of the aliphatic and indolic biosynthetic pathways is adapted from ref [11].

The synthesis of indolic GSL does not require side-chain elongation and only consists of core structure synthesis and side-chain modification, which can be divided into seven and two steps, respectively. All enzymes involved in indolic GSL synthesis have corresponding homologs in *B. nigra*, and 33 and 14 *BniGSL* genes encode 12 and 7 enzymes, respectively, to take part in the above two phases. Remarkably, four *BniGSL* genes encoding CYP79B2 and CYP79B3, the indolic cytochrome P450 enzymes that convert tryptophan derivatives into aldoximes, and all indolic *BniGSL* genes responsible for the side-chain modification, showed generally low expression, especially in stems and leaves (Figure 2B). In addition, 12 *BniGSL* genes encoding three cytochrome P450 members (i.e., CYP79A2, CYP79C1 and CYP79C2) were involved in the aromatic core structure GSL pathway but were almost not expressed in roots, stems, and leaves (Figure 2C). This finding might explain the few aromatic GSLs in *B. nigra*, and their content was extremely low.

3.4. Expression Patterns of BniGSL Genes Encoding TFs, Transporters and Proteins Involved in Co-Substrate Pathways

Thus far, a total of 10 TFs of the R2R3 domain MYB family have been characterized as key players in the regulation of GSL genes. MYB28, MYB29, and MYB76 play critical regulatory roles in aliphatic GSL biosynthesis. MYB34, MYB51, and MYB122 are essential regulators of indolic GSL biosynthesis. MYB115 and MYB118 are responsible for modulating the synthesis of aromatic GSLs. Similar to the enzymes involved in GSL biosynthesis, multiple aliphatic GSL-related *BniMYBs* showed high expression in stems and leaves, whereas all indolic and aromatic GSL-related *BniMYBs* had low expressions (Figure 3A). As expected, MYC2, MYC3, and MYC4, the common regulators of aliphatic and indolic GSL synthesis, were highly expressed in all three organs of *B. nigra* (Figure 3A), and had similar expression patterns of some *BniGSL* genes (i.e., *BniGGP1-1/2*, *BniSUR1-1*, *BniUGT74B1*, *BniSOT16-2*, and *BniSOT18-1*) involved in the core structure synthesis of aliphatic and indolic GSLs (Figure 2). Among other TFs that regulated GSL biosynthesis, except *IQD1* and *FRS 7*, at least one copy of a *BniGSL* gene was expressed in at least one organ (Figure 3B).

Five proteins have been experimentally characterized as transporters in GSL biosynthesis. BAT5s, which acts as chloroplast transporter in the side-chain elongation phase, exhibited a similar expression pattern to the GSL genes related to aliphatic GSL biosynthesis in *B. nigra*. SULTR1;1 and SULTR1;2 are two sulfate transporters that function in *Arabidopsis* roots, and their orthologs in *B. nigra* also showed a relatively high expression in roots. GTR1 and GTR2 are responsible for transporting synthesized GSLs from leaves to seeds and roots [50,51]. Expression data also showed that *BniGTR1s* and *BniGTR2s* were predominantly expressed in roots (Figure 3C).

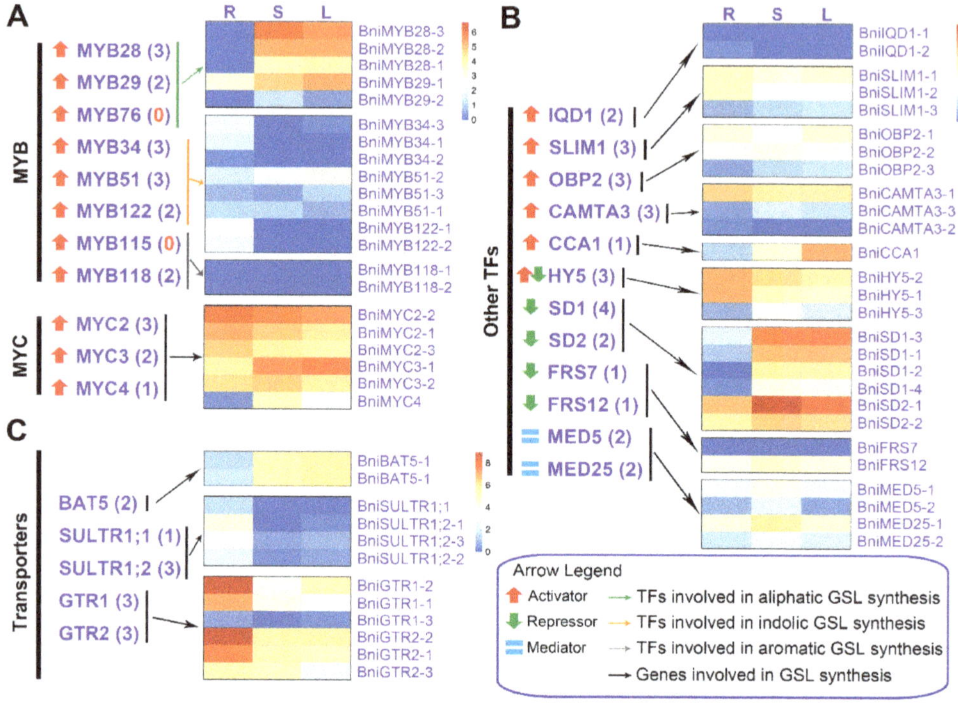

Figure 3. Heatmap of transcription factor (TFs) genes and transporter genes involved in GSL biosynthesis. Expression analysis of *BniMYBs* (**A**) and other TFs (**B**) involved in GSL biosynthesis. (**C**) Expression analysis of *BniGSLs* encoding transporters involved in GSL transportation. The number in parentheses represents the copy number of the gene. Abbreviations: BAT5, probable sodium/metabolite cotransporter BASS5; CAMTA3, calmodulin-binding transcription activator 3; CCA1, circadian clock-associated 1; FRS, Far1 related sequence; GTR, GSL transporter; HY5, long hypocotyl 5; IQD1, IQ domain 1; L, leaf; MED, mediator subunit; OBP2, UAS-tagged root patterning 3; R, root; S; stem; SD, protein sulfur deficiency-induced; SLIM1, sulfur limitation 1; SULTR, sulfate transporter.

The gene expression analysis showed that the expression patterns of the *BniGSL* genes encoding proteins involved in the co-substrate pathway were also different. For *BniGSL* genes that function in the sulfur assimilation process, except *BniAPK2-3*, *BniAPR1-2*, and *BniGSH1-4*, the remaining 17 genes were highly expressed in all three organs (Figure 4A), which were similar to the expression patterns of some core structure synthesis related *BniGSL* genes that play roles in aliphatic and indolic GSL synthesis (Figure 2). However, *BniGSL* genes involved in the synthesis of tryptophan and phenylalanine metabolism were all generally low-expressed in the three organs (Figure 4B,C), which indicated that the supplies of tryptophan for the synthesis of indolic GSL and benzoyl-coenzyme A (BzCoA) for the synthesis of benzoylated GSLs (BzGSLs) might not be sufficient. The low expression characteristics of these genes may be one of the reasons for the low indolic GSL content in *B. nigra* and the absence of BzGSLs and sinapoylated GSLs (SnGSLs).

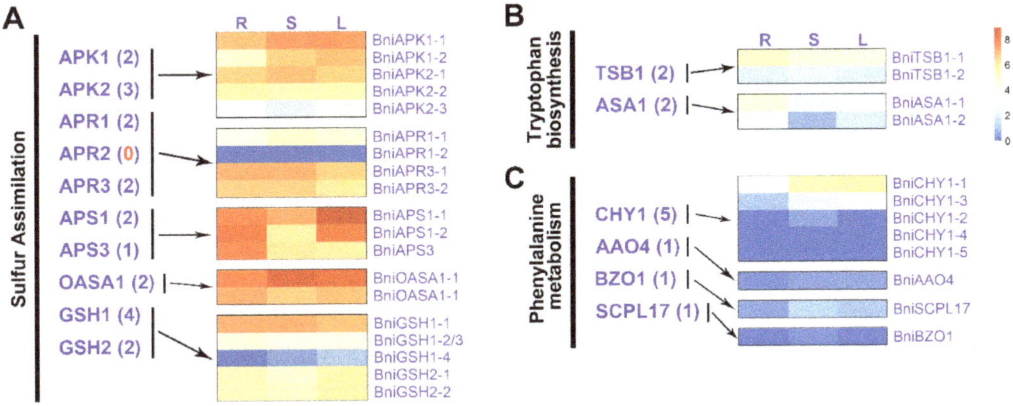

Figure 4. Heatmap of genes involved in the co-substrate pathways of GSL biosynthesis. Expression analysis of *BniGSLs* involved in sulfur assimilation (**A**), tryptophan biosynthesis (**B**), and phenylalanine metabolism (**C**). The number in parentheses represents the copy number of the gene. Abbreviations: AAO4, aldehyde oxidase 4; APK, adenylyl-sulfate kinase; APR, 5′-adenylylsulfate reductase; APS, ATP sulfurylase; BZO1, benzoyloxy GSL 1; ASA1, anthranilate synthase α subunit 1; CHY1, 3-hydroxyisobutyryl-CoA hydrolase 1; GSH1, glutamate−cysteine ligase; GSH2, glutathione synthetase; L, leaf; R, root; S, stem; SCPL17, serine carboxypeptidase-like 17; TSB1, tryptophan synthase β chain 1.

3.5. Expression Patterns of Candidate Key Genes Involved in the Synthesis of Aliphatic GSLs by qRT-PCR

qRT-PCR analysis was performed to reconfirm the expression patterns of 15 candidate key genes (including 12 structural genes and three TFs) involved in the biosynthesis of aliphatic GSLs. As shown in Figure 5, the results of qRT-PCR indicated that the expression levels of nine genes (i.e., *BniBCAT4-1*, *BniMAM1-2*, *BniIMD1*, *BniCYP79F1*, *BniCYP83A1-2*, *BniUGT74C1-1*, *BniFMO*$_{GS\text{-}OX1}$, *BniAOP2-1*, and *BniAOP2-2*) out of 12 structural genes in stems and leaves were five times or more than that in roots. The expression levels of the other three structural genes in three organs did not differ by more than 2.5 times. Nevertheless, in addition to *BniSOT17*, the expression levels of *BniIIL1-1/2* and *BniBCAT3-1* in stems and leaves were still higher than those in roots. For TFs, notably, as an orthologous gene of *MYB28*, which is the main regulatory gene of aliphatic GSL synthesis, *BniMYB28-2* was extremely highly expressed in stems and leaves. The expression of *BniMYC3-1* in stems and leaves was also higher than that in roots. In contrast, the expression of *BniMYB29-1* in three organs was not much different.

In short, the results of qRT-PCR were consistent with the results of RNA-Seq (Figures 3 and 4, Table S4), which together indicated that most of the candidate key genes involved in the synthesis of aliphatic GSLs in *B. nigra* were mainly highly expressed in stems and leaves, while relatively low in roots.

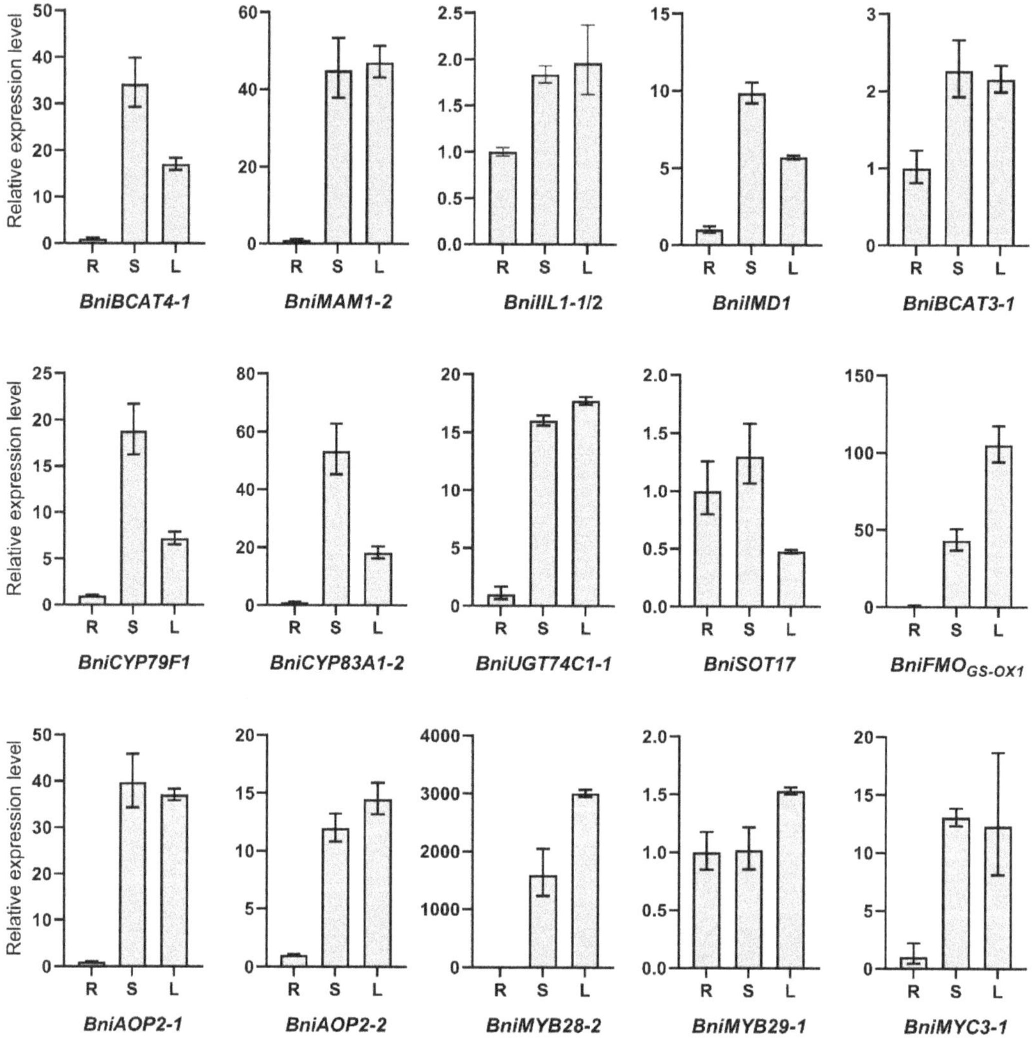

Figure 5. The qRT-PCR analysis of relative expression levels of 15 *BniGSL* genes in different organs of flowering plants. R, root; S, stem; L, leaf. *TIPS* was used as a reference gene, and the expression of *BniGSL* genes in root was set as 1. Error bars represent positive and negative deviations from three independent biological replicates.

3.6. Key MAM Genes Controlling Side-Chain Elongation in B. nigra

The aliphatic GSL biosynthesis initiated by methionine first needs to undergo a six-step side-chain elongation involving five types of enzymes. Among these enzymes, the MAM family remarkably contributes to the diversity of synthesized GSLs because different MAM members have different preferences for catalyzing the side-chain elongation process. In *Arabidopsis*, three tandemly duplicated *MAM* genes were identified, named *MAM1*, *MAM2* (absent in ecotype Columbia), and *MAM3*. The functional analysis revealed that MAM2 and MAM1 were correlated with the accumulation of 3 and 4 carbon (C) side-chain GSLs that had undergone the first and first two rounds of chain elongation, respectively [52,53], whereas MAM3 was able to catalyze the condensations in the first six elongation cycles [13].

To clarify the evolutionary relationships among MAM homologs, we performed a detailed phylogenetic analysis of predicted amino acid sequences of MAM members from Arabidopsis and three basic diploid species of *Brassica*. On the basis of syntenic and sequence similarity analysis, we identified nine, nine, and five MAM members from *B. rapa*, *B. oleracea*, and *B. nigra*, respectively. The resulting phylogenetic tree indicated that four, five, and two MAM members of the three basic diploid species of *Brassica* were phylogenetically close to AtMAM1/2, and three, two, and two MAM members seemed to be closely related to AtMAM3. In addition, there were two BraMAM, two BolMAM, and a BniMAM that were phylogenetically distant from AtMAM1/2 and AtMAM3 (Figure 6A).

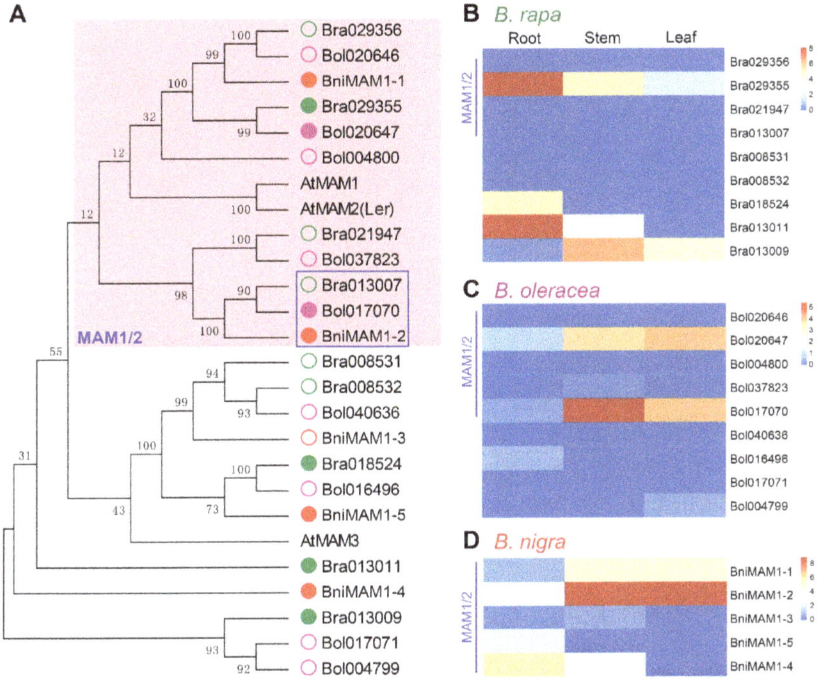

Figure 6. The phylogenetic tree and heatmap of *MAM* genes in *Brassica rapa*, *Brassica oleracea*, *Brassica nigra*, and *Arabidopsis thaliana*. (**A**) The phylogenetic tree of *MAM* genes. The *MAM* genes identified from *Brassica rapa*, *Brassica oleracea*, and *Brassica nigra* are indicated by green, purple, and red hollow circles or solid discs. The silenced genes are represented by a hollow circle, expressed functional genes are indicated by a solid disc. The heatmap of *MAM* genes in *Brassica rapa* (**B**), *Brassica oleracea* (**C**), and *Brassica nigra* (**D**). The expression data of *B. rapa* and *B. oleracea* were obtained from ref [48,49].

On the basis of expression data of the *BniGSL* genes obtained by RNA sequencing in this study, as well as the reported expression information of GSL genes in *B. rapa* and *B. oleracea* [48,49], we found that only four, two, and four *MAM* genes were expressed in *B. rapa*, *B. oleracea*, and *B. nigra*, respectively. Interestingly, a group of genes comprising Bra013007, Bol017070, and *BniMAM1-2*, was the only group of orthologs that showed high expression in *B. oleracea* and *B. nigra* but silenced in *B. rapa* (Figure 6B–D). This expression difference most likely explains why 3C and/or 4C GSLs can be synthesized and accumulated in large quantities in *B. nigra* (e.g., sinigrin) and *B. oleracea* (e.g., progoitrin, gluconapin, glucoraphanin, and sinigrin), but not in *B. rapa*. In addition, Bol020647 and *BniMAM1-1* may also contribute to the synthesis of 3C GSLs (Figure 6C,D). Meanwhile, it is likely the loss of *MAM3* leads to the low contents of 4C, 5C and long-chained aliphatic GSLs in *B. nigra* (Figure 6D).

3.7. *BniCYP79F1* Was Extremely Highly Expressed in *B. nigra*

The substrate-specific cytochrome P450s of the CYP79 family act as the entry point in GSL core structure synthesis by catalyzing the conversion of amino acid derivatives into the corresponding aldoximes. In Arabidopsis, seven CYP79s have been functionally characterized, and different members showed different preferences for amino acid-derived substrates. To reveal the possible connection between the expression of CYP79s and the GSL profiles, CYP79s involved in GSL synthesis in *B. rapa*, *B. oleracea*, and *B. nigra* were further analyzed (Figure 7). CYP79F1 and CYP79F2 are responsible for the biosynthesis of methionine-derived GSLs, and CYP79F2 only converts long-chained methionine derivatives into aldoximes [54–56]. Results showed that although there is only one ortholog of *CYP79F1* in each *Brassica* species, they were all highly expressed in at least one organ. The only difference was that Bra026058 was predominantly expressed in roots, Bol038222 was mainly expressed in stems, followed by leaves, while *BniCYP79F1* was extremely highly expressed in stems and leaves, while slightly lower in roots (Figure 7B–D). Surprisingly, there is no ortholog of *CYP79F2* in *Brassica* (Figure 7A). This fact may also connect to the low content of long-chained aliphatic GSLs in the GSL profiles of *Brassica*.

CYP79B2 and CYP79B3 take part in indolic GSL synthesis [57–59]. Genomic analysis showed that *CYP79B2* retained three copies in each of the three *Brassica* species, while *CYP79B3* retained only one copy (Figure 7). Moreover, these 12 *CYP79s* have been detected to be expressed in at least one organ. However, the expression patterns of *CYP79B2* and *CYP79B3* in the three species were different. In general, *CYP79B2* and *CYP79B3* were highly expressed in all three organs of *B. rapa*, while those in *B. oleracea* were predominantly expressed in leaves, followed by in roots. In *B. nigra*, their expression can be detected in roots but lower than those in stems and leaves (Figure 7B–D). These expression features may partly explain that the content of indolic GSL in *B. nigra* is lower than those in *B. rapa* and *B. oleracea*. In addition, homologous genes of *CYP79A2*, *CYP79C1*, and *CYP79C2* in *B. nigra* are more than those in *B. rapa* and *B. oleracea*. However, except for one copy of *CYP79A2* in *B. oleracea*, the expression of all homologs of these three genes in all three species was extremely low.

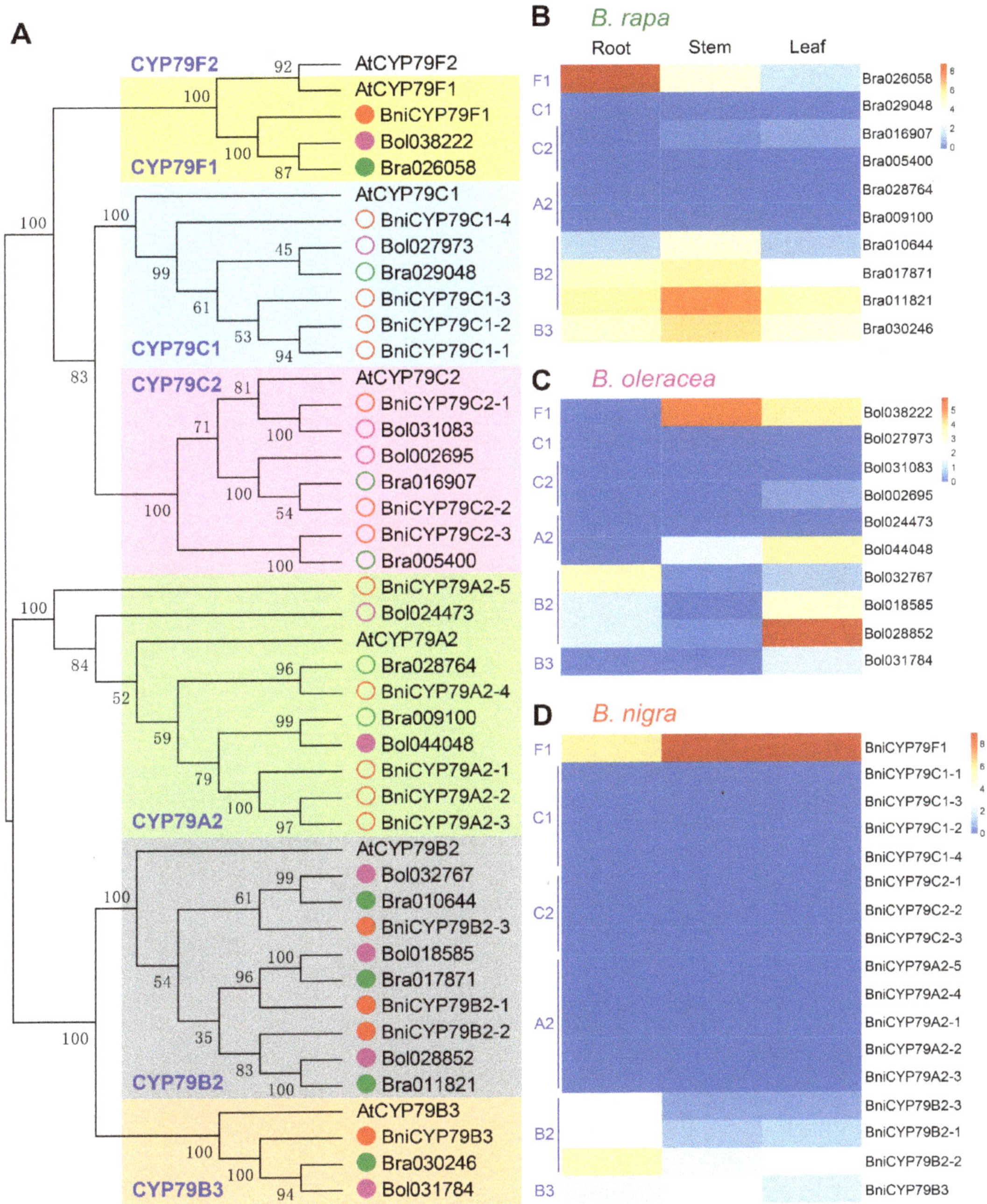

Figure 7. The phylogenetic tree and heatmap of *CYP79* genes in *Brassica rapa*, *Brassica oleracea*, *Brassica nigra*, and *Arabidopsis thaliana*. (**A**), The phylogenetic tree of *CYP79* genes. The *CYP79* genes identified from *Brassica rapa*, *Brassica oleracea*, and *Brassica nigra* are indicated by green, purple, and red hollow circles or solid discs. The silenced genes are represented by a hollow circle, expressed functional genes are indicated by a solid disc. The heatmap of *CYP79* genes in *Brassica rapa* (**B**), *Brassica oleracea* (**C**), and *Brassica nigra* (**D**). The expression data of *B. rapa* and *B. oleracea* were obtained from ref [48,49].

3.8. The Difference in AOP2 Genes Greatly Affect the Diversity of GSLs in Brassica

Side-chain modification is another pathway to enrich aliphatic GSL species in addition to the side-chain elongation. The *GS-AOP* locus is responsible for side-chain oxygenation and contains three genes encoding 2-oxoglutarate-dependent dioxygenases in Arabidopsis, namely *AOP1*, *AOP2*, and *AOP3* [60]. *AOP2* and *AOP3* are located within the *GSL-ALK* and *GSL-OHP* loci, respectively, and can convert methylsulfinylalkyl GSL into alkenyl and hydroxyalkyl GSL, respectively [23,61]. Although the functionality of *AOP1* is unknown, it is considered to be the ancestral gene of *AOP2* and *AOP3* by gene duplication events, suggesting that *AOP1* may also function in GSL biosynthesis [60].

AOP genes in *B. rapa*, *B. oleracea*, and *B. nigra* were identified and subjected to expression analysis to assess the contribution of *AOP* genes on aliphatic GSL diversity in *B. nigra*, as well as the difference in *AOP* genes in different *Brassica* species. Results showed that *B. rapa*, *B. oleracea*, and *B. nigra* contained three, two, and two *AOP1* genes, respectively; three, one, and two functional *AOP2* genes, respectively; and no *AOP3* homolog (Figure 8A). The absence of *AOP3* in these three *Brassica* species may be the key reason why they rarely contain hydroxyalkyl GSLs. The *AOP1* genes in *Brassica* were mainly expressed in roots (Figure 8B–D), suggesting that they may function in the side-chain modification of aliphatic GSLs in roots. Most notably, significant differences were observed in the gene function and expression of *AOP2* genes in *Brassica*. Although three *AOP2* copies in *B. oleracea* were identified, the presence of a premature stop codon resulted in two of them being nonfunctional [43], while the other functional *BolAOP2* had no expression data (Figure 8C), which might be caused by low expression. In contrast, all three *AOP2* genes in *B. rapa* were functional, and Bra00848 was highly expressed in both stems and leaves (Figure 8B). Similarly, two copies of *AOP2* in *B. nigra*, i.e., *BniAOP2-1* and *BniAOP2-1*, were also extremely highly expressed in stems and leaves (Figure 8D). The difference in *AOP2* genes in *Brassica* supports their unique GSL profiles and partially explains why *B. oleracea* is rich in glucoraphanin, but not in *B. rapa*, and why sinigrin is abundant in *B. nigra*.

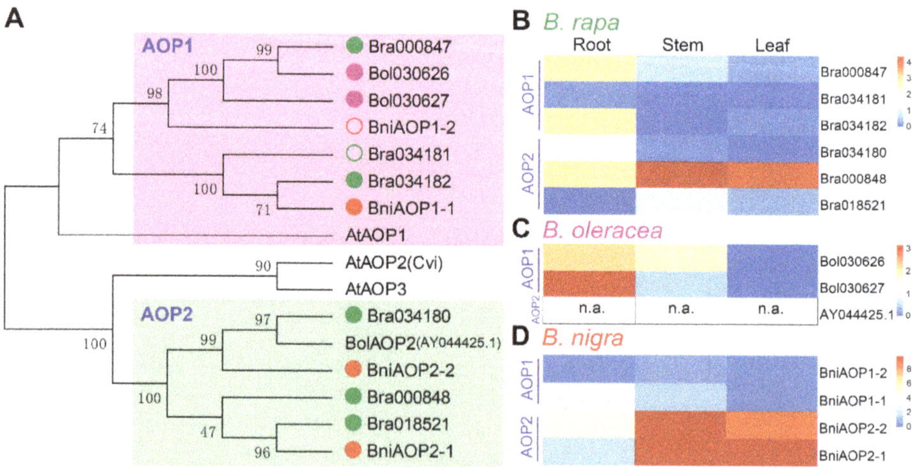

Figure 8. The phylogenetic tree and heatmap of *AOP* genes in *Brassica rapa*, *Brassica oleracea*, *Brassica nigra*, and *Arabidopsis thaliana*. (**A**) The phylogenetic tree of *AOP* genes. The *AOP* genes identified from *Brassica rapa*, *Brassica oleracea*, and *Brassica nigra* are indicated by green, purple, and red hollow circles or solid discs. The silenced genes are represented by a hollow circle, expressed functional genes are indicated by a solid disc. The heatmap of *AOP* genes in *Brassica rapa* (**B**), *Brassica oleracea* (**C**), and *Brassica nigra* (**D**). n.a., not available. The expression data of *B. rapa* and *B. oleracea* were obtained from ref [48,49].

4. Discussion

Despite being one of the three ancestral *Brassica* species in U's triangle model, studies on *B. nigra* always lag behind that on *B. rapa* and *B. oleracea*, including the study on the identification of GSL biosynthesis genes at the genome-wide level. Although the absolute content of various GSLs in *Brassica* can be strongly influenced by environmental factors, the patterns of GSL are mainly controlled genetically [62,63]. Moreover, it was reported that genetic factors were dominant in controlling the synthesis of aliphatic GSLs [64]. Therefore, the genome-wide and expression analyses of GSL genes can help delineate the dominant GSL synthesis pathway. Here, through these analyses, we identified the genes involved in the biosynthesis of GSLs in *B. nigra*, and proposed a sinigrin biosynthesis pathway involving multiple candidate key genes in *B. nigra* (Table S3, Figure 9).

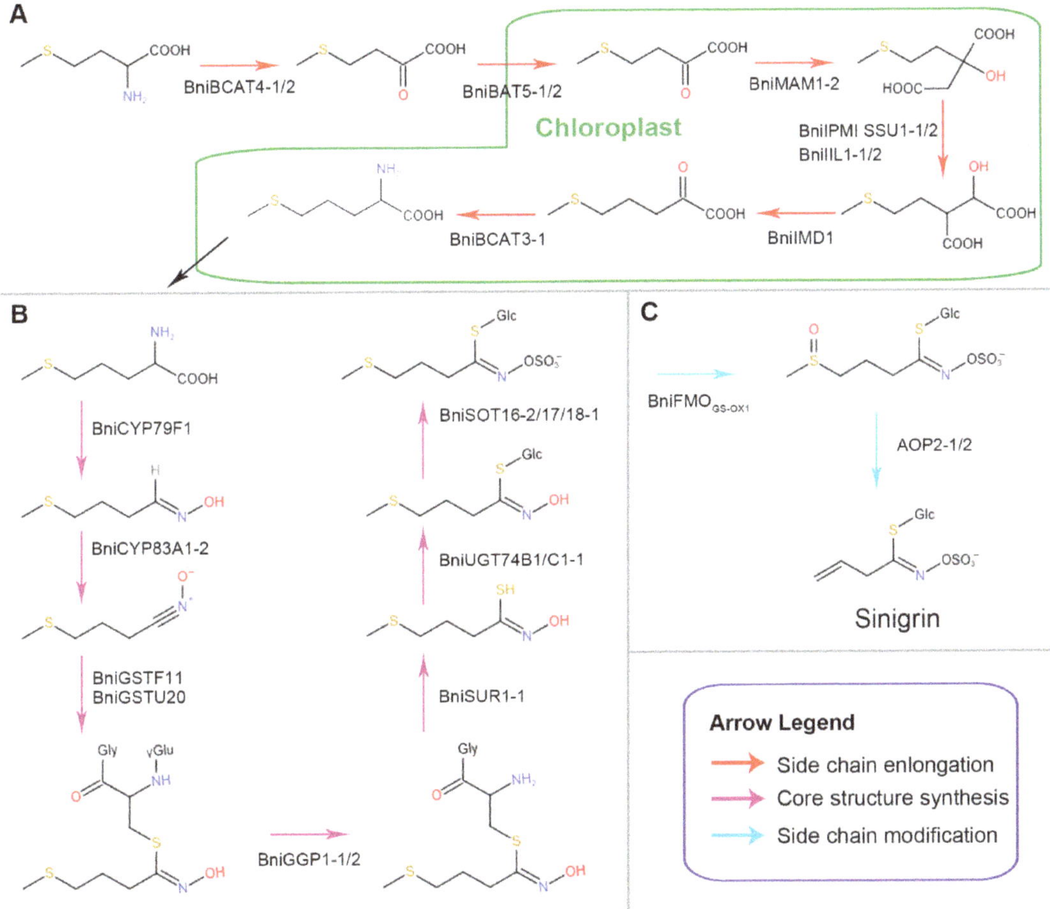

Figure 9. A pathway diagram of candidate key genes involved in sinigrin biosynthesis in *B. nigra*. (**A**) Side-chain elongation machinery. (**B**) Biosynthesis of core glucosinolate structure. (**C**) Secondary modification (alkenylations).

In this study, the GSL content survey once again confirmed that sinigrin was the most dominant GSL in *B. nigra* (Table 1), accounting for more than 90% of the total GSLs. In order to explore why *B. nigra* mainly synthesized and accumulated sinigrin from the genetic perspective. We searched out all GSL genes as much as possible on the basis of the latest whole-genome data of *B. nigra*. *BniGSL* genes were identified by comparing their related homologs in *A. thaliana*, *B. rapa*, and *B. oleracea*. A total of 184 *BniGSL* genes were identified, of which 182 could be labeled on eight chromosomes of *B. nigra* (Tables 2 and 3, Figure 1). Compared with the identified GSL genes in *B. rapa* and *B. oleracea*, more members of *BniGSL* genes were determined in this study, which might be due to the incomplete genome data of *B. rapa* and *B. oleracea* previously [42,43,65]. Nevertheless, the GSL genes in these three *Brassica* species were all amplified considerably in their process of evolution. Six modes of gene duplication have been demonstrated in previous research, including whole-genome duplication (WGD) and tandem duplication (TD). The *Brassica* genome is proven to have triplicated soon after its divergence from *Arabidopsis* [44,66,67]. A functional bias is observed in genes retained after WGD and TD, which may show positive or negative correlations in the expansion of different members in a certain gene family. Duplicated genes may enhance the potential for the quantitative variation of a particular trait [68–70]. Similar to the GSL genes in *B. rapa* and *B. oleracea*, WGD and TD are the main mechanisms accounting for the expansion of *BniGS* genes, and most of them are present in multiple copies (Figure 1, Tables 2 and 3). For instance, previous research showed that *SOT18* is a multigene subfamily with tandem arrays of genes in *B. rapa* (10 members) and *R. sativus* (11 members) [42,71]. Thirteen *BniSOT18s* were identified in the current *B. nigra* genome and there were four groups of members expanded through TD (Figure 1, Table 2), indicating that TD was another factor responsible for the expansion of GSL genes during the evolution of *B. nigra* genome similar to those of *B. rapa* and *R. sativus*.

In general, there is no substantial difference among the inventory of GSL genes in *B. nigra* and those in *B. rapa* and *B. oleracea*, since most GSL genes have successfully retained at least one copy during the evolution of the *Brassica*, despite the difference in copy number of some GSL genes (Tables 2 and 3). Moreover, the absence of some GSL genes exists in all three *Brassica* species. For example, no ortholog of *CYP79F2* and *AOP3* has been identified in *Brassica* [42,43,46], which basically explains the lack of long-chain aliphatic and hydroxyalkyl GSLs [36,38,72]. However, there are significant differences in the GSL profiles of the three *Brassica* species. For example, sinigrin is the main GSL in *B. nigra*, and the sinigrin content in *B. nigra* is much higher than those in *B. rapa* and *B. oleracea*.

By transcriptome sequencing, we believe that the specific expression patterns of *BniGSL* genes in *B. nigra* largely determine its unique GSL profile. In the aliphatic GSL biosynthesis pathway, except the oxidative decarboxylation that involves only one gene (i.e., *BniIMD1*), each step (until the alkenylation of basic GSL) involves two or more genes, and at least one gene is highly expressed in stems and leaves (Figure 2A), ensuring the synthesis of a large amount of aliphatic GSL. Furthermore, the loss of *MAM3* orthologs results in failure to synthesize aliphatic GSL with long side chain, while the absence of *AOP3* and *GSL-OH* prevented the hydroxyalkylation of the basic GSL and the conversion of alkenyl GSL into hydroxylated alkenyl GSL. Therefore, we speculated that the expression characteristics of the aliphatic *BniGSL* genes and the above-mentioned limitations are possibly the main reason for the synthesis of a large amount of sinigrin in *B. nigra*. Specifically, we concluded that the candidate key genes involved in the sinigrin synthesis pathway are *BniBCAT4-1/2*, *BniBAT5-1/2*, *BniMAM1-2*, *BniIPMI SSU1-1*/*BniIPMI SSU2*/*BniIIL1-1/2*, *BniIMD1*, *BniBCAT3-1*, *BniCYP79F1*, *BniCYP83A1-2*, *BniGSTF11*/*BniGSTU20*, *BniGGP1-1/2*, *BniSUR1-1*, *BniUGT74B1*/*BniUGT74C1-1*, *BniSOT16-2*/*BniSOT17*/*BniSOT18-1*, $BniFMO_{CG\text{-}OX1}$, and *BniAOP2-1/2* (Figure 9). Moreover, the results of expression analysis of *MAMs*, *CYP79s*, and *AOPs* in three *Brassica* species further indicate that the differences in gene expression have significant effects on GSL patterns and contents, and *BniMAM1-2*, *BniCYP79F1*, and *BniAOP2-1/2* may control the key nodes of the sinigrin synthesis pathway in *B. nigra*. In addition, the low expression of some key indolic and aromatic *BniGSL* genes in *B. nigra* is partially responsible for low indolic and aromatic GSL, respectively. These findings enriched our understanding of GSL biosynthesis patterns in *B. nigra* and provided guidance for changing the GSL profile in *B. nigra*, e.g., by regulating the expression of key *BniGSL* genes (e.g., overexpression/knockout) to change the GSL synthesis pathway in *B. nigra*, thereby constructing new varieties with different GSL profiles (e.g., high/low sinigrin). Moreover, the identification of GSL biosynthesis pathways in *B. nigra* also provided a reference for the regulation of GSL synthesis in other *Brassica* species.

In this study, we determined the GSL profile of *B. nigra*, made an inventory of *BniGSL* genes and characterized their expression patterns. This research contributes to the functional analysis of *BniGSL* genes and improvement of GSLs in *B. nigra*, and provides a new perspective for future research on GSL synthesis in other species.

Supplementary Materials: The following are available online at https://www.mdpi.com/article/10.3390/horticulturae7070173/s1; Figure S1: HPLC chromatograms of glucosinolates isolated from different organs of Brassica nigra. Figure S2: Correlation analysis of samples. Table S1: GSL genes reported in Brassica rapa and Brassica oleracea. Table S2: List of primers used for qRT-PCR. Table S3: The DNA, CDS and amino acid sequences of glucosinolate biosynthetic related genes (BniGSLs) in Brassica nigra. Table S4: Expression data of BniGSL genes in different organs of *Brassica nigra* (FPKM).

Author Contributions: Conceptualization, Y.Y., Y.H. and Z.Z.; methodology, Y.L., L.X. and E.G.; software, Y.L.; validation, Y.Y. and Y.H.; formal analysis, Y.L. and L.X.; investigation, Y.L. and E.G.; resources, Y.Y. and Y.Z.; data curation, Y.Y.; writing—original draft preparation, Y.Y. and L.X.; writing—review and editing, Y.Y., Y.H., and Z.Z.; visualization, L.X.; supervision, Y.Y. and Y.H.; project administration, Y.H. and Z.Z.; funding acquisition, Z.Z. All authors have read and agreed to the published version of the manuscript.

Funding: This work was supported by the National Natural Science Foundation of China (No. 32072557 and No. 31572115) and the Zhejiang Provincial Natural Science Foundation of China (No. LZ14C150001).

Institutional Review Board Statement: Not applicable.

Informed Consent Statement: Not applicable.

Data Availability Statement: The data used for the analysis in this study are available within the article and supplementary materials.

Conflicts of Interest: The authors declare no conflict of interest.

References

1. Blazevic, I.; Montaut, S.; Burcul, F.; Olsen, C.E.; Burow, M.; Rollin, P.; Agerbirk, N. Glucosinolate structural diversity, identification, chemical synthesis and metabolism in plants. *Phytochemistry* **2019**, *169*, 112100. [CrossRef]
2. Agerbirk, N.; Hansen, C.C.; Kiefer, C.; Hauser, T.P.; Ørgaard, M.; Asmussen Lange, C.B.; Cipollini, D.; Koch, M.A. Comparison of glucosinolate diversity in the crucifer tribe Cardamineae and the remaining order Brassicales highlights repetitive evolutionary loss and gain of biosynthetic steps. *Phytochemistry* **2021**, *185*, 112668. [CrossRef]
3. Halkier, B.A.; Gershenzon, J. Biology and Biochemistry of Glucosinolates. *Annu. Rev. Plant Biol.* **2006**, *57*, 303–333. [CrossRef] [PubMed]
4. Agerbirk, N.; Olsen, C.E. Glucosinolate structures in evolution. *Phytochemistry* **2012**, *77*, 16–45. [CrossRef]
5. Bones, A.M.; Rossiter, J.T. The enzymic and chemically induced decomposition of glucosinolates. *Phytochemistry* **2006**, *67*, 1053–1067. [CrossRef]
6. Padilla, G.; Cartea, M.E.; Velasco, P.; De Haro, A.; Ordás, A. Variation of glucosinolates in vegetable crops of *Brassica rapa*. *Phytochemistry* **2007**, *68*, 536–545. [CrossRef] [PubMed]
7. Hopkins, R.J.; van Dam, N.M.; van Loon, J.J.A. Role of glucosinolates in insect-plant relationships and multitrophic interactions. *Annu. Rev. Entomol.* **2009**, *54*, 57–83. [CrossRef]
8. Arumugam, A.; Abdull Razis, A.F. Apoptosis as a Mechanism of the Cancer Chemopreventive Activity of Glucosinolates: A Review. *Asian Pac. J. Cancer Prev.* **2018**, *19*, 1439–1448. [PubMed]
9. Mazumder, A.; Dwivedi, A.; Du Plessis, J. Sinigrin and Its Therapeutic Benefits. *Molecules* **2016**, *21*, 416. [CrossRef] [PubMed]
10. Sønderby, I.E.; Geu-Flores, F.; Halkier, B.A. Biosynthesis of glucosinolates—Gene discovery and beyond. *Trends Plant Sci.* **2010**, *15*, 283–290. [CrossRef]
11. Harun, S.; Abdullah-Zawawi, M.-R.; Goh, H.-H.; Mohamed-Hussein, Z.-A. A Comprehensive Gene Inventory for Glucosinolate Biosynthetic Pathway in *Arabidopsis thaliana*. *J. Agric. Food Chem.* **2020**, *68*, 7281–7297. [CrossRef] [PubMed]
12. Grubb, C.D.; Abel, S. Glucosinolate metabolism and its control. *Trends Plant Sci.* **2006**, *11*, 89–100. [CrossRef]
13. Textor, S.; Kraker, J.-W.; De Hause, B.; Gershenzon, J.; Tokuhisa, J.G. MAM3 catalyzes the formation of all aliphatic glucosinolate chain lengths in Arabidopsis. *Plant Physiol.* **2007**, *144*, 60–71. [CrossRef] [PubMed]

14. Gigolashvili, T.; Yatusevich, R.; Rollwitz, I.; Humphry, M.; Gershenzon, J.; Flügge, U.-I. The plastidic bile acid transporter 5 is required for the biosynthesis of methionine-derived glucosinolates in *Arabidopsis thaliana*. *Plant Cell* **2009**, *21*, 1813–1829. [CrossRef] [PubMed]
15. Zhang, J.; Wang, H.; Liu, Z.; Liang, J.; Wu, J.; Cheng, F.; Mei, S.; Wang, X. A naturally occurring variation in the *BrMAM-3* gene is associated with aliphatic glucosinolate accumulation in *Brassica rapa* leaves. *Hortic. Res.* **2018**, *5*, 69. [CrossRef] [PubMed]
16. Wang, C.; Dissing, M.M.; Agerbirk, N.; Crocoll, C.; Halkier, B.A. Characterization of Arabidopsis CYP79C1 and CYP79C2 by Glucosinolate Pathway Engineering in Nicotiana benthamiana Shows Substrate Specificity Toward a Range of Aliphatic and Aromatic Amino Acids. *Front. Recent Dev. Plant Sci.* **2020**, *11*, 57. [CrossRef] [PubMed]
17. Bak, S.; Feyereisen, R. The Involvement of Two P450 Enzymes, CYP83B1 and CYP83A1, in Auxin Homeostasis and Glucosinolate Biosynthesis. *Plant Physiol.* **2001**, *127*, 108–118. [CrossRef]
18. Naur, P.; Petersen, B.L.; Mikkelsen, M.D.; Bak, S.; Rasmussen, H.; Olsen, C.E.; Halkier, B.A. CYP83A1 and CYP83B1, two nonredundant cytochrome P450 enzymes metabolizing oximes in the biosynthesis of glucosinolates in Arabidopsis. *Plant Physiol.* **2003**, *133*, 63–72. [CrossRef]
19. Mikkelsen, M.D.; Naury, P.; Halkier, B.A. Arabidopsis mutants in the *C-S* lyase of glucosinolate biosynthesis establish a critical role for indole-3-acetaldoxime in auxin homeostasis. *Plant J.* **2004**, *37*, 770–777. [CrossRef]
20. Grubb, C.D.; Zipp, B.J.; Kopycki, J.; Schubert, M.; Quint, M.; Lim, E.-K.; Bowles, D.J.; Pedras, M.S.C.; Abel, S. Comparative analysis of Arabidopsis UGT74 glucosyltransferases reveals a special role of UGT74C1 in glucosinolate biosynthesis. *Plant J.* **2014**, *79*, 92–105. [CrossRef]
21. Piotrowski, M.; Schemenewitz, A.; Lopukhina, A.; Müller, A.; Janowitz, T.; Weiler, E.W.; Oecking, C. Desulfoglucosinolate sulfotransferases from *Arabidopsis thaliana* catalyze the final step in the biosynthesis of the glucosinolate core structure. *J. Biol. Chem.* **2004**, *279*, 50717–50725. [CrossRef]
22. Hirschmann, F.; Krause, F.; Baruch, P.; Chizhov, I.; Mueller, J.W.; Manstein, D.J.; Papenbrock, J.; Fedorov, R. Structural and biochemical studies of sulphotransferase 18 from *Arabidopsis thaliana* explain its substrate specificity and reaction mechanism. *Sci. Rep.* **2017**, *7*, 4160. [CrossRef] [PubMed]
23. Burow, M.; Atwell, S.; Francisco, M.; Kerwin, R.E.; Halkier, B.A.; Kliebenstein, D.J. The Glucosinolate Biosynthetic Gene AOP2 Mediates Feed-back Regulation of Jasmonic Acid Signaling in Arabidopsis. *Mol. Plant* **2015**, *8*, 1201–1212. [CrossRef] [PubMed]
24. Kong, W.; Li, J.; Yu, Q.; Cang, W.; Xu, R.; Wang, Y.; Ji, W. Two Novel Flavin-Containing Monooxygenases Involved in Biosynthesis of Aliphatic Glucosinolates. *Front. Plant Sci.* **2016**, *7*, 1292. [CrossRef] [PubMed]
25. Gigolashvili, T.; Engqvist, M.; Yatusevich, R.; Müller, C.; Flügge, U.-I. HAG2/MYB76 and HAG3/MYB29 exert a specific and coordinated control on the regulation of aliphatic glucosinolate biosynthesis in *Arabidopsis thaliana*. *New Phytol.* **2008**, *177*, 627–642. [CrossRef]
26. Sønderby, I.E.; Burow, M.; Rowe, H.C.; Kliebenstein, D.J.; Halkier, B.A. A complex interplay of three R2R3 MYB transcription factors determines the profile of aliphatic glucosinolates in Arabidopsis. *Plant Physiol.* **2010**, *153*, 348–363. [CrossRef]
27. Frerigmann, H.; Gigolashvili, T. MYB34, MYB51, and MYB122 distinctly regulate indolic glucosinolate biosynthesis in *Arabidopsis thaliana*. *Mol. Plant* **2014**, *7*, 814–828. [CrossRef]
28. Mitreiter, S.; Gigolashvili, T. Regulation of glucosinolate biosynthesis. *J. Exp. Bot.* **2021**, *72*, 70–91. [CrossRef]
29. Zhang, Y.; Li, B.; Huai, D.; Zhou, Y.; Kliebenstein, D.J. The conserved transcription factors, MYB115 and MYB118, control expression of the newly evolved benzoyloxy glucosinolate pathway in *Arabidopsis thaliana*. *Front. Plant Sci.* **2015**, *6*, 343. [CrossRef]
30. Bischoff, A.; Trémulot, S. Differentiation and adaptation in *Brassica nigra* populations: Interactions with related herbivores. *Oecologia* **2011**, *165*, 971–981. [CrossRef] [PubMed]
31. Obi, R.K.; Nwanebu, F.C.; Ndubuisi, U.U.; Orji, N.M. Antibacterial qualities and phytochemical screening of the oils of *Curcubita pepo* and *Brassica nigra*. *J. Med. Plants Res.* **2009**, *3*, 429–432.
32. Alam, M.B.; Hossain, M.S.; Haque, M.E. Antioxidant and anti-inflammatory activities of the leaf extract of *Brassica nigra*. *Int. J. Pharmaceut. Sci. Res.* **2011**, *2*, 303–310.
33. Song, X.; Wei, Y.; Xiao, D.; Gong, K.; Sun, P.; Ren, Y.; Yuan, J.; Wu, T.; Yang, Q.; Li, X.; et al. *Brassica carinata* genome characterization clarifies U's triangle model of evolution and polyploidy in *Brassica*. *Plant Physiol.* **2021**. [CrossRef] [PubMed]
34. Cheng, F.; Liang, J.; Cai, C.; Cai, X.; Wu, J.; Wang, X. Genome sequencing supports a multi-vertex model for Brassiceae species. *Curr. Opin. Plant Biol.* **2017**, *36*, 79–87. [CrossRef] [PubMed]
35. Sharma, S.; Padmaja, K.L.; Gupta, V.; Paritosh, K.; Pradhan, A.K.; Pental, D. Two plastid DNA lineages—Rapa/Oleracea and Nigra–within the tribe Brassiceae can be best explained by reciprocal crosses at hexaploidy: Evidence from divergence times of the plastid genomes and R-block genes of the A and B genomes of *Brassica juncea*. *PLoS ONE* **2014**, *9*, e93260. [CrossRef] [PubMed]
36. Volden, J.; Borge, G.I.A.; Hansen, M.; Wicklund, T.; Bengtsson, G.B. Processing (blanching, boiling, steaming) effects on the content of glucosinolates and antioxidant-related parameters in cauliflower (*Brassica oleracea* L. ssp. *botrytis*). *LWT Food Sci. Technol.* **2009**, *42*, 63–73. [CrossRef]
37. Wang, J.; Yu, H.; Zhao, Z.; Sheng, X.; Shen, Y.; Gu, H. Natural Variation of Glucosinolates and Their Breakdown Products in Broccoli (*Brassica oleracea* var. *italica*) Seeds. *J. Agric. Food Chem.* **2019**, *67*, 12528–12537. [CrossRef]
38. Yang, B.; Quiros, C.F. Survey of glucosinolate variation in leaves of *Brassica rapa* crops. *Genet. Resour. Crop Evol.* **2010**, *57*, 1079–1089. [CrossRef]

39. Mnzava, N.A.; Olson, K. Studies on tropical vegetables. Part 1: Seed amino, fatty acid and glucosinolate profile of Ethiopian mustards (*Brassica carinata* Braun). *Food Chem.* **1990**, *35*, 229–235. [CrossRef]
40. Rangkadilok, N.; Nicolas, M.E.; Richard, B.N.; Premier, R.R.; Eagling, D.R.; Taylor, P.W.J. Developmental changes of sinigrin and glucoraphanin in threee *Brassica* species (*Brassica nigra*, *Brassica juncea* and *Brassica oleracea* var. *italica*). *Sci. Hortic.* **2002**, *96*, 11–26. [CrossRef]
41. Bellostas, N.; Sørensen, J.C.; Sørensen, H. Profiling glucosinolates in vegetative and reproductive tissues of four Brassica species of the U-triangle for their biofumigation potential. *J. Sci. Food Agric.* **2007**, *87*, 1586–1594. [CrossRef]
42. Wang, H.; Wu, J.; Sun, S.; Liu, B.; Cheng, F.; Sun, R.; Wang, X. Glucosinolate biosynthetic genes in *Brassica rapa*. *Gene* **2011**, *487*, 135–142. [CrossRef]
43. Liu, S.; Liu, Y.; Yang, X.; Tong, C.; Edwards, D.; Parkin, I.A.P.; Zhao, M.; Ma, J.; Yu, J.; Huang, S.; et al. The *Brassica oleracea* genome reveals the asymmetrical evolution of polyploid genomes. *Nat. Commun.* **2014**, *5*, 3930. [CrossRef] [PubMed]
44. Yang, J.; Liu, D.; Wang, X.; Ji, C.; Cheng, F.; Liu, B.; Hu, Z.; Chen, S.; Pental, D.; Ju, Y.; et al. The genome sequence of allopolyploid Brassica juncea and analysis of differential homoeolog gene expression influencing selection. *Nat. Genet.* **2016**, *48*, 1225–1232. [CrossRef]
45. Zhu, B.; Yang, J.; He, Y.; Zang, Y.; Zhu, Z. Glucosinolate Accumulation and Related Gene Expression in Pak Choi (*Brassica rapa* L. ssp. *chinensis* var. *communis* N. Tsen & S.H. Lee Hanelt) in Response to Insecticide Application. *J. Agric. Food Chem.* **2015**, *63*, 9683–9689.
46. Yi, G.-E.; Robin, A.H.K.; Yang, K.; Park, J.-I.; Kang, J.-G.; Yang, T.-J.; Nou, I.-S. Identification and expression analysis of glucosinolate biosynthetic genes and estimation of glucosinolate contents in edible organs of *Brassica oleracea* subspecies. *Molecules* **2015**, *20*, 13089–13111. [CrossRef]
47. Roshan, K.; Arya, G.C.; Bisht, N.C. Differential expression and interaction specificity of the heterotrimeric G-protein family in Brassica nigra reveal their development- and condition-specific role. *Plant Cell Physiol.* **2014**, *11*, 1954–1968.
48. Tong, C.; Wang, X.; Yu, J.; Wu, J.; Li, W.; Huang, J.; Dong, C.; Hua, W.; Liu, S. Comprehensive analysis of RNA-seq data reveals the complexity of the transcriptome in *Brassica rapa*. *BMC Genomics* **2013**, *14*, 689. [CrossRef]
49. Yu, J.; Tehrim, S.; Zhang, F.; Tong, C.; Huang, J.; Cheng, X.; Dong, C.; Zhou, Y.; Qin, R.; Hua, W.; et al. Genome-wide comparative analysis of NBS-encoding genes between *Brassica* species and *Arabidopsis thaliana*. *BMC Genomics* **2014**, *15*, 3. [CrossRef]
50. Nour-Eldin, H.H.; Andersen, T.G.; Burow, M.; Madsen, S.R.; Jørgensen, M.E.; Olsen, C.E.; Dreyer, I.; Hedrich, R.; Geiger, D.; Halkier, B.A. NRT/PTR transporters are essential for translocation of glucosinolate defence compounds to seeds. *Nature* **2012**, *488*, 531–534. [CrossRef]
51. Nambiar, D.M.; Kumari, J.; Augustine, R.; Kumar, P.; Bajpai, P.K.; Bisht, N. GTR1 and GTR2 transporters differentially regulate tissue-specific glucosinolate contents and defence responses in the oilseed crop *Brassica juncea*. *Plant Cell Environ.* **2021**. [CrossRef] [PubMed]
52. Benderoth, M.; Textor, S.; Windsor, A.J.; Mitchell-Olds, T.; Gershenzon, J.; Kroymann, J. Positive selection driving diversification in plant secondary metabolism. *Proc. Natl. Acad. Sci. USA* **2006**, *103*, 9118–9123. [CrossRef] [PubMed]
53. Benderoth, M.; Pfalz, M.; Kroymann, J. Methylthioalkylmalate synthases: Genetics, ecology and evolution. *Phytochem. Rev.* **2009**, *8*, 255–268. [CrossRef]
54. Hansen, N.; Ostermeier, A. Completely Derandomized Self-Adaptation in Evolution Strategies. *Evol. Comput.* **2001**, *9*, 159–195. [CrossRef] [PubMed]
55. Chen, S.; Glawischnig, E.; Jørgensen, K.; Naur, P.; Jørgensen, B.; Olsen, C.-E.; Hansen, C.H.; Rasmussen, H.; Pickett, J.A.; Halkier, B.A. CYP79F1 and CYP79F2 have distinct functions in the biosynthesis of aliphatic glucosinolates in Arabidopsis. *Plant J.* **2003**, *33*, 923–937. [CrossRef]
56. Miao, H.; Wei, J.; Zhao, Y.; Yan, H.; Sun, B.; Huang, J.; Wang, Q. Glucose signalling positively regulates aliphatic glucosinolate biosynthesis. *J. Exp. Bot.* **2013**, *64*, 1097–1109. [CrossRef] [PubMed]
57. Mikkelsen, M.D.; Halkier, B.A. Metabolic engineering of valine- and isoleucine-derived glucosinolates in Arabidopsis expressing CYP79D2 from Cassava. *Plant Physiol.* **2003**, *131*, 773–779. [CrossRef]
58. Zhao, Y.; Hull, A.K.; Gupta, N.R.; Goss, K.A.; Alonso, J.; Ecker, J.R.; Normanly, J.; Chory, J.; Celenza, J.L. Trp-dependent auxin biosynthesis in *Arabidopsis*: Involvement of cytochrome P450s CYP79B2 and CYP79B3. *Genes Dev.* **2003**, *16*, 3100–3112. [CrossRef]
59. Zang, Y.-X.; Kim, D.-H.; Park, B.-S.; Hong, S.-B. Metabolic engineering of indole glucosinolates in Chinese cabbage hairy roots expressing Arabidopsis CYP79B2, CYP79B3, and CYP83B1. *Biotechnol. Bioprocess Eng.* **2009**, *14*, 467–473. [CrossRef]
60. Kliebenstein, D.J.; Kroymann, J.; Brown, P.; Figuth, A.; Pedersen, D.; Gershenzon, J.; Mitchell-Olds, T. Genetic Control of Natural Variation in Arabidopsis Glucosinolate Accumulation. *Plant Physiol.* **2001**, *126*, 811–825. [CrossRef]
61. Jensen, L.M.; Kliebenstein, D.J.; Burow, M. Investigation of the multifunctional gene *AOP3* expands the regulatory network fine-tuning glucosinolate production in *Arabidopsis*. *Front. Plant Sci.* **2015**, *6*, 762. [CrossRef]
62. Vallejo, F.; Tomás-Barberán, F.A.; Benavente-García, A.G.; García-Viguera, C. Total and individual glucosinolate contents in inflorescences of eight broccoli cultivars grown under various climatic and fertilisation conditions. *J. Sci. Food Agric.* **2003**, *83*, 307–313. [CrossRef]
63. Schonhof, I.; Krumbein, A.; Brückner, B. Genotypic effects on glucosinolates and sensory properties of broccoli and cauliflower. *Nahrung/Food* **2004**, *48*, 25–33. [CrossRef]

64. Farnham, M.W.; Wilson, P.E.; Stephenson, K.K.; Fahey, J.W. Genetic and environmental effects on glucosinolate content and chemoprotective potency of broccoli. *Plant Breed.* **2004**, *123*, 60–65. [CrossRef]
65. Zang, Y.-X.; Kim, H.U.; Kim, J.A.; Lim, M.-H.; Jin, M.; Lee, S.C.; Kwon, S.-J.; Lee, S.-I.; Hong, J.K.; Park, T.-H.; et al. Genome-wide identification of glucosinolate synthesis genes in *Brassica rapa*. *FEBS J.* **2010**, *276*, 3559–3574. [CrossRef]
66. Lysak, M.A.; Koch, M.A.; Pecinka, A.; Schubert, I. Chromosome triplication found across the tribe *Brassiceae*. *Genome Res.* **2005**, *15*, 516–525. [CrossRef]
67. Wang, X.; Wang, H.; Wang, J.; Sun, R.; Wu, J.; Liu, S.; Bai, Y.; Mun, J.-H.; Bancroft, I.; Cheng, F.; et al. The genome of the mesopolyploid crop species *Brassica rapa*. *Nat. Genet.* **2011**, *43*, 1035–1039. [CrossRef] [PubMed]
68. Li, J.; Hansen, B.G.; Ober, J.A.; Kliebenstein, D.J.; Halkier, B.A. Subclade of flavin-monooxygenases involved in aliphatic glucosinolate biosynthesis. *Plant Physiol.* **2008**, *148*, 1721–1733. [CrossRef] [PubMed]
69. Edger, P.P.; Pires, J.C. Gene and genome duplications: The impact of dosage-sensitivity on the fate of nuclear genes. *Chromosome Res.* **2009**, *17*, 699–717. [CrossRef] [PubMed]
70. Freeling, M. Bias in plant gene content following different sorts of duplication: Tandem, whole-genome, segmental, or by transposition. *Annu. Rev. Plant Biol.* **2009**, *60*, 433–453. [CrossRef]
71. Wang, J.; Qiu, Y.; Wang, X.; Yue, Z.; Yang, X.; Chen, X.; Zhang, X.; Shen, D.; Wang, H.; Song, J.; et al. Insights into the species-specific metabolic engineering of glucosinolates in radish (*Raphanus sativus* L.) based on comparative genomic analysis. *Sci. Rep.* **2017**, *7*, 16040. [CrossRef] [PubMed]
72. Chen, X. Glucosinolates in Chinese *Brassica campestris* Vegetables: Chinese Cabbage, Purple Cai-tai, Choysum, Pakchoi, and Turnip. *Hortscience* **2008**, *43*, 571–574. [CrossRef]

horticulturae

Article

Evaluation of Effect of Brassinolide in *Brassica juncea* Leaves under Drought Stress in Field Conditions

Naveen Naveen [1], Nisha Kumari [1,*], Ram Avtar [2], Minakshi Jattan [3], Sushil Ahlawat [4], Babita Rani [1], Kamla Malik [5], Anubhuti Sharma [6] and Manjeet Singh [2,*]

1 Department of Biochemistry, CCS Haryana Agricultural University, Hisar 125004, Haryana, India; jangranaveen4321@gmail.com (N.N.); babitachahalkharb@gmail.com (B.R.)
2 Oilseeds Section, Department of Genetics and Plant Breeding, CCS Haryana Agricultural University, Hisar 125004, Haryana, India; ramavtar0706@gmail.com
3 Cotton Section, Department of Genetics and Plant Breeding, CCS Haryana Agricultural University, Hisar 125004, Haryana, India; jattanmina@gmail.com
4 Department of Entomology, CCS Haryana Agricultural University, Hisar 125004, Haryana, India; sushilahlawat08@gmail.com
5 Department of Microbiology, CCS Haryana Agricultural University, Hisar 125004, Haryana, India; kamlamalik06@gmail.com
6 ICAR-Directorate of Rapeseed-Mustard Research, Sewar, Bharatpur 321303, Rajasthan, India; sharmaanubhuti98@gmail.com
* Correspondence: nishaahlawat211@gmail.com (N.K.); manjeetsingh125033@gmail.com (M.S.); Tel.: +91-9468117723 (N.K.); +91-9053127937 (M.S.)

Citation: Naveen, N.; Kumari, N.; Avtar, R.; Jattan, M.; Ahlawat, S.; Rani, B.; Malik, K.; Sharma, A.; Singh, M. Evaluation of Effect of Brassinolide in *Brassica juncea* Leaves under Drought Stress in Field Conditions. *Horticulturae* **2021**, *7*, 514. https://doi.org/10.3390/horticulturae7110514

Academic Editors: Xiaowu Wang, Jian Wu and Xu Cai

Received: 7 September 2021
Accepted: 17 November 2021
Published: 22 November 2021

Publisher's Note: MDPI stays neutral with regard to jurisdictional claims in published maps and institutional affiliations.

Copyright: © 2021 by the authors. Licensee MDPI, Basel, Switzerland. This article is an open access article distributed under the terms and conditions of the Creative Commons Attribution (CC BY) license (https://creativecommons.org/licenses/by/4.0/).

Abstract: Drought stress is considered to be a major factor responsible for reduced agricultural productivity, because it is often linked to other major abiotic stresses, such as salinity and heat stress. Understanding drought-tolerance mechanisms is important for crop improvement. Moreover, under drought conditions, it is possible that growth regulators are able to protect the plants. Brassinosteroids not only play a regulatory role in plant growth, but also organize defense mechanisms against various tresses. This study aimed to evaluate the effect of brassinolide on physio-biochemical amendment in two contrasting cultivars (drought-tolerant RH 725, and drought-sensitive RH 749) of *Brassica juncea* under drought stress. Two foliar sprayings with brassinolide (10 and 20 mg/L) were carried out in both cultivars (RH 725 and RH 749) at two stages—i.e., flower initiation, and 50% flowering—under stress conditions. The results clearly revealed that the activities of antioxidative enzymes and non-enzymatic antioxidants (carotenoids, ascorbic acid, and proline) increased significantly in RH 725 at 50% flowering, whereas 20 mg/L of brassinolide showed the most promising response. The different oxidative stress indicators (i.e., hydrogen peroxide, malondialdehyde, and electrolyte leakage) decreased to a significant extent at 20 mg/L of brassinolide spray in RH 725 at 50% flowering. This study indicates that brassinolide intensifies the physio-biochemical attributes by improving the antioxidant system and photosynthetic efficiency in RH 725 at 50% flowering. It is assumed that enhanced production of proline, improvement of the antioxidant system, and reduction in the amount of stress indicators impart strength to the plants to combat the stress conditions.

Keywords: antioxidants; *Brassica juncea*; brassinolide; drought stress; proline

1. Introduction

Rapeseed mustard comprises an important group of oilseed Brassica crops. In this group, Indian mustard [*Brassica juncea* (L.) Czern & Coss.] is an important edible, oil-yielding crop covering about 90% of the cultivated area under brassica oilseeds in India [1]. It is the third-largest source of vegetable oil in the world, after soybeans and palm oil. Indian mustard has the potential for quicker seed germination, high productivity, and heat and drought tolerance, along with enhanced insect and disease resistance if sown on time [2], whereas late sowing exposes the crop to abiotic and biotic stresses. There is a dire need to intensify in the production of food crops but, on the other hand, environmental stresses (biotic and abiotic) suppress the overall yield of agricultural crops. Drought stress is recognized as the main factor leading to the decline in agricultural productivity, because drought is persistently related to other major abiotic stresses, such as high-temperature stress and salinity [3]. It is estimated that by the end of the 21st century, the proportion of drought-prone areas will have doubled.

Reactive oxygen species (ROS) are continuously generated in plant mitochondria, plastids, peroxisomes, apoplasts, and cytosol as byproducts of different cellular metabolic pathways, and they hinder photosynthesis. The enhanced production of alkoxy radicals (RO^-), superoxide radicals (O^{2-}), perhydroxy radicals (HO^{2-}), hydrogen peroxide (H_2O_2), singlet oxygen ($_1O^2$), and hydroxyl radicals (OH^-) is a common end result of plant-rearing under different abiotic stresses [4,5]. The production of reactive oxygen species is the basis for oxidative stress, damaging plants by oxidizing membrane lipids, nucleic acids, proteins, and photosynthetic pigments [6]. The ROS hamper the plants' photosynthesis and enzymes of the Calvin cycle, altering chlorophyll components and causing damage to the photosynthetic apparatus [7]. To survive under such intense environmental conditions, and to increase their tolerance, plants have developed many intricate defense mechanisms. Stress tolerance in plants necessitates the activation of complex metabolic activities, including antioxidative pathways—especially ROS-scavenging systems within the cells that, in turn, can contribute to continued plant growth under stress conditions [8]. The different antioxidative enzymes in plants—such as superoxide dismutase (SOD), catalase (CAT), and peroxidase (POX)—scavenge these ROS molecules [9]; however, oxidative stress is generated in plants if there is an imbalance in ROS [10]. The resistance to oxidative stress relies on the overall balance between the fabrication of ROS and the antioxidant capability of the cells [11].

The drought-tolerance mechanism in plants also includes some plant growth regulators and secondary metabolites, such as auxin, abscisic acid (ABA), jasmonic acid, plant steroids, and ethylene. Among the variety of compounds used to alleviate plant stress, brassinosteroids (BRs) are considered to be plant hormones that regulate plant growth and productivity. Brassinosteroids (BRs) are polyhydroxylated steroidal plant hormones that play an essential role in the regulation of plant growth and development processes. Myriad studies have highlighted that these are crucial for regulating a range of physiological processes, such as cell proliferation, expansion, male fertility, senescence, leaf development, and vascular differentiation. These compounds have a wide range of biological activities, providing unique possibilities to increase crop yields by altering plant metabolism and protecting plants from environmental stresses [12]. The research conducted thusfar shows that BRs cause a wide range of morphological and physiological responses in plants [13,14]. In addition, BRs are known as regulators of transcription and translation mechanisms, by which they improve the levels of total proteins and enzymes [15], as well as increasing the seed yield at harvest [16]. BRs not only play a regulatory role in plant growth, but also participate in the establishment of defense mechanisms to deal with various biotic and abiotic stresses [13]. Several BRs with brassinolide as the main component have been evaluated in the field, and they have significantly increased crop yields. Exogenous application of BRs has improved tolerance to salinity [17], drought [16,18], high/low temperatures, and heavy metals [13]. There are few reports on the role of brassinosteroids in the unveiling of genes and metabolic pathways that confer drought resistance to Indian mustard [16]. However, the data that are currently available on the role of BRs in plant drought response, from the few studies that have been performed with genotypes of known drought sensitivity, are not very conclusive [19–21]. Any comparison of the impact of exogenously applied BRs on drought-tolerant/sensitive genotypes should reveal the BR-induced changes—particularly in the sensitive genotypes, because the tolerant genotypes should experience less intensive drought effects. This should be similar to the situation observed for BRs exogenously applied to plants exposed to drought ranging from mild/moderate to less intense; BRs always have a greater effect on more strongly stressed plants. There have also been some cases where the drought-tolerant genotype showed a more pronounced response to BRs than the drought-sensitive one, as has been reported in several previous studies. Thus, the situation is not so simple, and probably depends on plant species as well as on a mechanism that is responsible for the drought resistance/sensitivity of the respective genotype. In this context, the present study was designed with the objectives to address the following questions: (1) whether exogenous application of BRs could alleviate drought stress in Indian mustard, and (2) whether the drought-tolerant and drought-sensitive cultivars have similar responses to these treatments under drought stress.

2. Materials and Methods

2.1. Plant Materials

Two Indian mustard [*Brassica juncea* (L.) Czern & Coss.] cultivars—drought-tolerant (RH 725) and drought-sensitive (RH 749)—were used in this study. The two cultivars were sown in the Research Farm of Oilseeds Section, Department of Genetics and Plant Breeding, CCS Haryana Agricultural University, Hisar, in a randomized complete block design (RCBD) with three replications. The CCS HAU, Hisar, is situated at a latitude of 29°10′ N, longitude of 75°46′ E, and altitude 215.2 m above main sea level, and falls in the semi-tropical region of the western zone of India. Drought conditions were achieved by withholding irrigation from the crop. The weather data (rainfall) during crop season are presented in Supplementary Table S1, which shows that rainfall was negligible during the crop-growing period and the drought conditions were adequate for this study. Two foliar sprayings of brassinolide at 10 and 20 mg/L concentrations were carried out in both of the cultivars at two growth stages—i.e., flower initiation (42 days after sowing; DAS) and 50% flowering (52 DAS)—with water spray as a control. All physio-biochemical analysis was carried out on leaves, which were taken two days after each spray.

2.2. Physiological Parameters

Parameters such as photosynthetic rate, stomatal conductance, and transpiration rate were measured using an infrared gas analyzer (IRGA) system (LI-COR USA Model LI6400, LE, USA) as per the method employed by Silva et al. [22], and their corresponding units are as follows: photosynthetic rate (PR, μmol CO_2 m^{-2} s^{-1}), stomatal conductance (SC, mole H_2O m^{-2} s^{-1}), and transpiration rate (TR, mmol H_2O m^{-2} s^{-1}).

2.3. Enzyme Extraction and Assay

Leaves of both cultivars of *B. juncea*, after two days of each spray, were used for enzymatic studies. Extraction was carried out at 4 °C, and standardized extraction conditions with respect to the molarity and pH of the buffer were maintained in order to achieve maximum enzyme activity. One gram of leaf sample was macerated in a chilled pestle and mortar in the presence of 4 mL of 0.1 M potassium phosphate buffer (pH 7.0). The homogenate was centrifuged at 12,000× *g* rpm for 30 min in a refrigerated centrifuge at 4 °C. The supernatant was carefully decanted and used for the enzyme assay.

The SOD activity was determined by quantifying the ability of the enzyme to inhibit the photochemical reduction of nitro blue tetrazolium (NBT) to formazan [23]. One enzyme unit was defined as the amount of enzyme that could cause 50% inhibition of the photochemical reaction. The catalase activity was measured by following the method of Sinha [24]. One unit of enzyme activity was defined as the amount of enzyme required to consume 1 μmol H_2O_2 per minute under assay conditions. The POX activity was assayed by adopting the method of Shannon et al. [25]. One unit of enzyme activity was equivalent to one μmol of H_2O_2 oxidized per minute.

2.4. Non-Enzymatic Estimations

For extraction of carotenoids, 30 mg of the fresh leaves was cut into small discs and dipped in test tubes containing 3 mL of dimethyl sulfoxide (DMSO). The tubes were kept at room temperature overnight. The carotenoids extracted in the DMSO were estimated by the method of Hiscox and Israelstam [26]. Ascorbic acid was extracted from the leaves via homogenization in 5 mL of 5% (*w/v*) metaphosphoric acid in glacial acetic acid, and the homogenate was centrifuged at 10,000× *g* rpm for 25 min. The supernatant thus obtained was used for the estimation of ascorbic acid. Ascorbic acid content was estimated by the method of Roe [27]. For proline estimation, 1g of tissue was homogenized in 5 mL of sulfosalicylic acid (3%) and centrifuged at 10,000× *g* rpm for 25 min; the supernatant thus obtained was used for the estimation of proline content. Proline content was estimated by using the method of Bates et al. [28].

2.5. Extraction and Estimation of Oxidative Stress Indicators

For the extraction of H_2O_2 and MDA, 1.0 g of leaves from each treatment was taken and ground in 6 mL of chilled 0.8 N $HClO_4$ and centrifuged at 10,000× *g* rpm for 30 min. The clear supernatant thus obtained was used for further estimation. Hydrogen peroxide was estimated by the method of Sinha [24]. Malondialdehyde was estimated according to the method of Heath and Packer [29].The relative intactness of the plasma membrane was measured as the leakage percentage of electrolytes, as described by Gong et al. [30].

2.6. Statistical Analysis

Three-way ANOVA was applied to test the statistical significance of the treatments. Duncan'smultiple range test (DMRT) was applied for multiple comparisons of treatments' mean values. Pearson's product–moment correlation was used to test the relationships between the antioxidant parameters. All statistical analyses were performed using the OP STAT statistical software developed by CCS HAU, Hisar, India. Graphs were prepared using Microsoft Excel, 2013.

3. Results

3.1. Physiological Parameters

The stomatal conductance, photosynthetic rate, and transpiration rate were significantly influenced by cultivar, growth stage, brassinolide concentration, and their interactions (Table 1). The brassinolide foliar application (10 and 20 mg/L) increased the stomatal conductance, photosynthetic rate, and transpiration rate, and these increases had the greatest statistical significance in the drought-tolerant cultivar RH 725 as compared to drought-sensitive RH 749 at both the growth stage and the 50% flowering stage. The photosynthetic parameters at the 50% flowering stage were high with brassinolide (10 and 20 mg/L) spray in both cultivars (RH 725 and RH 749). Brassinolide (20 mg/L) enhanced the stomatal conductance, photosynthetic rate, and transpiration rate by 47.16, 40.92, and 31.06% at the flower initiation stage and 50.72, 46.04, and 33.76% at the 50% flowering stage, respectively, in RH 725. This enhancement was low in drought-sensitive RH 749 as comparison to drought-tolerant RH 725 (Figures 1–3).

Table 1. Analysis of variance in the effects of cultivars (C), sampling time points (ST), and different brassinolide concentrations (BC),as well as their interactions, on the superoxide dismutase activity (SOD), peroxidase activity (POX), catalase activity (CAT), ascorbic acid (ASA), proline (PRO), carotenoids (CC), hydrogen peroxide (H_2O_2), and malondialdehyde (MDA) content, as well as electrolyte leakage (EL), stomatal conductance (SC), photosynthetic rate (PR), and transpiration rate (TR).

SV	Df	Mean Squares											
		SOD	POX	CAT	ASA	PRO	CC	H_2O_2	MDA	EL	SC	PR	TR
C	1	1335.17 **	61.96 **	27,408.11 **	1333.49 **	26.87 **	22.14 **	38,037.56 **	59.50 **	225.91 **	0.65 **	147.87 **	5.48 **
ST	1	1400.01 **	18.30 **	6702.90 **	2295.71 **	46.10 **	18.16 **	24,628.80 **	44.94 **	2334.83 **	0.29 **	98.01 **	59.78 **
C × ST	1	190.35 **	0.40	88.59 **	19.90 **	0.61	0.79	150.79 **	8.05 **	8.17 **	0.01 **	12.39 **	0.08 **
BC	2	137.58 **	31.86 **	5914.39 **	1068.52 **	11.94 **	7.58 **	18,979.63 **	178.78 **	242.52 **	0.14 **	98.06 **	6.72 **
C × BC	2	14.73 **	1.79	320.08 **	33.10 **	0.51	0.47	178.68 **	0.84	4.07 *	0.02 **	8.20 **	0.20 **
ST × BC	2	15.35 **	0.48	192.53 **	6.57 **	0.63	0.48	124.60 **	5.13 **	1.26	0.01 **	2.13 **	0.54 **
C × ST × BC	2	1.95 *	0.01	2.95 *	0.47	0.00	0.12	94.83 **	1.09	0.29	0.00	0.46 **	0.06 **
Error	22	0.39	0.90	0.61	0.99	0.53	0.82	0.84	0.83	0.78	0.00	0.02 **	0.01 **

** Significant at $p \leq 0.01$; * significant at $p \leq 0.05$.

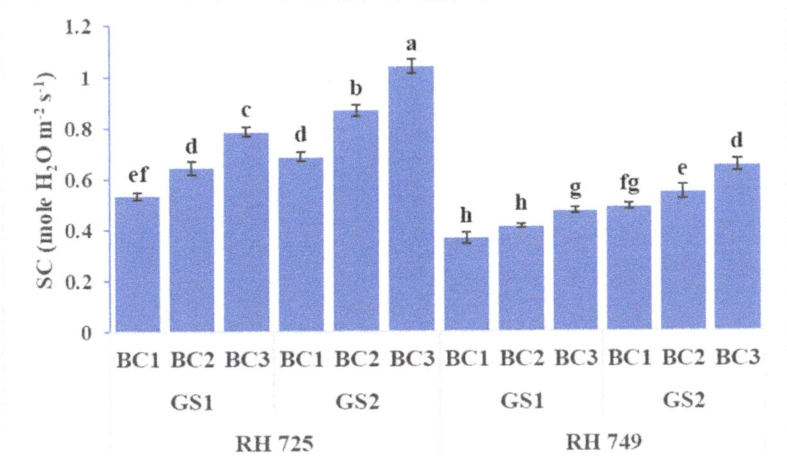

Figure 1. Comparisons of the effects of different concentrations of brassinolide sprays (BC1: control (water spray); BC2: 10 mg BRs/L water; BC3: 20 mg BRs/L water) on the stomatal conductance (SC) of drought-tolerant (RH 725) and drought-sensitive (RH 749) Indian mustard cultivars at the flower initiation (GS1) and 50% flowering (GS2) stages. Columns marked by different letters indicate significant differences ($p < 0.05$) between treatments based on Duncan's multiple range test. Error bars denote the standard errors of the mean.

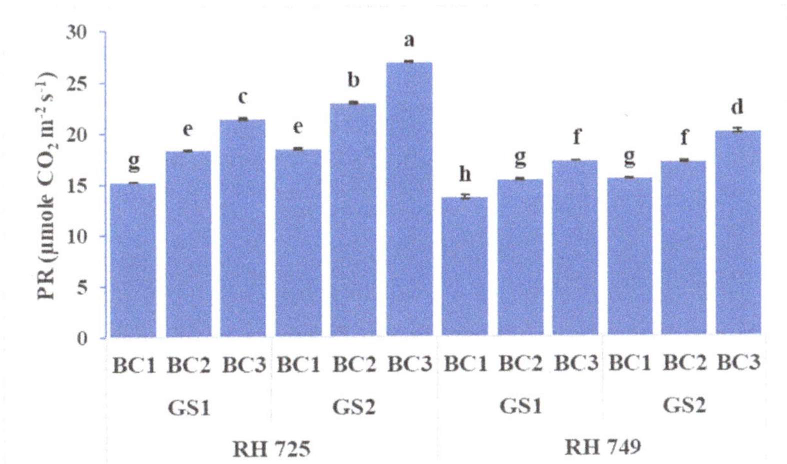

Figure 2. Comparisons of the effects of different concentrations of brassinolide sprays (BC1: control (water spray); BC2: 10 mg BRs/L water; BC3: 20 mg BRs/L water) on the photosynthetic rate (PR) of drought-tolerant (RH 725) and drought-sensitive (RH 749) Indian mustard cultivars at the flower initiation (GS1) and 50% flowering (GS2) stages. Columns marked by different letters indicate significant differences ($p < 0.05$) between treatments based on Duncan's multiple range test. Error bars denote the standard errors of the mean.

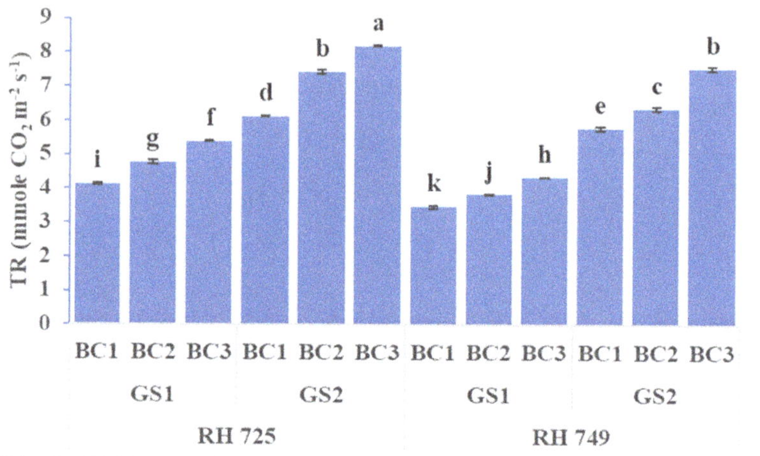

Figure 3. Comparisons of the effects of different concentrations of brassinolide sprays (BC1: control (water spray); BC2: 10 mg BRs/L water; BC3: 20 mg BRs/L water) on the transpiration rate (TR) of drought-tolerant (RH 725) and drought-sensitive (RH 749) Indian mustard cultivars at the flower initiation (GS1) and 50% flowering (GS2) stages. Columns marked by different letters indicate significant differences ($p < 0.05$) between treatments based on Duncan's multiple range test. Error bars denote the standard errors of the mean.

3.2. Oxidative Stress Indicators

Three-way ANOVA (Table 1) showed highly significant effects of all the three individual factors and their interactions on oxidative stress indicators, except that C × BC and C × GS × BC were insignificant for malondialdehyde, while GS × BC and C × GS × BC were insignificant for electrolyte leakage. Figures 4–6 show that brassinolide treatments significantly decreased the oxidative stress indicators—i.e., hydrogen peroxide (H_2O_2), malondialdehyde (MDA), and electrolyte leakage (EL)—in Indian mustard. Brassinolide foliar application at 20 mg/L showed a maximum decrease in H_2O_2, MDA, and EL over their respective controls at the 50% flowering stage of growth in RH 725. The percentage decrease in H_2O_2, MDA, and EL caused by brassinolide (20 mg/L) was 32.14, 60.37, and 37.94%, respectively, in RH 725 at the 50% flowering stage. On the other hand, at the same plant growth stage, this decrease was 18.56%, 54.98%, and 20.68% in H_2O_2, MDA, and EL, respectively, at 20 mg/L in RH 749, which was a sensitive cultivar. It is clear from these results that the tolerant cultivar RH 725 showed a significantly better response to brassinolide as compared to the sensitive cultivar RH 749.

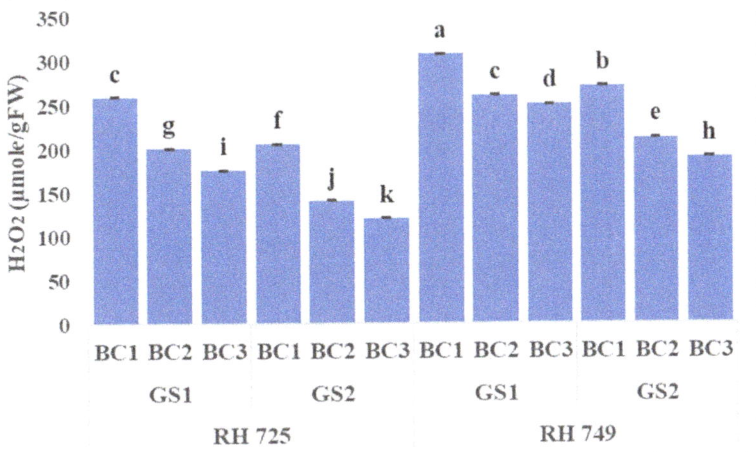

Figure 4. Comparisons of the effects of different concentrations of brassinolide sprays (BC1: control (water spray); BC2: 10 mg BRs/L water; BC3: 20 mg BRs/L water) on the hydrogen peroxidase (H_2O_2) concentration of drought-tolerant (RH 725) and drought-sensitive (RH 749) Indian mustard cultivars at the flower initiation (GS1) and 50% flowering (GS2) stages. Columns marked by different letters indicate significant differences ($p < 0.05$) between treatments based on Duncan's multiple range test. Error bars denote the standard errors of the mean.

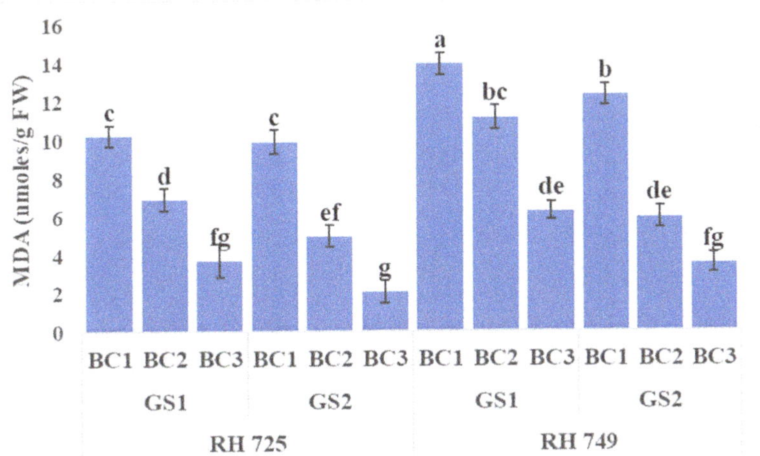

Figure 5. Comparisons of the effects of different concentrations of brassinolide sprays (BC1: control (water spray); BC2: 10 mg BRs/L water; BC3: 20 mg BRs/L water) on the malondialdehyde (MDA) concentration of drought-tolerant (RH 725) and drought-sensitive (RH 749) Indian mustard cultivars at the flower initiation (GS1) and 50% flowering (GS2) stages. Columns marked by different letters indicate significant differences ($p < 0.05$) between treatments based on Duncan's multiple range test. Error bars denote the standard errors of the mean.

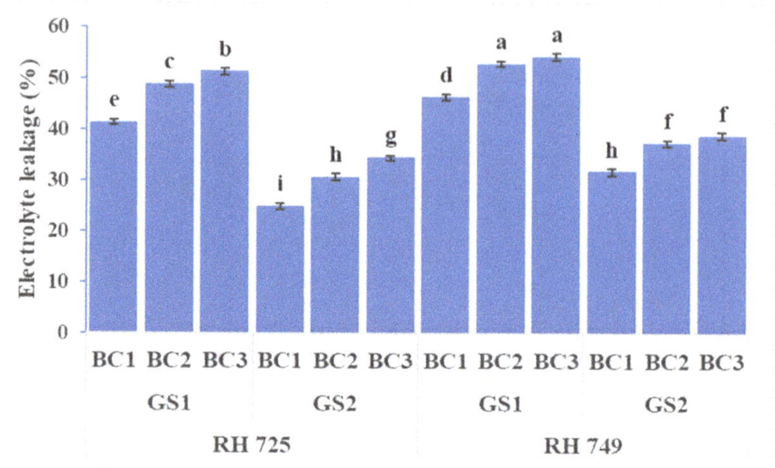

Figure 6. Comparisons of the effects of different concentrations of brassinolide sprays (BC1: control (water spray); BC2: 10 mg BRs/L water; BC3: 20 mg BRs/L water) on the electrolyte leakage (EL) concentration of drought-tolerant (RH 725) and drought-sensitive (RH 749) Indian mustard cultivars at the flower initiation (GS1) and 50% flowering (GS2) stages. Columns marked by different letters indicate significant differences ($p < 0.05$) between treatments based on Duncan's multiple range test. Error bars denote the standard errors of the mean.

3.3. *Enzymatic Antioxidants*

Three-way ANOVA revealed statistical significant effects of cultivar (C), growth stage (GS), brassinolide concentration (BC), and their interaction on SOD and CAT, whereas only the individual factors had significant effects on POX (Table 1). Brassinolide at two concentrations (10 and 20 mg/L), when applied to the plant, enhanced the activity of antioxidative enzymes (SOD, CAT, and POX) to a significant extent, but the tolerant cultivar (RH 725) showed significantly higher ($p < 0.05$) activities over the growth stages. However, all of the enzyme activities were increased significantly—particularly at the 50% flowering stage. The tolerant cultivar RH 725 exhibited the highest enzymatic activities (SOD, POX, and CAT) as compared to the sensitive cultivar RH 749 at the50% flowering stage. Brassinolide (20 mg/L) significantly enhanced the SOD, CAT, and POX activity by 31.73, 22.80, and 45.07% at the flower initiation stage and 36.92, 27.86, and 48.12% at the 50% flowering stage, respectively, in comparison to their controls in drought-tolerant RH 725, while this increase was less in the drought-sensitive cultivar RH 749 (Figures 7–9).

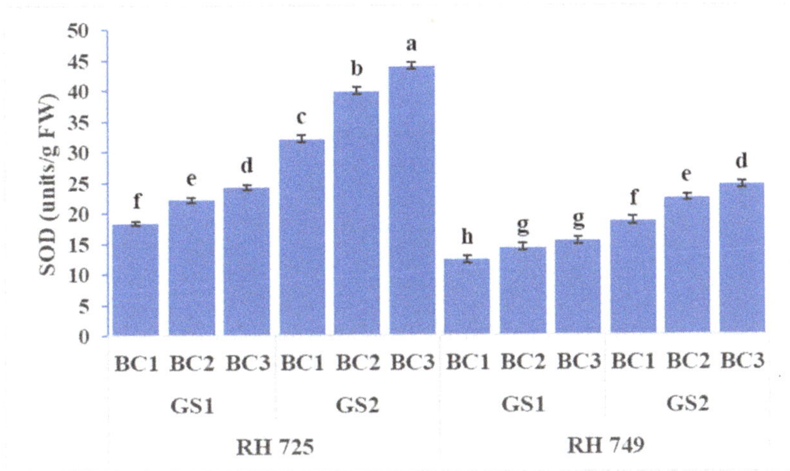

Figure 7. Comparisons of the effects of different concentrations of brassinolide sprays (BC1: control (water spray); BC2: 10 mg BRs/L water; BC3: 20 mg BRs/L water) on the superoxide dismutase (SOD) activity of drought-tolerant (RH 725) and drought-sensitive (RH 749) Indian mustard cultivars at the flower initiation (GS1) and 50% flowering (GS2) stages. Columns marked by different letters indicate significant differences ($p < 0.05$) between treatments based on Duncan's multiple range test. Error bars denote the standard errors of the mean.

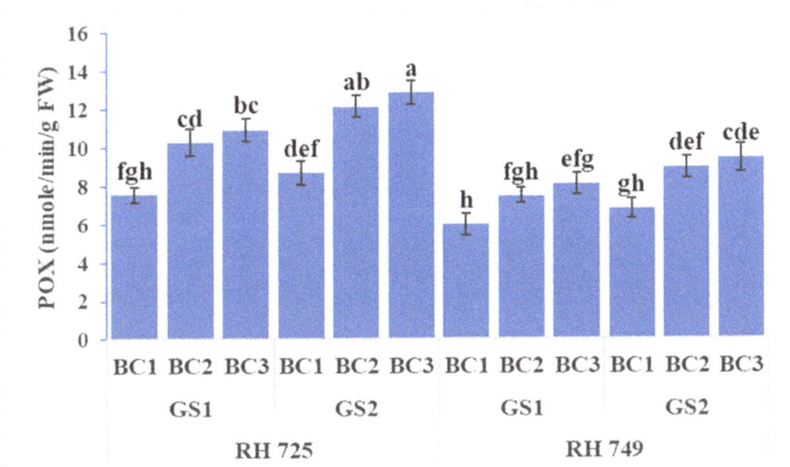

Figure 8. Comparisons of the effects of different concentrations of brassinolide sprays (BC1: control (water spray); BC2: 10 mg BRs/L water; BC3: 20 mg BRs/L water) on the peroxidase (POX) activity of drought-tolerant (RH 725) and drought-sensitive (RH 749) Indian mustard cultivars at the flower initiation (GS1) and 50% flowering (GS2) stages. Columns marked by different letters indicate significant differences ($p < 0.05$) between treatments based on Duncan's multiple range test. Error bars denote the standard errors of the mean.

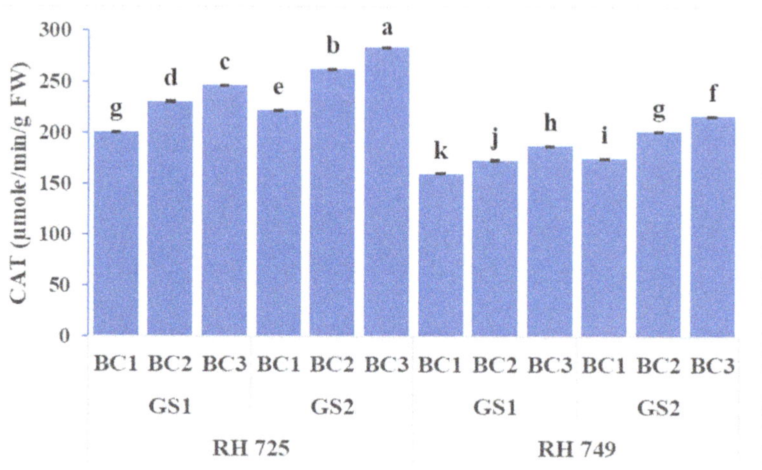

Figure 9. Comparisons of the effects of different concentrations of brassinolide sprays (BC1: control (water spray); BC2: 10 mg BRs/L water; BC3: 20 mg BRs/L water) on the catalase (CAT) activity of drought-tolerant (RH 725) and drought-sensitive (RH 749) Indian mustard cultivars at the flower initiation (GS1) and 50% flowering (GS2) stages. Columns marked by different letters indicate significant differences ($p < 0.05$) between treatments based on Duncan's multiple range test. Error bars denote the standard errors of the mean.

3.4. Non-Enzymatic Antioxidants

ANOVA showed significant effects of cultivar, growth stage, and brassinolide concentration on all three non-enzymatic antioxidants, while their interactions—viz.,C × GS, C × BC, and GS × BC—were significant for ascorbic acid only (Table 1). It is evident from Figures 10–12 that foliar application of brassinolide (10 and20 mg/L) significantly increased the non-enzymatic attributes in both cultivars, but this increase was more pronounced in the tolerant cultivar RH 725 as compared to RH 749,with increases in the levels of carotenoids, ascorbic acid, and proline (at 20 mg/L)of 27.04, 49.63, and 40.91% at the flower initiation stage and 41.55, 64.11, and 45.22% at the 50% flowering stage, respectively, over their respective controls in RH 725. Of the two concentrations of brassinolide and two stages of plant growth studied, 20 mg/L of brassinolide and the 50% flowering stage showed the greatest response in RH 725; this increase was less significant in the sensitive cultivar RH 749.

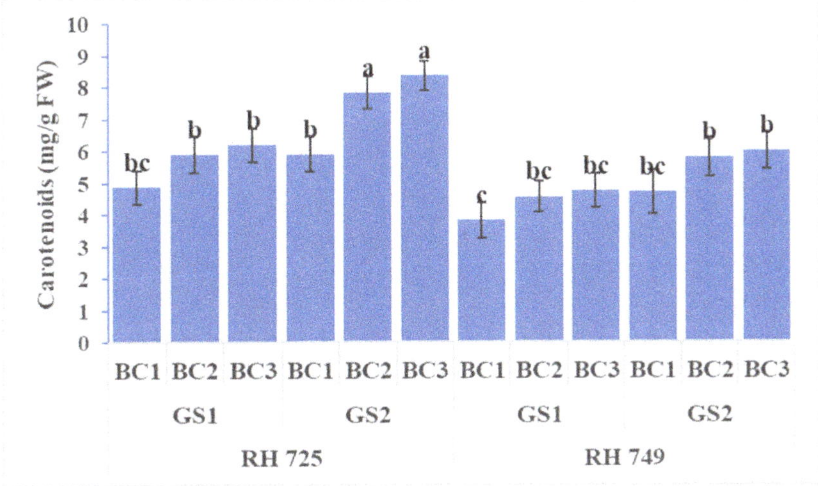

Figure 10. Comparisons of the effects of different concentrations of brassinolide sprays (BC1: control (water spray); BC2: 10 mg BRs/L water; BC3: 20 mg BRs/L water) on the carotenoids content (CC) of drought-tolerant (RH 725) and drought-sensitive (RH 749) Indian mustard cultivars at the flower initiation (GS1) and 50% flowering (GS2) stages. Columns marked by different letters indicate significant differences ($p < 0.05$) between treatments based on Duncan's multiple range test. Error bars denote the standard errors of the mean.

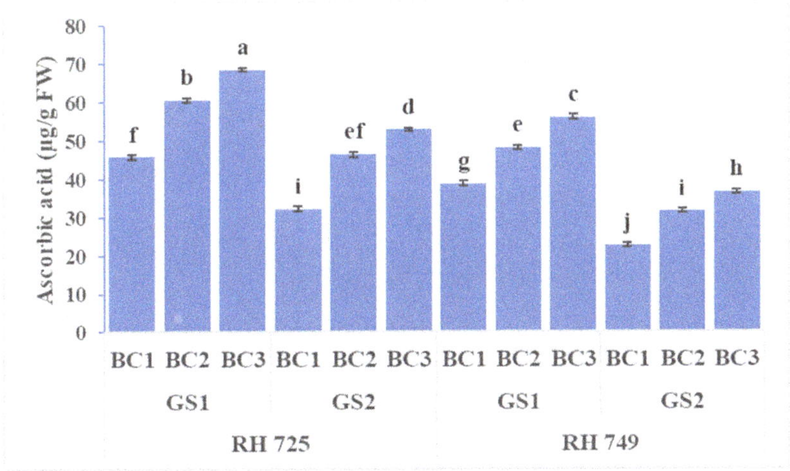

Figure 11. Comparisons of the effects of different concentrations of brassinolide sprays (BC1: control (water spray); BC2: 10 mg BRs/L water; BC3: 20 mg BRs/L water) on the ascorbic acid content (ASA) of drought-tolerant (RH 725) and drought-sensitive (RH 749) Indian mustard cultivars at the flower initiation (GS1) and 50% flowering (GS2) stages. Columns marked by different letters indicate significant differences ($p < 0.05$) between treatments based on Duncan's multiple range test. Error bars denote the standard errors of the mean.

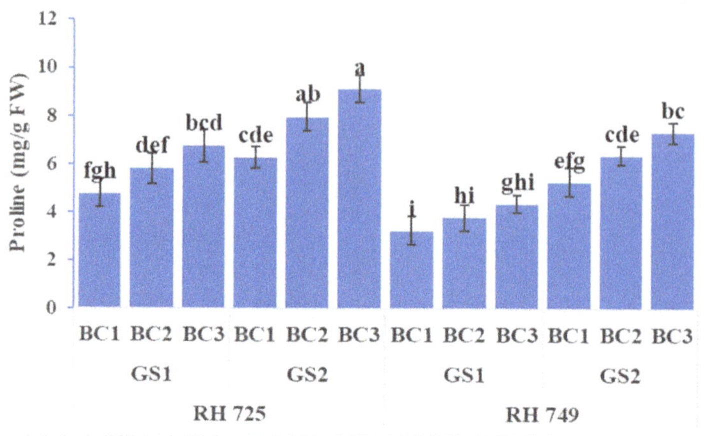

Figure 12. Comparisons of the effects of different concentrations of brassinolide sprays (BC1: control (water spray); BC2: 10 mg BRs/L water; BC3: 20 mg BRs/L water) on the proline content (PRO) of drought-tolerant (RH 725) and drought-sensitive (RH 749) Indian mustard cultivars at the flower initiation (GS1) and 50% flowering (GS2) stages. Columns marked by different letters indicate significant differences ($p < 0.05$) between treatments based on Duncan's multiple range test. Error bars denote the standard errors of the mean.

3.5. Correlation Analysis among Different Parameters

The Pearson's correlation coefficient matrix presented in Table 2 reveals the significant negative associations between both enzymatic and non-enzymatic antioxidants and oxidative stress indicators—*viz.*, H_2O_2 and MDA—except for ascorbic acid, which showed insignificant association between H_2O_2 and MDA, while it was significant and positively correlated with EL. Moreover, both of the oxidative stress indicators—H_2O_2 and MDA—showed a significant positive relationship with one another. The EL, which is the most important oxidative stress indicator, showed a significant negative association with SOD only. Proline content and all physiological parameters—i.e., stomatal conductance, photosynthetic rate, and transpiration rate—were positively correlated with all of the enzymatic antioxidants, while they were negatively associated with all oxidative stress indicators except for electrolyte leakage, which showed a significant negative association with transpiration rate only.

Table 2. Pearson's product–moment correlation matrix between different physio-biochemical parameters evaluated during the present study.

Variables	SOD	POX	CAT	CC	ASA	PRO	H_2O_2	MDA	EL	SC	PR	TR
SOD	1.000	0.864 **	0.911 **	0.960 **	0.091	0.920 **	−0.909 **	−0.644*	−0.640 *	0.948 **	0.903 **	0.857 **
POX		1.000	0.974 **	0.962 **	0.494	0.901 **	−0.981 **	−0.873 **	−0.220	0.945 **	0.955 **	0.739 **
CAT			1.000	0.963 **	0.442	0.916 **	−0.969 **	−0.811 **	−0.338	0.983 **	0.950 **	0.743 **
CC				1.000	0.263	0.957 **	−0.978 **	−0.801 **	−0.455	0.970 **	0.959 **	0.856 **
ASA					1.000	0.118	−0.355	−0.481	0.657 *	0.332	0.388 **	−0.172
PRO						1.000	−0.953 **	−0.824 **	−0.537	0.941 **	0.927 **	0.943 **
H_2O_2							1.000	0.874 **	0.353	−0.957 **	−0.960 **	−0.832 **
MDA								1.000	0.025	−0.774 **	−0.854 **	−0.711 **
EL									1.000	−0.434	−0.304	−0.666 **
SC										1.000	0.970 **	0.806 **
PR											1.000	0.817 **
TR												1.000

** Significant at $p \leq 0.01$; * significant at $p \leq 0.05$; SOD: superoxide dismutase; PX: peroxidase; CAT: catalase; CC: carotenoids content; ASA: ascorbic acid content; PRO: proline content; H_2O_2: hydrogen peroxide content; MDA: malondialdehyde content; EL: electrolyte leakage; SC: stomatal conductance; PR: photosynthetic rate; TR: transpiration rate.

4. Discussion

Analysis of variance showed significant effects of cultivar, growth stage, brassinolide, and their interactions on most of the studied traits, indicating that drought tolerance in Indian mustard is a cultivar- and growth-stage-specific but brassinolide-responsive trait. Similar patterns of results were obtained in many previous studies [31–35]. This indicates that the drought-tolerant cultivar RH 725 is more responsive to brassinolide, and has the capacity to cope with drought-induced oxidative stress and detoxify the oxidative stress indicators by significantly elevating both the non-enzymatic and enzymatic antioxidants. Nevertheless, drought-tolerant RH 725 also has the capacity to improve the physiological processes—*viz.*, stomatal conductance, photosynthetic rate, and transpiration rate—in order to maintain better physiology of the plant as compared to sensitive RH 749. The foliar application of brassinolide enhanced the levels of both enzymatic and non-enzymatic antioxidants, and caused decreased production of oxidative stress indicators (MDA, H_2O_2, and EL). However, we found that the physio-biochemical contents/activities in the leaves depend on the genotype of the cultivar. The drought-tolerant cultivar RH 725 had significantly higher levels of both enzymatic and non-enzymatic antioxidants, and higher physiological parameters, along with lower levels of oxidative stress indicators as compared to the drought-sensitive cultivar RH 749. This indicates that the drought-tolerant cultivar—particularly at the 50% flowering stage—is more responsive to exogenous application of BRs in terms of mitigating drought stress, compared to the sensitive cultivar. Similar results were also reported previously in many crops, including maize [36], sunflowers, [37]; tomatoes, [38], and chickpeas [39].

Brassinosteroids are attractive as original regulators in plants because of their ability to enhance cells in two ways: to provide defense, and to promote growth [40]. Tolerance provided by BR treatment is mediated via the provoked expression of genes involved in defense, regulation, antioxidant responses, and the production of high levels of H_2O_2,which results from enhanced activity of NADPH oxidase [41]. Brassinosteroids regulate the activity of antioxidative enzymes in the cells where ROS production is very high [42]. These results are consistent with the findings of Behnamnia et al. [18], who reported significant augmentation in SOD, CAT, and POX activity in *Lycopersicon esculentum* with the application of brassinolide under drought stress. Similar effects of brasinosteroids were also observed in maize [43], soybeans [44], wheat [45], and Indian mustard [16]. These findings consistent with those of Kumari and Thakur [46], who reported that BRs could regulate antioxidant enzymes such as superoxide dismutase, catalase, peroxidase, etc., in plants under different stress conditions.

Non-enzymatic antioxidants such as carotenoids, ascorbic acid, and proline play a vital rolein the metabolism of plants, by shielding them from stress conditions [47]. Plants produce the carotenoids, which are natural pigments and are involved in photoprotection and photosynthesis. Under drought conditions, carotenoids increase significantly. Ascorbic acid is one of the most powerful antioxidants, which scavenge harmful free radicals and other ROS [48]. Brassinolide was reported to increase the contents of ascorbic acid and total carotenoids in seedlings of drought-resistant (PAN 6043) and drought-sensitive (SC 701) cultivars of *Zea mays* under water stress [43]. Plants accrue low-molecular-mass compounds, such as proline [49], which acts as a non-enzymatic antioxidant that is well known to stabilize the sub-cellular structures of proteins and cell membranes, scavenging free radicals and buffering redox potential under various stress conditions. Proline also acts as a molecular chaperone that preserves the integrity of proteins and boosts the activity of various enzymes during stressful conditions [50]. Among the different compatible solutes, proline is the only molecule that protects the plants against singlet oxygen and damage induced by free radicals resulting from various stresses [51]. It has also been reported previously that BRs propel the expression of proline biosynthetic genes [52]. High proline content in plants under water stress is frequently observed in several plant species [53,54].

The product of membrane peroxidation is the thiobarbituric-acid-reactive substance malondialdehyde (MDA),which is used as a direct marker of membrane damage and lipid peroxidation. Reactive oxygen species attack the majority of the sensitive macromolecules in cells under various environmental stresses, interfering with their function. Drought stress resulted in an increase in MDA accumulation in the leaves of Indian mustard [53]. It was reported that level of lipid peroxidation induced by biotic stresses—such as oxygen deficiency [55], drought stress [56], and heat [57]—could be decreased by treatment with BRs. The results of the present study are consistent with earlier findings that the level of lipid peroxidation in *Brassica juncea* leaves was augmented during drought stress, and it was significantly minimized by BR application. Hydrogen peroxide is produced in the cells under normal as well as a wide range of stressful conditions, such as drought, chilling, UV irradiation, exposure to intense light, wounding, and intrusion by pathogens; it can generate singlet oxygen upon reaction with superoxide anions/HOCl, and it can degrade certain heme proteins to release iron ions, so it is considered to increase membrane permeability by degrading membrane lipids [58]. Therefore, it is important that H_2O_2 be scavenged rapidly by the antioxidative defense system.

Leaf membrane damage was determined by measurement of electrolyte leakage (EL), as described by Valentovic et al. [59]. Electrolyte leakage decreased considerably in *Curcuma alismatifolia* when subjected to water-deficit stress [60]. This is an indicator of a drought-tolerance mechanism in the species via the maintenance of membrane integrity and reduction in electrolyte leakage. The exposure of the plants to drought stress resulted in an increase in electrolyte leakage, which was mitigated by spraying with brassinolide. Houimli et al. [61] observed that exogenous application of brassinolide resulted in a significant reduction in electrolyte leakage under salt stress. Similarly, Coban and Baydar [62] reported a significant reduction in electrolyte leakage in maize, along with improved morphometric parameters, when brassinolide was applied.

Stomatal conductance, photosynthetic rate, and transpiration rate are important characteristics describing plants' water relations [63]. These are inter-related traits of plants that play major roles under stress conditions. In the present study, there were significant increases in stomatal conductance, photosynthetic rate, and transpiration rate with brassinolide spraying under drought stress. Similar increases in these parameters due to BR application have previously been observed in tomatoes, wheat, and cucumbers under both normal conditions and environmental stresses [41,45,57]. Brassinosteroids are also known to activate the key enzymes of photosynthesis, i.e., rubisco [64] and carbonic anhydrase [65]. The assimilation of CO_2 in the Calvin cycle is increased by high carbonic anhydrase activity, which is primarily ascribed to efficient functioning of rubisco [66], consequently improving the net photosynthetic rate and related attributes.

Furthermore, the significant negative association of both enzymatic and non-enzymatic antioxidants with oxidative stress indicators, along with their positive correlation with physiological parameters, again confirmed the role of the antioxidant defense system in mitigating the negative effects of drought stress. However, a further analysis of this phenomenon is certainly needed.

5. Conclusions

The present investigation found that both concentrations of brassinolide (10 and 20 mg/L) improved the plants' efficiency via different physio-biochemical amendments. However, with the 20 mg/L brassinolide spray at the 50% flowering stage, various physio-biochemical attributes showed a more emphatic response in RH 725 than in RH 749. Enhancement of the antioxidative system with improved antioxidative enzyme activity and accumulation of proline may strengthen the plants' ability to combat different stress conditions. Moreover, the drought-tolerant cultivar (RH 725) was superior in term of antioxidant defense system, as compared with the sensitive cultivar RH 749. Indeed, this could be one of the reasons for the former's higher drought tolerance. Understanding the mechanisms of drought tolerance in Indian mustard will make it possible for plant breeders and plant physiologists to develop specific techniques to mitigate the adverse effects of drought, and to maximize Indian mustard crop production.

Supplementary Materials: The following are available online at https://www.mdpi.com/article/10.3390/horticulturae7110514/s1, Supplementary Table S1: Weekly averaged rainfall data during the crop season of 2018-19 at CCS Haryana Agricultural University, Hisar.

Author Contributions: Conceptualization, N.K. and R.A.; methodology, N.K., R.A. and B.R.; validation, M.J., S.A. and N.K.; software, M.S., resources, R.A., investigation, N.N. and N.K.; data curation, M.S., N.N. and K.M.; writing—original draft, N.N.; writing—review and editing, M.S., N.K. and A.S. All authors have read and agreed to the published version of the manuscript.

Funding: This research was supported by Oilseeds Section, Department of Genetics and Plant Breeding, CCS Haryana Agricultural University, Hisar, India. [C (b) PB-3-ICAR].

Institutional Review Board Statement: Not applicable.

Informed Consent Statement: Not applicable.

Data Availability Statement: The data used for the analysis in this study are available within the article, while the datasets used or analyzed during the current study are available from the corresponding author upon reasonable request.

Acknowledgments: The first author is highly grateful to HOS, Oilseeds Section, Department of Genetics and Plant Breeding, CCS Haryana Agricultural University, Hisar, for providing the necessary facilities throughout her M.Sc. research work.

Conflicts of Interest: The authors declare no conflict of interest.

References

1. Singh, M.; Avtar, R.; Lakra, N.; Hooda, E.; Singh, V.K.; Bishnoi, M.; Kumari, N.; Punia, R.; Kumar, N.; Choudhary, R.R. Genetic and Proteomic Basis of Sclerotinia Stem Rot Resistance in Indian Mustard [*Brassica juncea* (L.) Czern & Coss.]. *Genes* **2021**, *12*, 1784.
2. Norton, R.; Burton, W.; Salisbury, P. Canola quality *Brassica juncea* for Australia. In Proceeding of 4th International Crop Science Congress, Gosford, Australia, 26 September–1 October 2004; p. 5.
3. Mahmood, T.; Ashraf, M.; Shahbaz, M. Does exogenous application of glycine-betaine as a pre-sowing seed treatment improve growth and regulate some key physiological attributes in wheat plants grown under water deficit conditions? *Pak. J. Bot.* **2009**, *41*, 1291–1302.
4. Gill, S.S.; Tuteja, N. Reactive oxygen species and antioxidant machinery in abiotic stress tolerance in crop plants. *Plant Physiol. Biochem.* **2010**, *48*, 909–930. [CrossRef]
5. Anjum, N.A.; Sofo, A.; Scopa, A.; Roychoudhury, A.; Gill, S.S.; Iqbal, M.; Lukatkin, A.S.; Pereira, E.; Duarte, A.C.; Ahmad, I. Lipids and proteins—Major targets of oxidative modifications in abiotic stressed plants. *Environ. Sci. Pollut. Res.* **2014**, *22*, 4099–4121. [CrossRef] [PubMed]
6. Singh, M.; Avtar, R.; Pal, A.; Punia, R.; Singh, V.K.; Bishnoi, M.; Singh, A.; Choudhary, R.R.; Mandhania, S. Genotype-Specific Antioxidant Responses and Assessment of Resistance Against *Sclerotinia sclerotiorum* Causing Sclerotinia Rot in Indian Mustard. *Pathogens* **2020**, *9*, 892. [CrossRef] [PubMed]
7. Nayyar, H.; Gupta, D. Differential sensitivity of C3 and C4 plants to water deficit stress: Association with oxidative stress and antioxidants. *Environ. Exp. Bot.* **2006**, *58*, 106–113. [CrossRef]
8. El-Mashad, A.A.A.; Mohamed, H.I. Brassinolide alleviates salt stress and increases antioxidant activity of cowpea plants (*Vigna sinensis*). *Protoplasma* **2012**, *249*, 625–635. [CrossRef] [PubMed]
9. Foyer, C.; Noctor, G. Oxygen processing in photosynthesis: Regulation and signaling. *New Phytol.* **2000**, *146*, 359–388. [CrossRef]
10. Tripathy, B.C.; Oelmüller, R. Reactive oxygen species generation and signaling in plants. *Plant Signal. Behav.* **2012**, *7*, 1621–1633. [CrossRef] [PubMed]
11. Mittler, R. Oxidative stress, antioxidants and stress tolerance. *Trends Plant Sci.* **2002**, *7*, 405–410. [CrossRef]
12. Arora, N.; Bhardwaj, R.; Sharma, P.; Arora, H.K. 28-Homobrassinolide alleviates oxidative stress in salt-treated maize (*Zea mays* L.) plants. *Braz. J. Plant Physiol.* **2008**, *20*, 153–157. [CrossRef]
13. Hayat, S.; Hasan, S.A.; Hayat, Q.; Ahmad, A. Brassinosteroids protect *Lycopersicon esculentum* from cadmium toxicity applied as shotgun approach. *Protoplasma* **2010**, *239*, 3–14. [CrossRef]
14. Bajguz, A.; Hayat, S. Effects of brassinosteroids on the plant responses to environmental stresses. *Plant Physiol. Biochem.* **2009**, *47*, 1–8. [CrossRef] [PubMed]
15. Hasan, S.A.; Hayat, S.; Ali, B.; Ahmad, A. 28-homobrassinolide protects chickpea (*Cicer arietinum*) from cadmium toxicity by stimulating antioxidants. *Environ. Pollut.* **2008**, *151*, 60–66. [CrossRef] [PubMed]
16. Fariduddin, Q.; Khanam, S.; Hasan, S.A.; Ali, B.; Hayat, S.; Ahmad, A. Effect of 28-homobrassinolide on the drought stress-induced changes in photosynthesis and antioxidant system of *Brassica juncea* L. *Acta Physiol.Plant* **2009**, *31*, 889–897. [CrossRef]
17. Ali, B.; Hayat, S.; Ahmad, A. 28-Homobrassinolide ameliorates the saline stress in chickpea (*Cicer arietinum* L.). *Environ. Exp. Bot.* **2007**, *59*, 217–223. [CrossRef]

18. Behnamnia, M.; Kalantari, K.M.; Ziaie, J. The effects of brassinosteroid on the induction of biochemical changes in *Lycopersicon esculentum* under drought stress. *Turk. J. Botany* **2009**, *33*, 417–428.
19. Zhang, M.; Zhai, Z.; Tian, X.; Duan, L.; Li, Z. Brassinolide alleviated the adverse effect of water deficits on photosynthesis and the antioxidant of soybean (*Glycine max* L.). *Plant Growth Regul.* **2008**, *56*, 257–264. [CrossRef]
20. Shen, X.Y.; Dai, J.Y.; Hu, A.C.; Gu, W.L.; He, R.Y.; Zheng, B. Studies on physiological effects of brassinolide on drought resistance in maize. *J. Shenyang Agric. Univ.* **1990**, *21*, 191–195.
21. Gill, M.B.; Cai, K.; Zhang, G.; Zeng, F. Brassinolide alleviates the drought-induced adverse effects in barley by modulation of enzymatic antioxidants and ultrastructure. *Plant Growth Regul.* **2017**, *82*, 447–455. [CrossRef]
22. Silva, F.V.D.F.; Mendes, B.D.S.; Rocha, M.D.S.; Brito, J.F.D.; Beltrão, N.E.D.M.; Sofiatti, V. Photosynthetic pigments and gas exchange in castor bean under conditions of above the optimal temperature and high CO_2. *Acta Sci. Agron.* **2015**, *37*, 331–337. [CrossRef]
23. Beauchamp, C.; Fridovich, I. Superoxide dismutase: Improved assays and an assay applicable to acrylamide gels. *Anal. Chem.* **1971**, *44*, 276–287. [CrossRef]
24. Sinha, A.K. Calorimetric assay of catalase. *Anal. Chem.* **1972**, *47*, 389–395.
25. Shannon, L.M.; Key, E.; Law, J.Y. Peroxidase isoenzymes from horse reddish roots: Isolation and physical properties. *J. Biol. Chem.* **1966**, *241*, 2166–2172. [CrossRef]
26. Hiscox, J.D.; Israelstam, G.F. A method for the extraction of chlorophyll from leaf tissue without maceration. *Can. J. Bot.* **1979**, *57*, 1332–1334. [CrossRef]
27. Roe, J.H. Chemical determination of ascorbic dehydroascorbic and diketogluconic acids. *Methods Biochem. Anal.* **1954**, *1*, 115–139.
28. Bates, L.S.; Waldren, R.P.; Teare, I.D. Rapid determination of free proline for water-stress studies. *Plant Soil* **1973**, *39*, 205–207. [CrossRef]
29. Heath, R.L.; Packer, L. Photoperoxidation in isolated chloroplasts: I. Kinetics and stoichiometry of fatty acid peroxidation. *Arch. Biochem. Biophys.* **1968**, *125*, 189–198. [CrossRef]
30. Gong, J.R.; Zhao, A.F.; Huang, Y.M.; Zhang, X.S.; Zhang, C.L. Water relations, gas exchange, photochemical efficiency, and peroxidative stress of four plant species in the Heihe drainage basin of northern China. *Photosynthetica* **2006**, *44*, 355–364. [CrossRef]
31. Ali, M.A.; Jabran, K.; Awan, S.I.; Abbas, A.; Zulkiffal, M.; Acet, T.; Farooq, J.; Rehman, A. Morpho-physiological diversity and its implications for improving drought tolerance in grain sorghum at different growth stages. *Aust. J. Crop Sci.* **2011**, *5*, 311–320.
32. Kumari, A.; Avtar, R.; Kumari, N.; Jattan, M.; Rani, B. Screening for drought tolerance in Indian mustard (*Brassica juncea* L.) genotypes based on yield contributing characters and physiological parameters. *J. Oilseed Brassica* **2019**, *10*, 1–7.
33. Anjum, S.A.; Wang, L.C.; Farooq, M.; Hussain, M.; Xue, L.L.; Zou, C.M. Brassinolide application improves the drought tolerance in maize through modulation of enzymatic antioxidants and leaf gas exchange. *J. Agron. Crop Sci.* **2011**, *197*, 177–185. [CrossRef]
34. Dehghan, M.; Balouchi, H.; Yadavi, A.; Zare, E. Improve wheat (*Triticum aestivum*) performance by brassinolide application under different irrigation regimes. *S. Afr. J. Bot.* **2020**, *130*, 259–267. [CrossRef]
35. Chen, L.; Yang, H.; Fang, Y.; Guo, W.; Chen, H.; Zhang, X.; Dai, W.; Chen, S.; Hao, Q.; Yuan, S.; et al. Overexpression of GmMYB14 improves high-density yield and drought tolerance of soybean through regulating plant architecture mediated by the brassinosteroid pathway. *Plant Biotechnol. J.* **2021**, *19*, 702–716. [CrossRef] [PubMed]
36. Tůmová, L.; Tarkowská, D.; Řehořová, K.; Marková, H.; Kočová, M.; Rothová, O.; Čečetka, P.; Holá, D. Drought-tolerant and drought-sensitive genotypes of maize (*Zea mays* L.) differ in contents of endogenous brassinosteroids and their drought-induced changes. *PLoS ONE* **2018**, *13*, e0197870. [CrossRef]
37. Ghasemi, M.; Jahanbin, S.; Latifmanesh, H.; Farajee, H.; Mirshekari, A. Effect of brassinolide foliar application on some physiological and agronomic characteristics of sunflower (*Helianthus annuus* L.) under drought stress conditions. *J. Crop Prod.* **2021**, *14*, 31–48.
38. Jangid, K.K.; Dwivedi, P. Physiological and biochemical changes by nitric oxide and brassinosteroid in tomato (*Lycopersicon esculentum* Mill.) under drought stress. *Acta Physiol Plant.* **2017**, *39*, 73. [CrossRef]
39. Verma, J.; Kakralya, B.L.; Jakhar, M.L. Effect of Brassinolide on physiological aspects of Chick pea (*Cicer arietinum* L.) under drought conditions. *J. Plant Sci. Res.* **2012**, *28*, 151–155.
40. Sun, Y.; Fan, X.Y.; Cao, D.M.; Tang, W.; He, K.; Zhu, J.Y. Integration of brassinosteroid signal transduction with the transcription network for plant growth regulation in Arabidopsis. *Dev. Cell.* **2010**, *19*, 765–777. [CrossRef] [PubMed]
41. Xia, X.J.; Huang, L.F.; Zhou, Y.H.; Mao, W.H.; Shi, K.; Wu, J.X.; Asami, T.; Chen, Z.; Yu, J.Q. Brassinosteroids promote photosynthesis and growth by enhancing activation of Rubisco and expression of photosynthetic genes in *Cucumis sativus*. *Planta* **2009**, *230*, 1185–1196. [CrossRef] [PubMed]
42. Ashraf, M.; Akram, N.A.; Arteca, R.N.; Foolad, M.R. The physiological, biochemical and molecular roles of brassinosteroids and salicylic acid in plant processes and salt tolerance. *Crit. Rev. Plant Sci.* **2010**, *29*, 162–190. [CrossRef]
43. Li, L.; Van Staden, J.; Jager, A.K. Effects of plant growth regulators on the antioxidant system in seedlings of two maize cultivars subjected to water stress. *Plant Growth Regul.* **1998**, *25*, 81–87. [CrossRef]
44. Zhang, Y.; Luo, Y.X.; Hou, H.; Jiang, Q.; Chen and Tang, R.H. Chilling acclimation induced changes in the distribution of H_2O_2 and antioxidant system of strawberry leaves. *Agric. J.* **2008**, *3*, 286–291.

45. Shahbaz, M.; Ashraf, M.; Athar, H. Dose exogenous application of 24-epibrassinolide ameliorate salt induced growth inhibition in wheat (*Triticum aestivum* L.)? *Plant Growth Regul.* **2008**, *55*, 51–64. [CrossRef]
46. Kumari, S.; Thakur, A. The Effects of Water Stress and Brassinosteroid on Apple Varieties. *Int. J. Econ. Plants* **2019**, *6*, 1–6. [CrossRef]
47. Tyagi, A.; Santha, I.M.; Mehta, S.L. Effect of water stress on proline content and transcript levels in *Lathyrus sativus*. *Indian J. Biochem. Biophys.* **1999**, *36*, 207–210.
48. Smirnoff, N. Ascorbate biosynthesis and function in photoprotection. *Philos. Trans. R Soc. Lond. B Biol. Sci.* **2000**, *355*, 1455–1464. [CrossRef]
49. Bajji, M.; Lutts, S.; Kinet, J.M. Water deficit effects on solute contribution to osmotic adjustment as a function of leaf ageing in three durum wheat (*Triticum durum Desf.*) cultivars performing differently in arid conditions. *Plant Sci. J.* **2001**, *160*, 669–681. [CrossRef]
50. Szabados, L.; Savoure, A. Proline: A multifunctional amino acid. *Trends Plant Sci.* **2010**, *15*, 89–97. [CrossRef] [PubMed]
51. Alyemeni, M.N.; Al-Quwaiz, S.M. Effect of 28-homobrassinolide on the performance of sensitive and resistant varieties of *Vigna radiata*. *Saudi J. Biol. Sci.* **2016**, *23*, 698–705. [CrossRef]
52. Ozdemir, F.; Bor, M.; Demiral, T.; Turkan, I. Effects of 24- epibrassinolide on seed germination, seedling growth, lipid peroxidation, proline content and antioxidative system of rice (*Oryza sativa* L.) under salinity stress. *Plant Growth Regul.* **2004**, *42*, 203–211. [CrossRef]
53. Kumari, N.; Avtar, R.; Kumari, A.; Sharma, B.; Rani, B.; Sheoran, R.K. Antioxidative response of Indian mustard subjected to drought stress. *J. Oilseed Brassica* **2018**, *9*, 40–44.
54. Rani, B.; Madan, S.; Pooja, K.S.; Sharma, K.D.; Kumari, N.; Kumar, A. Mitigating the effect of drought stress on yield in wheat (*Triticum aestivum*) using arbuscular mycorrhiza fungi (*Glomus mosseae*). *Indian J. Agric. Sci.* **2018**, *88*, 95–100.
55. Ershova, A.N.; Khripach, V.A. Effect of epibrassinolide on lipid peroxidation in *Pisum sativum* at normal aeration and under oxygen deficiency. *Russ. J. Plant Physiol.* **1996**, *43*, 750–752.
56. Robinson, J.M.; Bunce, J.A. Influence of Drought-Induced Water Stress on Soybean and Spinach Leaf Ascorbate-Dehydro ascorbate level and Redox Status. *Int. J. Plant Sci.* **2000**, *161*, 271–279. [CrossRef]
57. Ogweno, J.O.; Song, X.S.; Shi, K.; Hu, W.H.; Mao, W.H.; Zhou, Y.H.; Nogués, S. Brassinosteroids alleviate heat-induced inhibition of photosynthesis by increasing carboxylation efficiency and enhancing antioxidant systems in *Lycopersicon esculentum*. *J. Plant Growth Regul.* **2008**, *27*, 49–57. [CrossRef]
58. Krieger-Liszkay, A. Singlet oxygen production in photosynthesis. *J. Exp. Bot.* **2005**, *56*, 337–346. [CrossRef]
59. Valentovic, P.; Luxova, M.; Kolarovic, L.; Gasparikova, O. Effect of osmotic stress on compatible solutes content, membrane stability and water relations in two maize cultivars. *Plant Soil Environ.* **2006**, *52*, 186–191. [CrossRef]
60. Jungklang, J.; Saengnil, K.; Uthaibutra, J. Effects of water-deficit stress and paclobutrazol on growth, relative water content, electrolyte leakage, proline content and some antioxidant changes in *Curcuma alismatifolia* Gagnep. cv. Chiang Mai Pink. *Saudi J. Biol. Sci.* **2017**, *24*, 1505–1512. [CrossRef]
61. Houimli, S.I.M.; Denden, M.; Mouhandes, B.D. Effects of 24-epibrassinolide on growth, chlorophyll, electrolyte leakage and proline by pepper plants under NaCl-stress. *Eur. Asian J. Biosci.* **2010**, *4*, 96–104. [CrossRef]
62. Coban, O.; Baydar, N.G. Brassinosteroid effects on some physical and biochemical properties and secondary metabolite accumulation in peppermint (*Mentha piperita* L.) under salt stress. *Ind. Crop. Prod.* **2016**, *86*, 251–258. [CrossRef]
63. Farooq, M.; Aziz, T.; Wahid, A.; Lee, D.J.; Siddique, K.H. Chilling tolerance in maize: Agronomic and physiological approaches. *Crop Pasture Sci.* **2009**, *60*, 501–516. [CrossRef]
64. Yu, J.Q.; Huang, L.F.; Hu, W.H.; Zhou, Y.H.; Mao, W.H.; Ye, S.F.; Nogue's, S. A role for brassinosteroids in the regulation of photosynthesis in Cucumis sativus. *J. Exp. Bot.* **2004**, *55*, 1135–1143. [CrossRef] [PubMed]
65. Yusuf, M.; Fariduddin, Q.; Hayat, S.; Hasan, S.A.; Ahmad, A. Protective responses of 28-homobrssinolide in cultivars of Triticumaestivum with different levels of nickel. *Arch. Environ. Contam. Toxicol.* **2011**, *60*, 68–76. [CrossRef] [PubMed]
66. Bajguz, A.; Asami, T. Suppression of Wolffiaarrhiza growth by brassinazole, an inhibitor of brassinosteroid biosynthesis and its restoration by endogenous 24-epibrassinolide. *Phytochemistry* **2005**, *66*, 1787–1796. [CrossRef] [PubMed]

horticulturae

MDPI

Article

BnA.JAZ5 Attenuates Drought Tolerance in Rapeseed through Mediation of ABA–JA Crosstalk

Biting Cao [1,2], Jinjuan Bai [2], Xuan Wang [2], Yanfeng Zhang [3], Xiang Yu [4,*], Shengwu Hu [1,*] and Yuke He [2,*]

1 State Key Laboratory of Crop Stress Biology in Arid Areas and College of Agronomy, Northwest A&F University, Xianyang 712100, China; caobis@163.com
2 National Key Laboratory of Plant Molecular Genetics, Center for Excellence in Molecular Plant Science, Shanghai Institute of Plant Physiology and Ecology, Chinese Academy of Sciences, Fenglin Road 300, Shanghai 200032, China; jjbai@cemps.ac.cn (J.B.); wangxuan@cemps.ac.cn (X.W.)
3 Hybrid Rape Research Center of Shaanxi Province, Xianyang 712100, China; zhangyfcl@126.com
4 Joint International Research Laboratory of Metabolic & Developmental Sciences, School of Life Sciences and Biotechnology, Shanghai Jiao Tong University, Shanghai 200240, China
* Correspondence: yuxiang2021@sjtu.edu.cn (X.Y.); swhu83251@nwafu.edu.cn (S.H.); heyk@sippe.ac.cn (Y.H.); Tel.: +86-29-8708-2604 (S.H.); +86-21-5492-4111 (Y.H.); Fax: +86-21-5492-4111 (Y.H.)

Abstract: Drought stress reduces water availability in plant cells and influences rapeseed yield. Currently, key genetic regulators that contribute to rapeseed response to drought remain largely unexplored, which limits breeding of drought-resistant rapeseed. In this study, we found that *Brassica napus JASMONATE ZIM-DOMAIN 5* (*BnA.JAZ5*), one of the transcriptional repressors functioning in the jasmonate (JA) signaling pathway, was triggered by drought treatment in rapeseed, and drought-susceptibility increased in *BnA.JAZ5*-overexpressing rapeseed plants as compared to wild-type plants, resulting in a lower survival rate after recovery from dehydration. After recovery for 3 days, 22–40% of *p35S::BnA.JAZ5* transgenic plants survived, while approximately 61% of wild-type plants survived. Additionally, seed germination of *BnA.JAZ5*-overexpressing rapeseed was hyposensitive to abscisic acid (ABA). The germination rate of five transgenic lines was 32~42% under 9 μM ABA treatment, while the germination rate of wild-type plants was 14%. We also found that the average stomatal density of five overexpressing lines was 371~446/mm^2, which is higher than that of wild-type (232/mm^2) plants under normal conditions. These results indicate that *BnA.JAZ5* regulated drought response in an ABA-dependent manner, possibly by affecting stomatal density. Interestingly, methyl jasmonate (MeJA) treatment rescued the ABA-hyposensitive seed germination, revealing crosstalk between JAZ5-meidated JA and the ABA signaling pathway. Taken together, our results suggest that *BnA.JAZ5* attenuated drought resistance through the ABA-dependent pathway, which could represent important genetic loci for drought-resistant rapeseed breeding.

Keywords: *BnA.JAZ5*; drought stress; jasmonates; ABA; rapeseed

Citation: Cao, B.; Bai, J.; Wang, X.; Zhang, Y.; Yu, X.; Hu, S.; He, Y. *BnA.JAZ5* Attenuates Drought Tolerance in Rapeseed through Mediation of ABA–JA Crosstalk. *Horticulturae* **2022**, *8*, 131. https://doi.org/10.3390/horticulturae8020131

Academic Editor: Young-Doo Park

Received: 9 December 2021
Accepted: 29 January 2022
Published: 31 January 2022

Publisher's Note: MDPI stays neutral with regard to jurisdictional claims in published maps and institutional affiliations.

Copyright: © 2022 by the authors. Licensee MDPI, Basel, Switzerland. This article is an open access article distributed under the terms and conditions of the Creative Commons Attribution (CC BY) license (https://creativecommons.org/licenses/by/4.0/).

1. Introduction

Various environmental stresses affect plant growth and yield. Water availability is the most important abiotic factor contributing to plant evolution [1]. Establishment of seedlings is directly inhibited by drought stress, which reduces plant densities and yields [2]. With the reduction in arable land worldwide, global food production needs to be increased to feed an ever-growing population [3]. Thereby, genetic engineering must be integrated with breeding technologies to develop climate-resilient crops adaptable to environmental changes leading to stress conditions. Rapeseed (*Brassica napus*, AACC, 2n = 36), as an important oilseed crop used for animal fodder and human consumption, is very sensitive to environmental stresses during its growth and reproductive stages [4,5].

Plants have evolved different strategies for protection against drought. A total of 90% of soil water used by plants is lost through transpiration via the opening of stomata [5], and drought stress causes insufficient uptake of water from soil by roots to meet the requirement of plant transpiration [6,7]. Stomatal density can modulate plant transpiration rates [7,8]. Stomatal density is controlled by both genetic factors and environmental cues [8–10]. Overexpression of *Arabidopsis thaliana GT-2 LIKE 1* (*AtGTL1*), *AtERECTA*, *STOMATAL DENSITY AND DISTRIBUTION1* (*AtSDD1*) and *Oryza sativa stress-induced protein kinase gene 1* (*OsSIK1*) enhances drought tolerance, which is associated with a reduction in stomatal density [7,8,11]. Drought stress induces transcription of dehydration-responsive element-binding proteins (DREBs) and activates genes involved in water movement and chaperone functions, such as late embryogenesis abundant (LEA) proteins [12–15]. RESPONSIVE TO DESICCATION 29A (RD29A), a stress-responsive marker [16], can be used as a control in stress treatments. Overexpression of the gene encoding Δ^1-pyrroline-5-carboxylate synthetase 1 (P5CS1), an enzyme involved in proline biosynthesis, enhances osmotic stress tolerance [17]. Phospholipase C (PLC) works upstream of DREB2 [18].

Jasmonates (JAs) are crucial hormones that regulate plant response to abiotic and biotic stresses [19]. JASMONATE ZIM-DOMAIN (JAZ) proteins are inhibitors of the jasmonic acid signaling pathway [20]. Low jasmonoyl–isoleucine (JA-Ile) levels permit the accumulation of JAZ proteins, which can interact with several bHLH-type transcription factors, including MYC2, MYC3, and MYC4, to activate the transcription of some early JA-responsive genes [21–23]. JAZ proteins can also repress MYC activity to recruit NINJA and TOPLESS [24–26], which participate in JA-Ile perception and induce JAZ degradation [20,27,28]. Some JAZ members interact with other proteins, such as MYB75, that inhibit trichome initiation and anthocyanin accumulation [29]. Some JAZ members play roles in increasing plant tolerance to abiotic stress, while *OsJAZ1* negatively regulates drought resistance by modulating JA and abscisic acid (ABA) signaling in rice [30].

Although JAZ proteins were clearly identified as transcriptional repressors of JA responses in *Arabidopsis* [20,21,31,32], their functions in rapeseed resistance to abiotic stress are unknown. In this study, we explore the function of *BnA.JAZ5*, a rapeseed (*Brassica napus*) homolog of JAZ5, in drought resistance, which could be a potential genetic recourse for stress-resistant plant breeding.

2. Materials and Methods

2.1. Plant Material and Growth Conditions

The rapeseed accession K407 was acquired from the Hybrid Rape Research Center of Shaanxi Province [33]. Seeds of K407 and *p35S::BnA.JAZ5* plants were sterilized according to the method of Li et al. (2020) [34] and grown on Murashige and Skoog (MS) plates with 1% sucrose in darkness for 3 days at 4 °C. Then, seedings were transferred to a greenhouse for growth at 22 °C under long-day conditions (16 h light/8 h dark). The seedlings were finally transplanted to Songjiang Farm Station of the Shanghai Institute of Plant Physiology and Ecology, China in early September 2017 and 2018 [35]. To preserve the phenotypic uniformity of individual plants of the wild-type accession, potential aberrant forms were eliminated at several developmental stages.

2.2. Plasmid Construction and Transformation

The coding sequences of *BnA.JAZ5* genes were amplified from cDNA of *B. napus* using appropriate primers (Supplementary Table S1). The sequences were inserted into a chimeric binary vector. We followed the protocol described by Moloney et al. (1989) and Verma et al. (2012) for rapeseed hypocotyl transformation [30,36], and *Agrobacterium tumefaciens strain* GV3101 pMP90RK was selected [34]. *p35S::BnA.JAZ5* was transferred into *B. napus* accession K407, and transgenic plants were selected on kanamycin after *Agrobacterium*-mediated transformation. Positive seedlings (T0) were transplanted to the greenhouse. T1-generation *p35S::BnA.JAZ5* plants were identified by PCR genotyping using transgene-specific oligos and maintained in a greenhouse as described above.

2.3. RNA Extraction and Real-Time PCR

All plant material was sampled and frozen immediately in liquid nitrogen, and total RNA was extracted with TRIzol (Invitrogen). Total RNA was treated with DNase I (Takara, Shanghai, China), and PrimeScript reverse transcriptase (Takara) was used for cDNA synthesis. Real-time PCR was performed in a 303 MyiQ2 two-color real-time PCR detection system (Bio-Rad, Richmond, CA, USA) [35]. At least three biological replicates per gene were carried out. Transcript levels were normalized relative to those of *UBC21* cDNA [37]. PCR amplifications were performed in 20 mL reaction volumes containing SYBR Green PCR Master Mix (Applied Biosystems), cDNA, and the primers (listed in Supplementary Table S1). Quantification of relative expression levels followed a previously reported method [38].

2.4. MeJA and ABA Treatments

The *B. napus* accession K407 and transgenic lines were subjected to MeJA and ABA treatments. Initial ABA and MeJA concentrations were set according to [39,40]. Seeds were sown on solid MS plates containing MeJA (50 μM), ABA (9 μM), or both MeJA (50 μM) and ABA (9 μM) and treated at 4 °C for 4 d before moving to 22 °C for seed germination. Seed germination was recorded with images taken on the third day after the seed was grown at 22 °C. The emergence of radicle was scored as germination. Cotyledon opening and cotyledon expansion were also recorded [41–43]. The experiment was repeated using different batches of seeds, and at least 150 seeds were used per line. At least three biological replicates were carried out. For every biological replicate, we tested the seeds from the same batch at least three times as technical replicates.

2.5. Stress Treatment at Seedling Stage

For drought-stress treatment of seedlings, the seeds were sown in MS solid medium containing 400 mM mannitol, and then seeds were stratified at 4 °C in the dark for 4 d. The number of germinated seeds was recorded after time intervals of 1, 2, 3, 4, and 5 days. Cotyledon opening and cotyledon expansion were also recorded [41–43]. For determination of root length, the sterilized seeds were first sown on half-strength MS medium for 3 days, transplanted to fresh medium containing 400 mM mannitol and grown vertically. At least three biological replicates were carried out, and at least 150 seeds were used per line. For every biological replicate, we tested the seeds from the same batch at least three times as technical replicates.

For PEG treatment, the seedlings grown on 1/2 MS medium for 7 days were moved to fresh medium with 20% PEG-6000. Seedling samples were harvested after 0, 12, and 48 h according to the methods of Verslues et al. [44,45]. At least three biological replicates were carried out, and at least 150 seeds were used per line. For every biological replicate, we tested the seeds from the same batch at least three times as technical replicates

2.6. Dehydration Treatments

For seedling dehydration, the seedlings grown on half-strength MS medium for 7 days were transplanted to the surface of a Petri dish, maintained in a greenhouse at 22 °C under dark conditions, and then collected for an assay of dehydration effects after 0, 2, 6, and 12 h. At least three biological replicates were carried out, and at least 150 seeds were used per line.

For leaf dehydration, leaves were cut from the five-leaf-stage plants, maintaining the same dehydration conditions as seedlings. Leaves were harvested after time intervals of 0, 6, 12, 24, 36, 48, and 60 h [46]. At least three biological replicates were carried out, and at least 30 seeds were used per line.

For the whole plant dehydration treatment, soil was weighed to ensure that there was equal weight of soil in each pot and the same volume of water was poured into each pot. Seven-day-old seedlings were transplanted into the pots and grown until the five-leaf stage. Water was then withheld for 2 weeks. Plant survival rates were recorded based on the number of plants surviving 3 days after rewatering [47]. At least three biological replicates were carried out, and at least 30 seeds were used per line.

2.7. Measurement of Leaf Stomatal Density

Plants at the five-leaf stage were treated with drought stress (no watering for 7 days). Third leaves were collected, cut into leaf sections (0.5 cm in length), starting from the middle of a leaf, and fixed overnight in FAA at 25 °C. The leaf sections were rinsed three times using distilled water, dehydrated in ethanol series (30%, 50%, 70%, 80%, and 95%), and rinsed three times in 100% ethanol. The dehydrated samples were sputter-coated with gold, and stomatal observation was performed with a Hitachi S-2300 electron microscope. The number of stomata per mm^2 in middle regions of leaves of five-leaf-stage plants was recorded [48]. At least three biological replicates were carried out, and at least 30 seeds were used per line.

2.8. Water Loss Assay

Leaves from plants at the five-leaf stage were detached and immediately weighed. The leaves were then placed on a laboratory bench and weighed according to the schedule (designated as Wi; Wi represents the weight of leaves at time *i*). Fresh weight (FW) loss was calculated according to the initial weight of the detached leaves. The water-loss rate (WLR) was also calculated: WLR = (FW − Wt)/FW. At least three biological replicates were carried out, and at least 20 seeds were used per line.

2.9. Measurement of Relative Water Content

Relative water content (RWC) was measured as described by Kumar et al. [49], with minor modification. Leaves were detached from plants at the five-leaf stage and immediately weighed to record their fresh weight (FW). The detached leaves were placed in distilled water for 12 h, blotted dry, and weighed to record their turgid weight (TW). To determine the dry weight (DW), the turgid leaves were subjected to oven drying at 70 °C for 36 h. RWC was calculated using the following equation: RWC = ((FW − DW) × 100)/(TW − DW). At least three biological replicates were carried out, and at least 20 seeds were used per line.

2.10. Sequence Alignment and Phylogenetic Analysis

Arabidopsis JAZ protein sequences were obtained from the Arabidopsis Information Resource (https://www.arabidopsis.org/ (accessed on 3 November 2021). BnA.JAZ5 and BnC.JAZ5 were selected based on their high similarity with AtJAZ5. Full-length amino-acid sequence multiple alignments were performed in ClustalW. Unrooted phylogenetic trees were constructed from the aligned amino-acid sequences in MEGA 6.0, and bootstrapping was carried out with 1000 iterations [50].

2.11. Statistical Analysis

Student's *t*-tests were used for statistical analyses, and statistical significance was determined with p value < 0.05.

3. Results

3.1. Expression Patterns of JA Signaling Regulator BnA.JAZ5 Revealed Its Potential Function in Rapeseed Response to Drought Stress

JAs are involved in plant development, reproduction, and defense. Within JA signaling cascades, JAZ proteins play a central role. *BnA.JAZ5* and *BnC.JAZ5* are two of the *JAZ5* homoeologous genes present in polyploid rapeseed (Figures 1A and S1). To investigate whether JAZ5 in rapeseed is involved in drought stress response, we checked the expression profiles of *BnA.JAZ5* and *BnC.JAZ5* in 18-day-old rapeseed treated with 20% high-molecular-weight polyethylene glycol (PEG-6000) for 48 h using real-time PCR. We found that *BnA.JAZ5* was strongly induced by 20% PEG-6000, whereas *BnC.JAZ5* was not (Figures 1B and S4E). In addition, the expression level of *BnA.JAZ5* was further increased under continued dehydration stress (Figures 1C and S4E). Taken together, these results suggest that *BnA.JAZ5* may be involved in plant drought response.

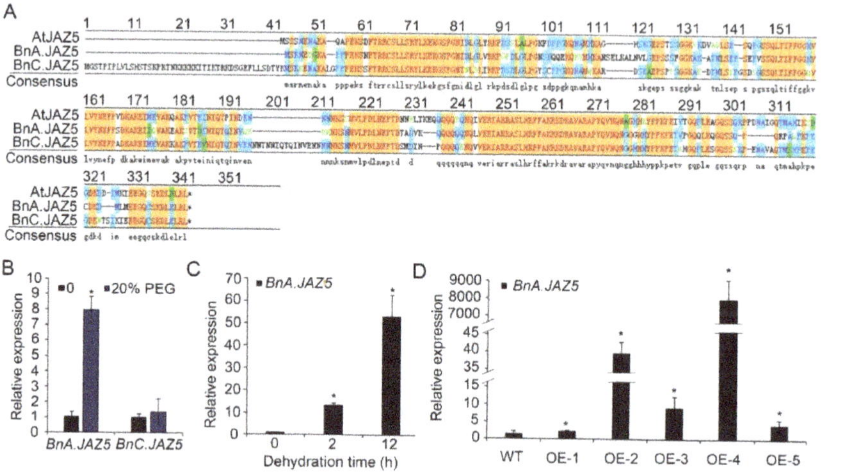

Figure 1. Homologous alignment of AtJAZ5 in rapeseed and expression levels of *BnA.JAZ5* and *BnC.JAZ5*. (**A**) Alignment of the protein sequences of AtJAZ5 in *Arabidopsis* and BnA.JAZ5 and BnC.JAZ5 in *B. napus*. (**B**) Expression levels of *BnA.JAZ5* and *BnC.JAZ5* under 20% PEG-6000. (**C**) Expression levels of *BnA.JAZ5* under dehydration stress. Detached leaves of five-leaf-stage plants were subjected to dehydration treatment for 2 and 12 h. (**D**) The expression levels of *BnA.JAZ5* in *p35S::BnA.JAZ5* plants (OE-1, OE-2, OE-3, OE-4 and OE-5) and wild-type plants. Error bars indicate standard errors. Asterisks indicate significant differences ($p < 0.05$). WT, accession K407, OE-1, OE-2, OE-3, OE-4, OE-5, and *p35S::BnA.JAZ5* lines.

3.2. Overexpression of BnA.JAZ5 Reduced Drought Tolerance

To study the functions of *BnA.JAZ5* in response to drought, we constructed *BnA.JAZ5*-overexpressing transgenic plants by introducing a *p35S::BnA.JAZ5* construct into the K407 cultivar of rapeseed (designated as *p35S::BnA.JAZ5*), which was self-fertilized and further selected to obtain homozygous lines for subsequent experiments. In total, five independent lines (OE-1, OE-2, OE-3 OE-4, and OE-5) were generated, and we found that *BnA.JAZ5* was overexpressed from 3-fold to 7000-fold compared to the wild-type plants (Figure 1D).

To study the tolerance of *p35S::BnA.JAZ5*-overexpressing plants to drought stress, seeds of the transgenic plants and K407 were sown on MS medium containing 400 mM mannitol (Figure 2). Compared with the control (0 mM mannitol), germination rates of all five of these transgenic lines were significantly reduced under 400 mM mannitol (Figures 2A,B and S4A) and exhibited much lower percentages of expanded cotyledons on the fifth day after the seed was sown (Figure 2C). Root elongation of *p35S::BnA.JAZ5* plants was inhibited compared with wild-type plants (Figure 2D,E). These results indicate that *BnA.JAZ5* decreased drought tolerance of *B. napus*.

Compared to the wild-type plants, the leaves of *p35S::BnA.JAZ5* plants rolled and wilted earlier (Figure 3A). Relative water content (RWC) and water-loss rate (WLR) are two indicators of cell and tissue hydration. Therefore, we tested the WLR and found the WLR of *BnA.JAZ5*-overexpressing rapeseed plants was significantly higher than that of wild-type (Figures 3B and S4B). Besides, the RWC of *BnA.JAZ5*-overexpressing rapeseed plants was significantly lower than that of K407 under drought conditions (Figures 3C and S4C). These results suggest that *BnA.JAZ5* played a negative role in drought resistance.

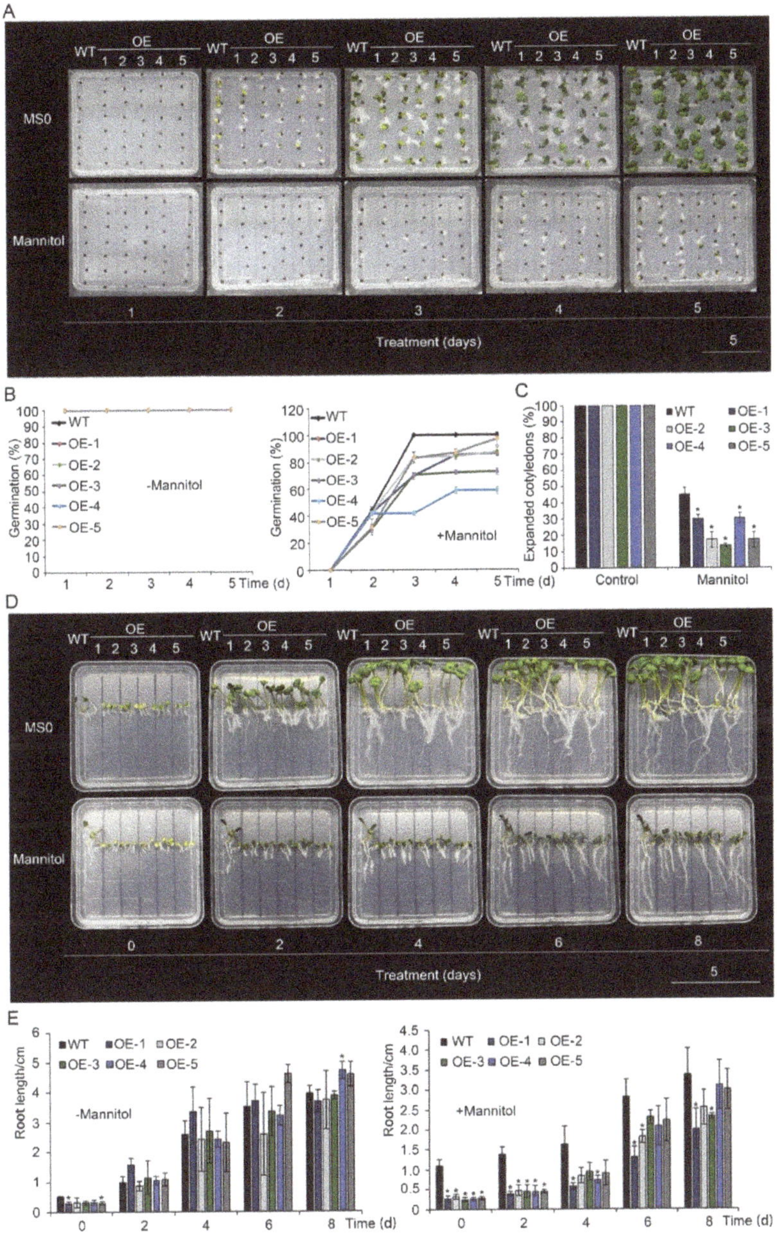

Figure 2. Drought-stress responses of *p35S::BnA.JAZ5* rapeseed plants. (**A**) Phenotypes of germinated six-day-old overexpressing lines under 400 mM mannitol treatment. (**B**) Seed germination rate of the five overexpressing lines under 400 mM mannitol treatment. (**C**) The percentage of wild-type and *p35S::BnA.JAZ5* rapeseed seedlings with expanded cotyledons was scored 5 d after stratification on MS medium supplemented with 400 mM mannitol. (**D**) Phenotypes of root elongation of five overexpressing lines under 400 mM mannitol treatment. (**E**) Root length of five overexpressing lines under 400 mM mannitol treatment. Error bars represent standard errors. Asterisks indicate significant differences between wild-type and *p35S::BnA.JAZ5* lines ($p < 0.05$).

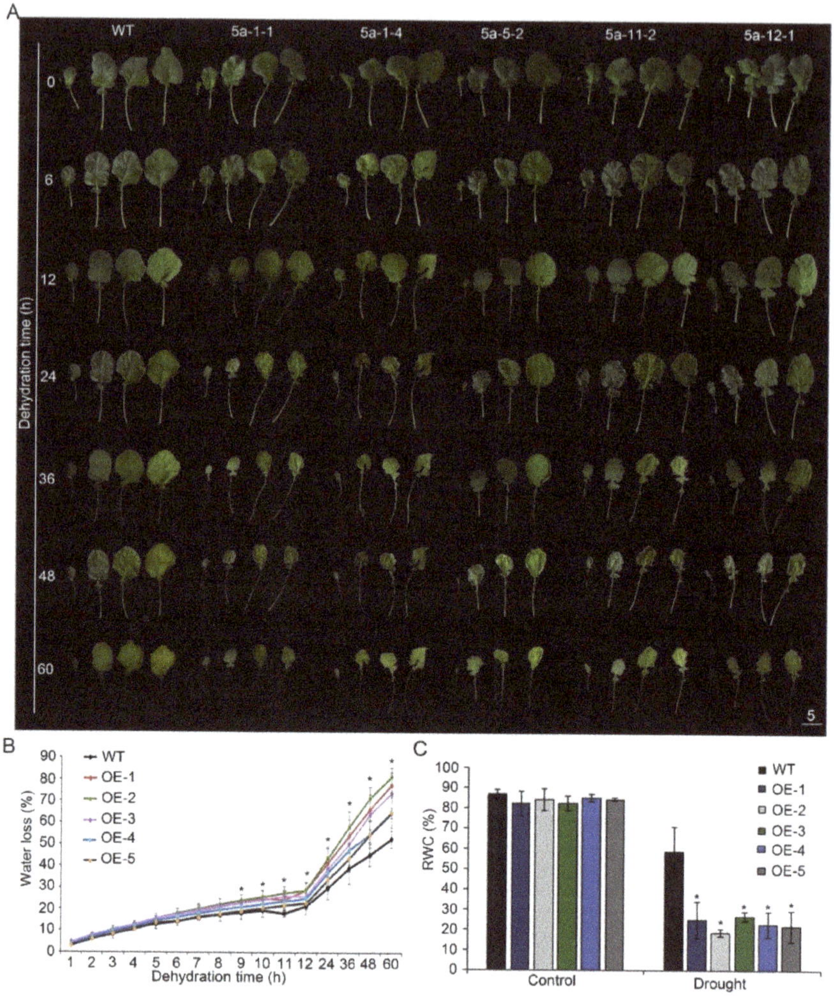

Figure 3. Dehydration test of *p35S::BnA.JAZ5* plants at the five-leaf stage. (**A**) Phenotypes of detached leaves of *p35S::BnA.JAZ5* plants. (**B**) WLR of five overexpressing lines. Water loss was measured at the indicated time points. Asterisks indicate significant differences between wild-type and *p35S::BnA.JAZ5* lines ($p < 0.05$, Student's t-test). (**C**) RWC of five overexpressing lines under drought-stress treatment. Error bars represent standard errors. Asterisks indicate significant differences between control and drought stress ($p < 0.05$, Student's t-test). WT, accession K407. OE-1, OE-2, OE-3, OE-4, OE-5, and *p35S::BnA.JAZ5* lines. Scale bar, 5 cm.

To further investigate the response of plants under drought conditions, five overexpressing lines were planted in the greenhouse for 5 weeks and then subjected to drought stress by withholding water for 14 days (Figure 4A). Drought-associated symptoms, such as leaf rolling and wilting, appeared earlier in *p35S::BnA.JAZ5* plants than in the wild-type during drought treatment. After recovery for 3 days, only 22–40% of *p35S::BnA.JAZ5* transgenic plants survived, when approximately 61% of wild-type ones survived (Figures 4B and S4D). Overall, these results revealed that *BnA.JAZ5* was a negative regulator of drought resistance.

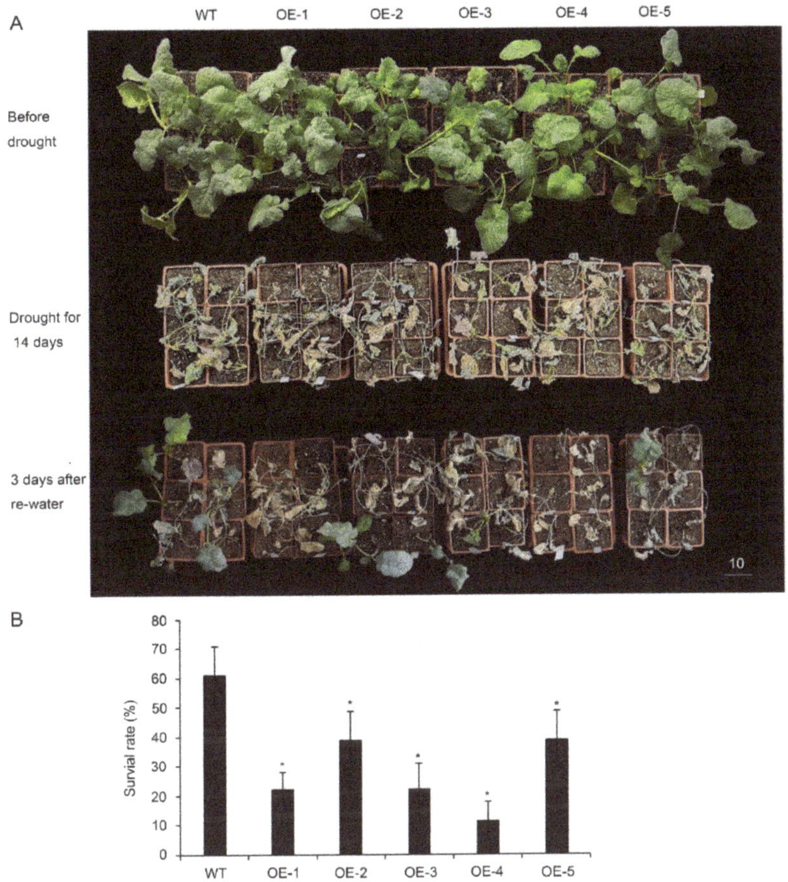

Figure 4. Drought resistance testing of *p35S::BnA.JAZ5* plants at the five-leaf stage. (**A**) Drought resistance assay of five overexpressing lines. (**B**) Survival rates of five overexpressing lines after 3-day recovery from drought stress treatment. WT, accession K407. OE-1, OE-2, OE-3, OE-4 and OE-5, *p35S::BnA.JAZ5* plants. Asterisks indicate significant differences between wild-type and *p35S::BnA.JAZ5* lines ($p < 0.05$, Student's *t*-test). Scale bar, 10 cm.

3.3. *BnA.JAZ5* Overexpression Altered Plant Responses to JA and ABA

Given that *JAZ5* belongs to the *JAZ* family, whose members are key inhibitors of the JA signaling pathway, we examined the function of JA on seed germination of *BnA.JAZ5*-overexpressing rapeseed. Overall, methyl jasmonate (MeJA) inhibited the seed germination of both wide-type and *BnA.JAZ5*-overexpressing rapeseed (Figure 5A–E). However, the germination rate of these five overexpressing lines showed no significant change under treatment with 50 μM MeJA (Figures 5A,B and S4A), and the cotyledon expansion of the overexpressing lines was no significant difference on media containing 50 μM MeJA on the 4th day after seed was sown (Figure 5A,C). In fact, we found that, in wide-type plants, with 50 μM MeJA treatment, *BnA.JAZ5* expression increased up to 46-fold as compared with the control (Figure 5E). These results suggested that *BnA.JAZ5* overexpression did not affect the seed germination sensitivity to JA.

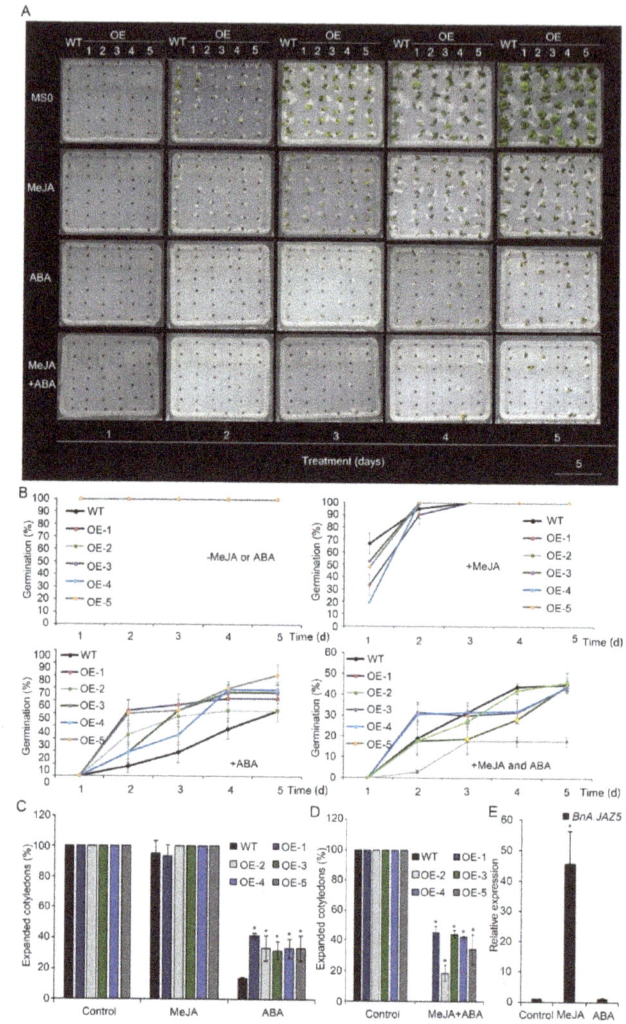

Figure 5. Seed germination of *p35S::BnA.JAZ5* plants on MS medium supplemented with 50 μM MeJA, 9 μM ABA or both 50 μM MeJA and 9 μM ABA, respectively. (**A**) Phenotypes of seed germination of five overexpressing lines under 50 μM MeJA, 9 μM ABA or both 50 μM MeJA and 9 μM ABA treatments. (**B**) Seed germination rate of five overexpressing lines under 50 μM MeJA, 9 μM ABA or both 50 μM MeJA and 9 μM ABA treatments. (**C**) The percentage of wild-type and *p35S::BnA.JAZ5* rapeseed seedlings with expanded cotyledons was scored 4 d after stratification on MS medium supplemented with 50 μM MeJA or 9 μM ABA. (**D**) The percentage of wild-type and *p35S::BnA.JAZ5* rapeseed seedlings with expanded cotyledons was scored 9 d after stratification on MS medium supplemented with both 50 μM MeJA and 9 μM ABA. (**E**) Expression levels of *BnA.JAZ5* under 50 μM MeJA or 9 μM ABA treatments. Error bars represent standard errors. Asterisks indicate significant differences between wild-type and *p35S::BnA.JAZ5* lines ($p < 0.05$). WT, accession K407. OE-1, OE-2, OE-3, OE-4 and OE-5, *p35S::BnA.JAZ5* lines. Scale bar, 5 cm.

In rapeseed seeds, exogenous application of ABA inhibits germination [51]. To test the possible roles of *BnA.JAZ5* in ABA signaling, we treated the wild-type and five overexpressing lines with 9 μM ABA, as suggested by previous studies [52,53]. From this analysis, we found the seed germination of five overexpressing lines exhibited hyposensitivity to ABA as compared to wild-type plants (Figures 5A,B and S4A). Additionally, compared with the wild-type line, the five overexpressing lines exhibited much higher percentages of expanded cotyledons on media containing 9 μM ABA on the 5th day after the seed was sown (Figure 5A,C). In addition, we found no change in the expression levels of *BnA.JAZ5* as compared to the control with 9 μM ABA treatment (Figure 5E). These findings indicate that *BnA.JAZ5* may regulate drought response in an ABA-dependent manner.

Next, to test the JAZ5 function in JA–ABA crosstalk, we treated the plants with a combination of 50 μM MeJA and 9 μM ABA. As expected, the seed germination of wild-type and *p35S::BnA.JAZ5*-overexpressing plants was severely inhibited (Figures 5A,B,D and S4A). Interestingly, while the seed germination rate of *p35S::BnA.JAZ5*-overexpressing plants was higher as compared to wild-type plants under ABA treatment, the seed germination rate showed no difference with the combination of ABA and JA treatment, suggesting that JAZ5 plays a role in crosstalk between JA and ABA. Taken together, these results suggest that *BnA.JAZ5* regulated rapeseed plant responses to drought stress through ABA and JA signaling.

3.4. *BnA.JAZ5* Regulates ABA-Dependent Stress-Responsive Genes

To explore the role of *BnA.JAZ5* in drought stress through the ABA signaling pathway, we examined JA- and ABA-responsive marker genes in plants under 400 mM mannitol treatment (Figure 6A). Our real-time PCR results revealed that *BnA.JAZ5* overexpression led to downregulation of the JA-responsive gene *BnMYC2* and upregulation of the ABA-responsive gene *BnABF3* under 400 mM mannitol treatment (Figure 6A). In addition, we examined stress-related marker genes in plants under dehydration treatment (Figure S2A). *BnA.JAZ5* overexpression led to downregulation of the stress-related genes *BnP5CS*, *BnCT-STM1*, and *BnLEA76* (Figure S2B), and the expression of *BnPLC* was repressed in *BnA.JAZ5*-overexpressing lines after 6 h and 12 h drought treatment as compared to wide-types. This finding suggests that *BnA.JAZ5* attenuated the drought tolerance of the transgenic plants through downregulation of ABA-dependent stress-responsive genes.

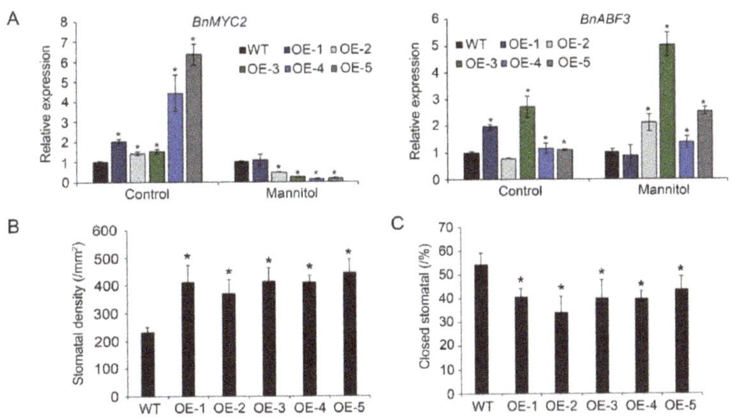

Figure 6. Changes in stomatal density and expression levels of stress-related marker genes. (**A**) Expression patterns of JA- and ABA-responsive genes under 400 mM mannitol stress. Error bars represent standard errors. Asterisks indicate significant differences ($p < 0.05$). (**B**) Stomatal density of OE-1, OE-2, OE-3, OE-4, and OE-5 lines. (**C**) Percentage of closed stomata under drought stress. Asterisks indicate significant differences between WT and *p35S::BnA.JAZ5* lines ($p < 0.05$, Student's *t*-test). WT, accession K407; OE-1, OE-2, OE-3, OE-4, OE-5, and *p35S::BnA.JAZ5* transgenic lines.

3.5. BnA.JAZ5 Overexpression Increased Stomatal Density and Reduced Stomatal Closure

Guard-cell signaling plays a critical role in plant drought response, which is regulated by ABA [1]. We measured stomatal density under normal and drought-stress conditions. Under normal conditions, the average stomatal density of five overexpressing lines was higher than that of wild-type plants (Figures 6B and S3). After drought stress, the percentage of closed stomata of five overexpressing lines was lower than that of the wild-type plants (Figures 6C and S3). These results suggest that *BnA.JAZ5* overexpression increased stomatal density under normal conditions and reduced stomatal closure when transgenic plants were subjected to drought stress.

4. Discussion

Rapeseed (*B. napus*), soybean, and oil palm are the most extensively grown oil crop species worldwide. Rapeseed, however, is very sensitive to water stress, a main factor of crop failure in rapeseed [5]. Identifying genes related to dehydration stress or markers linked to these genes is a key step in genomics-assisted breeding in rapeseed [54]. JAZ proteins are key components of the JA signaling pathway. The *OsJAZ* genes were induced upon abiotic stress in *OsbHLH148*-overexpressing plants [55]. *OsJAZ9* can increase salt and cold tolerance in rice by inhibiting the expression of *OsbHLH062* and *OsMYB30* [56,57]. *GsJAZ2* overexpression in *Arabidopsis* reduces plant sensitivity to salt stress [58]. In rice, interaction of OsJAZ proteins with a basic helix–loop–helix protein leads to drought tolerance [55–57], whereas *OsJAZ1* negatively regulates drought resistance by modulating JA and ABA signaling [52]. The varied roles of JAZ members indicate that JAZ protein family members regulate abiotic stresses differentially. However, little is known about JAZ functional roles and their mechanisms in rapeseed. We collected cotyledon, first true leaf, root (five-leaf stage), third leaf (five-leaf stage), bud, flower, and young silique samples from the rapeseed accession K407 for real-time PCR analysis. We found that *BnA.JAZ5* was generally expressed in all these tissues of rapeseed plants (Figure S2C). Furthermore, both 20% PEG-6000 treatment and dehydration stress significantly induced the expression of *BnA.JAZ5*. The germination rate of *p35S::BnA.JAZ5* transgenic seeds was weaker than that of wild-type seeds on medium containing 400 M mannitol. Fewer *p35S::BnA.JAZ5* plants survived than wild-type plants following withholding of water for 14 days and recovery for 3 days. Thus, we thought that *BnA.JAZ5* may have a negative role in drought resistance, and this conclusion is in agreement with those concerning *OsJAZ1* [52].

Most plant transpiration occurs in stomata, and stomatal density and/or movement can adjust the transpiration rate [59,60]. Leaves can respond to water status by changing stomata density and guard-cell size under water stress [61]. We found that stomatal density was significantly increased in *p35S::BnA.JAZ5* plants, and a lower proportion of closed stomata was observed in *p35S::BnA.JAZ5* plants under drought stress. This change caused the *p35S::BnA.JAZ5* plants to lose water more rapidly under drought stress, thereby accelerating death and reducing drought resistance. Consistently, *p35S::BnA.JAZ5* plants had lower RWC under drought stress and higher WLR than wild-type plants. Stomatal closure is one of the ABA-regulated pathways activated by water-deficit conditions [62,63]. A smaller stomatal aperture and stomatal density contribute to reduced water loss from plant cells, thereby enhancing drought or osmotic stress tolerance [8–10,64]. *BnA.JAZ5* gene did not respond to exogenous ABA treatment, but *p35S::BnA.JAZ5* plants were ABA-hyposensitive, and the germination rate of *p35S::BnA.JAZ5* seeds on MS medium containing ABA was higher than that of wild-type seeds. Studies have shown that JAZ proteins interact with ABA-responsive transcription factors in *Arabidopsis* [15,41]. The expression of JA-responsive gene *BnMYC2* was downregulated, and the expression of ABA-responsive gene *BnABF3* was upregulated in *p35S::BnA.JAZ5* plants under 400 mM mannitol treatment. At the same time, *BnA.JAZ5* could be induced by MeJA. In addition, the proline synthesis gene *BnP5CS* and stress tolerance-related genes *BnRD29A*, *BnLEA*, and *BnCYSTM1* tend to be downregulated in *p35S::BnA.JAZ5* plants under drought-stress conditions. These results suggest that *BnA.JAZ5* negatively regulates drought resistance in rapeseed through JA–ABA crosstalk, and this process affects stomatal development. *BnA.JAZ5* can be used as a candidate gene for improving drought resistance of rapeseed. We can design sgRNA to edit these *BnA.JAZ5* homologs using the CRISPR/Cas9 system to obtain rape germplasm with higher drought resistance.

5. Conclusions

In this study, the function of rapeseed *BnA.JAZ5* in drought response was characterized. *BnA.JAZ5*-overexpressing rapeseed plants were further investigated under drought conditions and plant hormone (JA and ABA) treatment, revealing the role of *BnA.JAZ5* in attenuating rapeseed drought resistance in an ABA-dependent manner. Molecular study showed that *BnA.JAZ5* regulated ABA-mediated stress-responsive genes *MYC2* and *ABF3*. Finally, significantly increased stomatal density and reduced stomatal closure were observed in *BnA.JAZ5*-overexpressing rapeseed plants. Taken together, these results suggest that *BnA.JAZ5*-mediated crosstalk between JA and ABA signaling pathways contributed to the rapeseed response to drought stress. *BnA.JAZ5* could be used as CRISPR-editing genetic loci for drought-resistant rapeseed breeding.

Supplementary Materials: The following supporting information can be downloaded at: https://www.mdpi.com/article/10.3390/horticulturae8020131/s1, Table S1. List of primers used in this study. Figure S1. Phylogeny tree of rapeseed BnA.JAZ5 (GSBRNA2T00103000001) and BnC.JAZ5 (GSBRNA2T00089260001) with *Arabidopsis* JAZ genes. Figure S2. Changes in plant phenotype and gene expression under dehydration stress. Figure S3. SEM images showing adaxial epidermal cells of OE-1, OE-2, OE-3, OE-4, and OE-5 lines at 350× magnification. Figure S4. Molecular data of another two biological replicates per experiment.

Author Contributions: Y.H. and S.H. designed the project. B.C., X.W., J.B. and Y.Z. performed the experiments. B.C., X.Y., S.H. and Y.H. wrote the manuscript. All authors have read and agreed to the published version of the manuscript.

Funding: This work was supported by grants from the National Key Research and Development Program of China (grant Nos. 2016YFD0101900 and 2016YFD0100500) and the Natural Science Foundation of China (grant Nos. 31771442 and 31571261).

Institutional Review Board Statement: Not applicable.

Informed Consent Statement: Not applicable.

Data Availability Statement: Gene sequences generated in this study were uploaded to NCBI (*AtJAZ5*: MW233040; *BnA8.JAZ5*: MW233041; *BnC8.JAZ5*: MW233042; *BnP5CS*: MW233043; *BnPLC*: MW233044; *BnCYSTM1*: MW233045; and *BnLEA76*: MW233046).

Conflicts of Interest: The authors declare no conflict of interest.

References

1. Zhu, J.K. Salt and drought stress signal transduction in plants. *Annu. Rev. Plant Biol.* **2002**, *53*, 247–273. [CrossRef] [PubMed]
2. Cook, E.R.; Seager, R.; Cane, M.A.; Stahle, D.W. North American drought: Reconstructions, causes, and consequences. *Earth-Sci. Rev.* **2007**, *81*, 93–134. [CrossRef]
3. Tester, M.; Langridge, P. Breeding technologies to increase crop production in a changing world. *Science* **2010**, *327*, 818–822. [CrossRef] [PubMed]
4. Allender, C.J.; King, G.J. Origins of the amphiploid species *Brassica napus* L. investigated by chloroplast and nuclear molecular markers. *BMC Plant Biol.* **2010**, *10*, 54. [CrossRef]
5. Wan, J.X.; Griffiths, R.; Ying, J.F.; McCourt, P.; Huang, Y.F. Development of drought-tolerant Canola (*Brassica napus* L.) through genetic modulation of ABA-mediated stomatal responses. *Crop Sci.* **2009**, *49*, 1539–1554. [CrossRef]
6. Jeong, J.S.; Kim, Y.S.; Baek, K.H.; Jung, H.; Ha, S.H.; Do Choi, Y.; Kim, M.; Reuzeau, C.; Kim, J.K. Root-specific expression of *OsNAC10* improves drought tolerance and grain yield in rice under field drought conditions. *Plant Physiol.* **2010**, *153*, 185–197. [CrossRef]
7. Yoo, C.Y.; Pence, H.E.; Jin, J.B.; Miura, K.; Gosney, M.J.; Hasegawa, P.M.; Mickelbart, M.V. The *Arabidopsis* GTL1 transcription factor regulates water use efficiency and drought tolerance by modulating stomatal density via transrepression of *SDD1*. *Plant Cell* **2010**, *22*, 4128–4141. [CrossRef]

8. Masle, J.; Gilmore, S.R.; Farquhar, G.D. The *ERECTA* gene regulates plant transpiration efficiency in *Arabidopsis*. *Nature* **2005**, *436*, 866–870. [CrossRef]
9. Berger, D.; Altmann, T. A subtilisin-like serine protease involved in the regulation of stomatal density and distribution in *Arabidopsis thaliana*. *Genes Dev.* **2000**, *14*, 1119–1131. [CrossRef]
10. Shpak, E.D.; McAbee, J.M.; Pillitteri, L.J.; Torii, K.U. Stomatal patterning and differentiation by synergistic interactions of receptor kinases. *Science* **2005**, *309*, 290–293. [CrossRef]
11. Ouyang, S.Q.; Liu, Y.F.; Liu, P.; Lei, G.; He, S.J.; Ma, B.; Zhang, W.K.; Zhang, J.S.; Chen, S.Y. Receptor-like kinase OsSIK1 improves drought and salt stress tolerance in rice (*Oryza sativa*) plants. *Plant J.* **2010**, *62*, 316–329. [CrossRef]
12. Shinozaki, K.; Yamaguchi-Shinozaki, K. Gene networks involved in drought stress response and tolerance. *J. Exp. Bot.* **2007**, *58*, 221–227. [CrossRef] [PubMed]
13. Wang, W.; Vinocur, B.; Altman, A. Plant responses to drought, salinity and extreme temperatures: Towards genetic engineering for stress tolerance. *Planta* **2003**, *218*, 1–14. [CrossRef] [PubMed]
14. Xiong, L.; Zhu, J.K. Molecular and genetic aspects of plant responses to osmotic stress. *Plant Cell Environ.* **2002**, *25*, 131–139. [CrossRef] [PubMed]
15. Yoshida, T.; Mogami, J.; Yamaguchi-Shinozaki, K. ABA-dependent and ABA-independent signaling in response to osmotic stress in plants. *Curr. Opin. Plant Biol.* **2014**, *21*, 133–139. [CrossRef] [PubMed]
16. Yamaguchi-Shinozaki, K.; Shinozaki, K. Characterization of the expression of a desiccation-responsive *rd29* gene of *Arabidopsis thaliana* and analysis of its promoter in transgenic plants. *Mol. Gen. Genet.* **1993**, *236*, 331–340. [CrossRef]
17. Kishor, P.; Hong, Z.; Miao, G.H.; Hu, C.; Verma, D. Overexpression of Δ^1-Pyrroline-5-Carboxylate Synthetase Increases Proline Production and Confers Osmotolerance in Transgenic Plants. *Plant Physiol.* **1995**, *108*, 1387–1394. [CrossRef]
18. Zhang, Q.; van Wijk, R.; Zarza, X.; Shahbaz, M.; van Hooren, M.; Guardia, A.; Scuffi, D.; Garcia-Mata, C.; Van den Ende, W.; Hoffmann-Benning, S.; et al. Knock-down of Arabidopsis *PLC5* reduces primary root growth and secondary root formation while overexpression improves drought tolerance and causes stunted root hair growth. *Plant Cell Physiol.* **2018**, *59*, 2004–2019. [CrossRef]
19. Devoto, A.; Turner, J.G. Jasmonate-regulated Arabidopsis stress signalling network. *Physiol. Plant* **2005**, *123*, 161–172. [CrossRef]
20. Thines, B.; Katsir, L.; Melotto, M.; Niu, Y.; Mandaokar, A.; Liu, G.; Nomura, K.; He, S.Y.; Howe, G.A.; Browse, J. JAZ repressor proteins are targets of the SCF^{COI1} complex during jasmonate signalling. *Nature* **2007**, *448*, 661–665. [CrossRef]
21. Fernandez-Calvo, P.; Chini, A.; Fernandez-Barbero, G.; Chico, J.M.; Gimenez-Ibanez, S.; Geerinck, J.; Eeckhout, D.; Schweizer, F.; Godoy, M.; Franco-Zorrilla, J.M.; et al. The *Arabidopsis* bHLH transcription factors MYC3 and MYC4 are targets of JAZ repressors and act additively with MYC2 in the activation of jasmonate responses. *Plant Cell* **2011**, *23*, 701–715. [CrossRef] [PubMed]
22. Lorenzo, O.; Chico, J.M.; Sanchez-Serrano, J.J.; Solano, R. *JASMONATE-INSENSITIVE1* encodes a MYC transcription factor essential to discriminate between different jasmonate-regulated defense responses in Arabidopsis. *Plant Cell* **2004**, *16*, 1938–1950. [CrossRef] [PubMed]
23. Niu, Y.; Figueroa, P.; Browse, J. Characterization of JAZ-interacting bHLH transcription factors that regulate jasmonate responses in *Arabidopsis*. *J. Exp. Bot.* **2011**, *62*, 2143–2154. [CrossRef] [PubMed]
24. An, C.; Li, L.; Zhai, Q.; You, Y.; Deng, L.; Wu, F.; Chen, R.; Jiang, H.; Wang, H.; Chen, Q.; et al. Mediator subunit MED25 links the jasmonate receptor to transcriptionally active chromatin. *Proc. Natl. Acad. Sci. USA* **2017**, *114*, E8930–E8939. [CrossRef]
25. Pauwels, L.; Barbero, G.F.; Geerinck, J.; Tilleman, S.; Grunewald, W.; Perez, A.C.; Chico, J.M.; Bossche, R.V.; Sewell, J.; Gil, E.; et al. NINJA connects the co-repressor TOPLESS to jasmonate signalling. *Nature* **2010**, *464*, 788–791. [CrossRef]
26. Shyu, C.; Figueroa, P.; Depew, C.L.; Cooke, T.F.; Sheard, L.B.; Moreno, J.E.; Katsir, L.; Zheng, N.; Browse, J.; Howe, G.A. JAZ8 lacks a canonical degron and has an EAR motif that mediates transcriptional repression of jasmonate responses in Arabidopsis. *Plant Cell* **2012**, *24*, 536–550. [CrossRef]
27. Katsir, L.; Schilmiller, A.L.; Staswick, P.E.; He, S.Y.; Howe, G.A. COI1 is a critical component of a receptor for jasmonate and the bacterial virulence factor coronatine. *Proc. Natl. Acad. Sci. USA* **2008**, *105*, 7100–7105. [CrossRef]
28. Yan, J.; Yao, R.; Chen, L.; Li, S.; Gu, M.; Nan, F.; Xie, D. Dynamic perception of jasmonates by the F-Box protein COI1. *Mol. Plant* **2018**, *11*, 1237–1247. [CrossRef]
29. Qi, T.; Song, S.; Ren, Q.; Wu, D.; Huang, H.; Chen, Y.; Fan, M.; Peng, W.; Ren, C.; Xie, D. The Jasmonate-ZIM-domain proteins interact with the WD-Repeat/bHLH/MYB complexes to regulate Jasmonate-mediated anthocyanin accumulation and trichome initiation in *Arabidopsis thaliana*. *Plant Cell* **2011**, *23*, 1795–1814. [CrossRef]
30. Chini, A.; Fonseca, S.; Fernandez, G.; Adie, B.; Chico, J.M.; Lorenzo, O.; Garcia-Casado, G.; Lopez-Vidriero, I.; Lozano, F.M.; Ponce, M.R.; et al. The JAZ family of repressors is the missing link in jasmonate signalling. *Nature* **2007**, *448*, 666–671. [CrossRef]
31. Song, S.; Qi, T.; Huang, H.; Ren, Q.; Wu, D.; Chang, C.; Peng, W.; Liu, Y.; Peng, J.; Xie, D. The Jasmonate-ZIM domain proteins interact with the R2R3-MYB transcription factors MYB21 and MYB24 to affect Jasmonate-regulated stamen development in *Arabidopsis*. *Plant Cell* **2011**, *23*, 1000–1013. [CrossRef] [PubMed]
32. Zhang, Y.; Wang, X.; Zhang, W.; Yu, F.; Tian, J.; Li, D.; Guo, A. Functional analysis of the two *Brassica AP3* genes involved in apetalous and stamen carpelloid phenotypes. *PLoS ONE* **2011**, *6*, e20930. [CrossRef] [PubMed]
33. Li, Y.; Li, X.; Yang, J.; He, Y. Natural antisense transcripts of *MIR398* genes suppress microR398 processing and attenuate plant thermotolerance. *Nat. Commun.* **2020**, *11*, 5351. [CrossRef] [PubMed]

34. Ren, W.; Wu, F.; Bai, J.; Li, X.; Yang, X.; Xue, W.; Liu, H.; He, Y. BcpLH organizes a specific subset of microRNAs to form a leafy head in Chinese cabbage (*Brassica rapa* ssp. *pekinensis*). *Hortic. Res.* **2020**, *7*, 1. [CrossRef] [PubMed]
35. Verma, S.S.; Yajima, W.R.; Rahman, M.H.; Shah, S.; Liu, J.J.; Ekramoddoullah, A.K.; Kav, N.N. A cysteine-rich antimicrobial peptide from *Pinus monticola* (PmAMP1) confers resistance to multiple fungal pathogens in canola (*Brassica napus*). *Plant Mol. Biol.* **2012**, *79*, 61–74. [CrossRef] [PubMed]
36. Moloney, M.M.; Walker, J.M.; Sharma, K.K. High-efficiency transformation of *Brassica napus* using *Agrobacterium* vectors. *Plant Cell Rep.* **1989**, *8*, 238–242. [CrossRef]
37. Lv, J.; Huang, Q.; Sun, Y.; Qu, G.; Guo, Y.; Zhang, X.; Zhao, H.; Hu, S. Male sterility of an *AHAS*-mutant induced by tribenuron-methyl solution correlated with the decrease of AHAS activity in *Brassica napus* L. *Front. Plant Sci.* **2018**, *9*, 1014. [CrossRef]
38. Livak, K.J.; Schmittgen, T.D. Analysis of relative gene expression data using real-time quantitative PCR and the $2^{-\Delta\Delta CT}$ Method. *Methods* **2001**, *25*, 402–408. [CrossRef]
39. Zhu, Z.; Sun, B.; Xu, X.; Chen, H.; Zou, L.; Chen, G.; Cao, B.; Chen, C.; Lei, J. Overexpression of *AtEDT1/HDG11* in Chinese Kale (*Brassica oleracea* var. *alboglabra*) enhances drought and osmotic stress tolerance. *Front. Plant Sci.* **2016**, *7*, 1285. [CrossRef]
40. Ma, Q.; Xia, Z.; Cai, Z.; Li, L.; Cheng, Y.; Liu, J.; Nian, H. *GmWRKY16* enhances drought and salt tolerance through an ABA-mediated pathway in *Arabidopsis thaliana*. *Front. Plant Sci.* **2018**, *9*, 1979. [CrossRef]
41. Pan, J.; Hu, Y.; Wang, H.; Guo, Q.; Chen, Y.; Howe, G.A.; Yu, D. Molecular mechanism underlying the synergetic effect of jasmonate on abscisic acid signaling during seed germination in Arabidopsis. *Plant Cell* **2020**, *32*, 3846–3865. [CrossRef] [PubMed]
42. Dave, A.; Hernandez, M.L.; He, Z.; Andriotis, V.M.; Vaistij, F.E.; Larson, T.R.; Graham, I.A. 12-oxo-phytodienoic acid accumulation during seed development represses seed germination in *Arabidopsis*. *Plant Cell* **2011**, *23*, 583–599. [CrossRef] [PubMed]
43. Piskurewicz, U.; Lopez-Molina, L. Basic techniques to assess seed germination responses to abiotic stress in *Arabidopsis thaliana*. *Methods Mol. Biol.* **2016**, *1398*, 183–196.
44. Verslues, P.E.; Agarwal, M.; Katiyar-Agarwal, S.; Zhu, J.; Zhu, J.K. Methods and concepts in quantifying resistance to drought, salt and freezing, abiotic stresses that affect plant water status. *Plant J.* **2006**, *45*, 523–539. [CrossRef] [PubMed]
45. Van der Weele, C.M.; Spollen, W.G.; Sharp, R.E.; Baskin, T.I. Growth of Arabidopsis thaliana seedlings under water deficit studied by control of water potential in nutrient-agar media. *J. Exp. Bot.* **2000**, *51*, 1555–1562. [CrossRef]
46. Sato, H.; Suzuki, T.; Takahashi, F.; Shinozaki, K.; Yamaguchi-Shinozaki, K. NF-YB2 and NF-YB3 have functionally diverged and differentially induce drought and heat stress-specific genes. *Plant Physiol.* **2019**, *180*, 1677–1690. [CrossRef]
47. Chen, J.; Nolan, T.M.; Ye, H.; Zhang, M.; Tong, H.; Xin, P.; Chu, J.; Chu, C.; Li, Z.; Yin, Y. Arabidopsis WRKY46, WRKY54, and WRKY70 transcription factors are involved in Brassinosteroid-regulated plant growth and drought responses. *Plant Cell* **2017**, *29*, 1425–1439. [CrossRef]
48. Cao, W.; Cao, B.; Wang, X.; Bai, J.; Xu, Y.Z.; Zhao, J.; Li, X.; He, Y.; Hu, S. Alternatively spliced *BobCAL* transcripts alter curd morphotypes in a collection of Chinese cauliflower accessions. *Hortic. Res.* **2020**, *7*, 160. [CrossRef]
49. Kumar, D.; Yusuf, M.A.; Singh, P.; Sardar, M.; Sarin, N.B. Modulation of antioxidant machinery in α-tocopherol-enriched transgenic *Brassica juncea* plants tolerant to abiotic stress conditions. *Protoplasma* **2013**, *250*, 1079–1089. [CrossRef]
50. Danisman, S.; van Dijk, A.D.; Bimbo, A.; van der Wal, F.; Hennig, L.; de Folter, S.; Angenent, G.C.; Immink, R.G. Analysis of functional redundancies within the Arabidopsis *TCP* transcription factor family. *J. Exp. Bot.* **2013**, *64*, 5673–5685. [CrossRef]
51. Schopfer, P.; Plachy, C. Control of seed germination by abscisic acid. II. Effect on embryo water uptake in *Brassica napus* L. *Plant Physiol.* **1984**, *76*, 155–160. [CrossRef]
52. Fu, J.; Wu, H.; Ma, S.; Xiang, D.; Liu, R.; Xiong, L. OsJAZ1 attenuates drought resistance by regulating JA and ABA signaling in rice. *Front. Plant Sci.* **2017**, *8*, 2108. [CrossRef]
53. Ju, L.; Jing, Y.; Shi, P.; Liu, J.; Chen, J.; Yan, J.; Chu, J.; Chen, K.M.; Sun, J. JAZ proteins modulate seed germination through interaction with ABI5 in bread wheat and *Arabidopsis*. *New Phytol.* **2019**, *223*, 246–260. [CrossRef]
54. Xue, Y.; Warburton, M.L.; Sawkins, M.; Zhang, X.; Setter, T.; Xu, Y.; Grudloyma, P.; Gethi, J.; Ribaut, J.M.; Li, W.; et al. Genome-wide association analysis for nine agronomic traits in maize under well-watered and water-stressed conditions. *Theor. Appl. Genet.* **2013**, *126*, 2587–2596. [CrossRef]
55. Seo, J.S.; Joo, J.; Kim, M.J.; Kim, Y.K.; Nahm, B.H.; Song, S.I.; Cheong, J.J.; Lee, J.S.; Kim, J.K.; Choi, Y.D. OsbHLH148, a basic helix-loop-helix protein, interacts with OsJAZ proteins in a jasmonate signaling pathway leading to drought tolerance in rice. *Plant J.* **2011**, *65*, 907–921. [CrossRef]
56. Wu, H.; Ye, H.; Yao, R.; Zhang, T.; Xiong, L. OsJAZ9 acts as a transcriptional regulator in jasmonate signaling and modulates salt stress tolerance in rice. *Plant Sci.* **2015**, *232*, 1–12. [CrossRef]
57. Lv, Y.; Yang, M.; Hu, D.; Yang, Z.; Ma, S.; Li, X.; Xiong, L. The OsMYB30 transcription factor suppresses cold tolerance by interacting with a JAZ protein and suppressing β-amylase expression. *Plant Physiol.* **2017**, *173*, 1475–1491. [CrossRef]
58. Zhu, D.; Cai, H.; Luo, X.; Bai, X.; Deyholos, M.K.; Chen, Q.; Chen, C.; Ji, W.; Zhu, Y. Over-expression of a novel JAZ family gene from *Glycine soja*, increases salt and alkali stress tolerance. *Biochem. Biophys. Res. Commun.* **2012**, *426*, 273–279. [CrossRef]
59. Kim, T.H.; Bohmer, M.; Hu, H.; Nishimura, N.; Schroeder, J.I. Guard cell signal transduction network: Advances in understanding abscisic acid, CO_2, and Ca^{2+} signaling. *Annu. Rev. Plant Biol.* **2010**, *61*, 561–591. [CrossRef]
60. Yu, H.; Chen, X.; Hong, Y.Y.; Wang, Y.; Xu, P.; Ke, S.D.; Liu, H.Y.; Zhu, J.K.; Oliver, D.J.; Xiang, C.B. Activated expression of an *Arabidopsis* HD-START protein confers drought tolerance with improved root system and reduced stomatal density. *Plant Cell* **2008**, *20*, 1134–1151. [CrossRef]

61. Xu, Z.; Zhou, G. Responses of leaf stomatal density to water status and its relationship with photosynthesis in a grass. *J. Exp. Bot.* **2008**, *59*, 3317–3325. [CrossRef]
62. Gonzalez-Guzman, M.; Pizzio, G.A.; Antoni, R.; Vera-Sirera, F.; Merilo, E.; Bassel, G.W.; Fernandez, M.A.; Holdsworth, M.J.; Perez-Amador, M.A.; Kollist, H.; et al. *Arabidopsis* PYR/PYL/RCAR receptors play a major role in quantitative regulation of stomatal aperture and transcriptional response to abscisic acid. *Plant Cell* **2012**, *24*, 2483–2496. [CrossRef]
63. Leung, J.; Giraudat, J. Aabscisic acid signal transduction. *Annu. Rev. Plant Physiol. Plant Mol. Biol.* **1998**, *49*, 199–222. [CrossRef]
64. Huang, X.Y.; Chao, D.Y.; Gao, J.P.; Zhu, M.Z.; Shi, M.; Lin, H.X. A previously unknown zinc finger protein, DST, regulates drought and salt tolerance in rice via stomatal aperture control. *Genes Dev.* **2009**, *23*, 1805–1817. [CrossRef]

horticulturae

MDPI

Review

Rapeseed as an Ornamental

Meili Xiao [1], Huadong Wang [1], Xiaonan Li [2], Annaliese S. Mason [3,4] and Donghui Fu [1,*]

1 Key Laboratory of Crop Physiology, Ecology and Genetic Breeding, Ministry of Education, Agronomy College, Jiangxi Agricultural University, Nanchang 330045, China; xml846@163.com (M.X.); wanghuadong_jxau@163.com (H.W.)
2 Zaojiao Agricultural Science Park, Shifang 618400, China; lixina4956@sina.com
3 Plant Breeding Department, Institute of Crop Science and Resource Conservation, University of Bonn, 53115 Bonn, Germany; annaliese.mason@uni-bonn.de
4 Department of Plant Breeding, IFZ Research Centre for Biosystems, Land Use and Nutrition, Justus Liebig University, 35392 Giessen, Germany
* Correspondence: fudhui@163.com; Tel.: +86-0791-8381-3142

Abstract: Rapeseed (*Brassica napus*) is one of the most important oil crops worldwide. However, an intriguing new use for rapeseed has recently developed: as an ornamental. Tourism based on blossoming fields of these yellow flowers has become a new economic growth opportunity in China. From a breeding perspective, two main problems currently limit the potential of rapeseed as an ornamental. First, the flowering period is quite short (30 days on average), which limits economic income; second, the flower color in commercial cultivars is currently limited to bright yellow, which may pall quickly for sightseers. This review summarizes the possible problems of using rapeseed as an ornamental, and details factors affecting the flowering period, how the flowering period can be prolonged by integrating optimal cultivation measures or/and spraying with chemical reagents, and ways of creating and breeding rapeseed with diverse flower colors.

Keywords: tourism rapeseed; prolonging flowering period; flower color; chemical regulation; *Brassica napus*

Citation: Xiao, M.; Wang, H.; Li, X.; Mason, A.S.; Fu, D. Rapeseed as an Ornamental. *Horticulturae* **2022**, *8*, 27. https://doi.org/10.3390/horticulturae8010027

Academic Editors: Xiaowu Wang, Jian Wu and Xu Cai

Received: 23 November 2021
Accepted: 23 December 2021
Published: 28 December 2021

Publisher's Note: MDPI stays neutral with regard to jurisdictional claims in published maps and institutional affiliations.

Copyright: © 2021 by the authors. Licensee MDPI, Basel, Switzerland. This article is an open access article distributed under the terms and conditions of the Creative Commons Attribution (CC BY) license (https://creativecommons.org/licenses/by/4.0/).

1. Introduction

Rapeseed (*Brassica napus*) is widely planted worldwide as an important source of edible oil, forage, condiments and vegetables. In China, rapeseed oil comprised 22.3% of edible oil production and accounted for 23.4% of the domestic consumption of vegetable oil in 2017–2018 [1]. However, with rapid urbanization and development and increasing labor costs, the relative economic benefits of planting rapeseed in China are progressively declining. Moreover, mechanized harvesting is often not feasible or easy to implement, such as in terrace cultivation. These factors have contributed to a recent major decline in the cultivation area of rapeseed in China [2]. Concurrently, the average income is increasing, along with work pressures and demand for recreational activities. Agricultural tourism is a new concept that is developing rapidly in China and around the world as a result of progressive urbanization, and rapeseed cultivation is perfectly suited to meet this demand.

Increasingly, rapeseed is becoming a well-known tourist attraction worldwide. Famous scenic spots include Jeju Island in Korea, and Takikawa, Tokyo and Yokohama in Japan, and Cambridgeshire in England. In China, there are at least 20 scenic spots which include rapeseed fields as a point of interest, and ten larger scenic spots which are well-known due to large-scale rapeseed cultivation: Hanzhong in Shanxi Province, Xinghua in Jiangsu Province, Jingmen in Hubei Province, Luoping in Yunnan Province, Tongnan in Chongqing, Menyuan in Qinghai Province, Rui'an in Zhejiang Province, Fengxian in Shanghai, Wuyuan in Jiangxi Province, and Guiding in Guizhou Province. The use of rapeseed as an ornamental is increasing across China. Each year, these scenic spots can attract over 5 million tourists. As an example, Wuyuan County, in Jiangxi Province, China, is considered the most beautiful village in China, due to its natural environmental surroundings and terrace agriculture, which includes rapeseed as a scenic attraction. Tourist income accounts for 51% of the yearly GDP of Wuyuan County (US$145 million each year), demonstrating the strong economic stimulus underlying the rapid development of rapeseed as an ornamental in China.

Rapeseed has a number of desirable qualities for exploitation as an ornamental. First of all, rapeseed is easy to grow and can be used for both agriculture and tourism at the same time. Compared with other tourism projects, the initial investment has a lower commercial risk and can promote farmers' income and economic development. Second, rapeseed fields are impressive and eye-catching, particularly in terraces: the bright yellow color is striking and signals prosperity in China. Thirdly, rapeseed flowers are also in bloom for a relatively long period (30 days on average) compared to other plants, such as peach blossom, cherry blossoms, pear blossom and apricot flower (10–20 day bloom period), improving potential tourist income.

The desirable features for rapeseed as an ornamental include bright flower color, large petals, lodging resistance, strong pest/disease resistance, longer flowering period, more flowers per unit area, and increased height (preferably over one meter). Currently, there are several problems which are restricting development of ornamental rapeseed, each of which is worth investigating and resolving.

2. Relatively Short Flowering Period

2.1. Problem

The first important problem is that the flowering period of rapeseed is relatively short for an ornamental (approximately 30 days in a normal season). According to incomplete statistics, in China, a tourist attraction that uses rapeseed as an ornamental plant can create about US$1.57 million in ticket revenue each day. Here, the flowering period is defined as the time between the first flower opening and flowering finishing.

2.2. Factors Affecting Flowering Period and the Corresponding Countermeasures

In order to solve this problem and maximize the flowering period of rapeseed, it is necessary to clarify the general factors affecting the length of the flowering period. Three major factors regulate the flowering period in rapeseed (Figure 1): the natural environment, cultivation practices, and hormonal control of flowering.

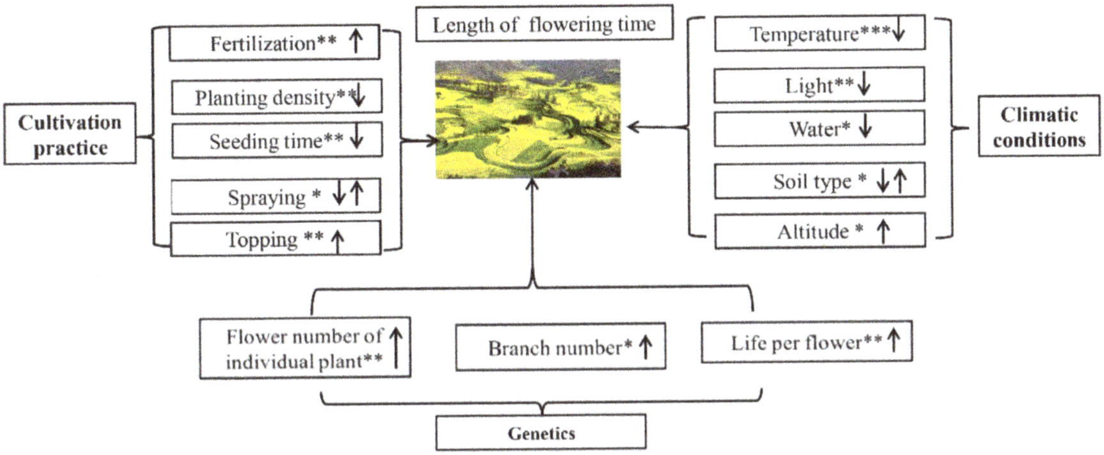

Figure 1. Summary of the factors affecting the length of the rapeseed flowering period. * represents the mild degree of importance, ** represents the moderate degree of importance, *** represents the severe degree of importance. Vertical up "↑" and down arrows "↓" represent positive and negative effects, respectively.

(1) Natural environmental conditions

Temperature variation greatly affects plant growth and development, and the circadian clock below the threshold of the activating temperature stresses regulation networks [3]. In addition, higher temperatures can promote metabolic activity, cell growth, photosynthesis, and plant growth and development [4,5], while lower temperatures (e.g., 20 °C) may decrease enzymatic activities and biochemical reactions, with complex effects on flowering phenotypes [3]. Several signaling pathways involving gibberellins (GAs), auxin, brassinosteroids and transcription factor phytochrome-interacting factor 4 (PIF4) are predicted to be involved in the regulation of growth in response to temperature [6]. While the cold response of rapeseed flowering is complex, it is clear that low temperatures may slow down the flowering process. Generally speaking, rapeseed blossoms faster at temperatures over 15 °C, while the flowering period may be reduced from 30 days to 25 days or less with increasing temperatures. The flowering period of rapeseed is negatively correlated with the length of the growing period across varieties: namely, early-flowering rapeseed blooms at lower temperatures, with a longer flowering period than late-flowering rapeseed [7]. However, varieties which flower too early are highly likely to suffer injury (fading, damage or wilting) caused by the cold, such that the rapeseed flowers generally do not look good when the temperature is below 12 °C [8].

Altitude is the second factor affecting the flowering period, although ultimately the effect of altitude on the flowering period is also associated with the effect of temperature. Higher altitudes have lower temperatures, with an average decrease of 0.6 °C per 100 m above sea level [9]. For terracing in particular, flowering on the top of mountains will be slower than at the bottom, which requires us to adjust sowing times according to local land altitudes to achieve a simultaneous flowering effect. Generally speaking, it is best to sow from higher elevation to lower elevation. Thus, the sowing date is determined by the local climate and altitude. In the Yangtze River Valley, China, it is generally better to adopt a direct seeding mode from October 1st to October 10th and to use a transplanting mode of sowing from October 20th to November 1st.

Light is related to terrain or fog, and light intensity may affect plant growth and development. For example, low light conditions are beneficial for the flowering and growth of *Arabidopsis thaliana* [5]. Moreover, water availability is co-determined by climate, terrain, and irrigation management. Many other environmental factors, such as nutrition and different stresses, can affect the flowering period [10]. Many unfavorable environmental factors, such as water deficiency (drought) and water excess (flooding), can trigger earlier flowering in plants [10]. However, these effects may not always be significant. For example, various drought and flooding treatments do not affect flowering time in wild genotypes of Arabidopsis [11–13]. This may be because plants which initiate the reproductive stage (e.g., flowering) in a timely fashion regardless of environmental stresses have greater reproductive success [10,14,15].

(2) Cultivation practices

Fertilization is critical to the flowering period, although effective utilization of fertilizer depends on the soil type. In general, under low mineral nutrition, Arabidopsis plants tend to flower later [12], but rapeseed tends to flower earlier (Fu et al., unpublished data). In addition, under the prerequisite of non-lodging, the application of more nitrogen fertilizer may delay flowering time, and the total growth period in rapeseed slightly increases with increasing fertilizer application by 1 to 2 days [16]. The reason for this is likely that N signaling regulates nuclear CRY1 (blue-light receptor cryptochrome 1, CRY1) protein abundance, and affects the normal flowering process of the central circadian clock (e.g., GI and CO), as was observed in A. thaliana [17]. We know that Boron-deficient fertilizer may also lead to flowering rather than fruiting. Moderate and full fertilizer application, including organic manure or farm manure, are probably likely to ensure rapeseed flowers have brighter colors and that the rapeseed has a longer flowering period.

Planting density is another factor affecting the flowering period. Higher planting density (e.g., over 300 thousand plants per hectare in the Yangtze Valley, China) is negatively correlated with the flowering period, possibly because it may decrease the number of flowers produced per individual plant or on the main inflorescence, resulting in faster completion of the flowering period. When planting density increases, the total rapeseed growth period has been observed to shorten by three to five days [18]. As a result of increasing planting density, the plant height, the stem diameter, the numbers of effective branches and the length of the main inflorescence decrease, while the effective branching position increases [19,20]. Hence, increasing planting density could promote faster rapeseed growth, especially in early sowing, while sparser planting (e.g., less than 105,000 plants per hectare in the Yangtze Valley, China) may increase the flowering period. In addition, we know that in rapeseed, early seeding results in earlier flowering with a longer flowering period, due to blooming in lower temperatures. Therefore, late sowing times (e.g., sowing after October 25th in the Yangtze River Valley, China) require a higher planting density (e.g., over 300,000 plants per hectare) to make up for the reduction in branches per individual plant.

Moreover, topping after rapeseed bolting (when the plant reaches 30 cm, mechanically remove the top 2–3 cm of the stem) can delay the early flowering period by 7–10 days, and can delay the final flowering period by 3–8 days [21]. Partial topping (topping a certain percentage of plants, e.g., every second plant) may delay the whole flowering period for three to eight days, as the early flowering stage does not change. However, this is a suboptimal strategy for prolonging rapeseed flowering in terms of the use of rapeseed as an ornamental, because fewer flowers will be in bloom across the flowering period in fields that have been topped (as a matter of personal observation). Hence, the optimal time (sunny morning), stage (plant height reaches 30 cm) and proportion (30–50%) of topping should be carefully chosen in order to prolong flowering time, but not to cause a detrimental ornamental effect.

Finally, changing the planting mode also can be a means to prolong the flowering period. In many scenic spots, rapeseed can potentially be used as green manure after the tourism period is over, rather than being harvested as an oil crop, because the tourism income is far higher than the profits of rapeseed oil. To prolong the flowering period, a mixed sowing model of early, middle and later flowering varieties in a ratio of 1:1:1 or 1:3:1 has been applied, and has been shown to achieve higher economic value by prolonging the total rapeseed flowering time (over 5 days, even over 7 days) in Wuyuan, China

(Figure 2) [22]. Hence, planting different cultivars with diverse flowering periods should be considered a viable strategy for prolonging the total flowering period.

(3) Hormonal control of flowering

Targeted breeding of flowering traits should provide an efficient and economic means of prolonging the flowering period in rapeseed. The three main factors impacting the flowering period which should be amenable to trait improvement are the number of flowers per plant, the number of branches per plant and lifetime per flower. Lifetime per flower is the most important factor, and is associated with regulation of flower senescence and abscission. Flower senescence and abscission are affected by several environmental factors, such as seasonal changes (e.g., dark and low-light conditions), insect-mediated pollination, and stresses such as water scarcity, salt, cold and high temperature, wounding and pathogen attack [23–25]. However, the timing of flower senescence is mostly regulated by individual or multiple plant hormones, such as ethylene, cytokinins and abscisic acid.

Flower abscission in almost all monocotyledonous and eudicotyledonous plant species is highly sensitive to ethylene [26]. Therefore, mutating or silencing the key genes related to flower abscission may prolong flower longevity. Little research has been done in this area on rapeseed. However, in Japanese morning glory, an NAC transcription factor, designated *EPHEMERAL1* (*EPH1*), was induced to expression independently by ethylene signaling, and it positively controlled programmed cell death during petal senescence to cause flowers to bloom for a second day [27]. Similarly, in *petunia*, silencing of *PhFBH4* (a basic helix-loop-helix transcription factor) using virus-induced gene silencing or an antisense approach prolonged flower longevity by modulating the ethylene biosynthesis pathway [28]. Also, in Campanula, a naturally occurring 7 bp frameshift of a key ethylene insensitive homologous gene (*Cmeil2*) in the ethylene signaling pathway may be used to screen for flower longevity [29]. Similarly, transgenic Torenia plants incorporating a fragment of a *1-aminocyclopropane-1-carboxylic acid* (*ACC*) oxidase gene had a two day extension of flower longevity [30]. In addition, *Petunia hybrida* plants transformed with boers, a mutated ethylene receptor sensor gene of Brassica oleracea, showed extended flower longevity and apparently larger flowers [31].

Figure 2. Prolonging the flowering period by mixing early-flowering cultivars with a late-flowering cultivar. The circle to the left of the red line indicates mixed sowing of rapeseed from early flowering cultivars and late flowering cultivar, which shows longer flowering period by at least five days than the normal cultivar shown in the circle to the right of the red line.

Aside from ethylene-related genes, the mutation or silencing of other related genes in the same flowering pathway can have similar effects. For instance, ectopically-expressed *TERMINAL FLOWER 1* in transgenic Arabidopsis plants can greatly prolong the vegetative and reproductive phases [32,33]. As well, the anthocyanin regulator *ANTHOCYANIN1* mutant an1 in petunia led to an increase of flower longevity compared to wild-type flowers [34]. Ethylene is thought to primarily regulate flower senescence in ethylene-sensitive flowers, while abscisic acid (ABA) is considered to be the primary regulator in ethylene-insensitive flowers [25]. ABA (abscisic acid) negatively regulated the ethylene biosynthetic pathway in all flower tissues of *Hibiscus rosa-sinensis* L. [35]. In addition, cytokinins can delay petal senescence [36]. Both an increase in ethylene and a reduction in cytokinin determine the initiation of senescence, which directly or indirectly leads to an increase in levels of reactive oxygen species [37]. Furthermore, signal transduction during the process of floral senescence may be associated with G-proteins, changes in calcium activity and the adjustment of protein phosphorylation and dephosphorylation [23]. The flowering period is comprehensively regulated by multiple hormones, and thus, the flowering period may be regulated by spraying with hormone reagents (as described separately below) or by genetic means.

In addition, compatible pollination triggers a series of post-pollination events such as petal senescence [38]. Therefore, lack of pollination or reduced pollination may hypothetically prolong the flowering period. In fact, using male sterile lines with normal flower organs (e.g., *Ogu* cytoplasmic male sterility in *B. napus* with both the sterile plant rate and sterility reaching 100%), grown at a distance from fertile lines which may act as pollinators, has recently been demonstrated to delay the flowering period by five to seven days [39].

Certainly, lines or cultivars with long flowering periods can also be screened for or explored, because different varieties exhibit diverse flowering periods, particularly in low temperatures [7]. This approach by the authors yielded a particularly desirable accession (Figure 3) which can continuously flower for about two months: new flowers are in bloom while old flowers on some branches wither away or pods grow. Although the yield of this accession is 15% less than common varieties because the different maturity times for pods on different branches results in loss of some seeds prior to harvest, this is not a consideration in some areas where rapeseed is used as an ornamental.

Figure 3. A rapeseed variety, LF9, with a longer flowering period than common *B. napus* by at least ten days. This variety can grow and bloom in batches; new branches grow up after old branches flower. Red triangles indicate old siliques.

2.3. Prolonging the Flowering Period by Chemical Reagents

There are three main approaches and reagents for prolonging the flowering period in rapeseed (Figure 4). Firstly, the final flowering stage can be delayed; secondly, the early flowering stage can be extended, and thirdly, both the first and second approaches can be applied simultaneously. A number of reagents are used to adjust the flowering period in horticultural plants, but these are not frequently reported in agronomic crops. Some chemicals are applied to regulate the life of fresh flowers by interfering with the floral development process via hormones: for instance, the ingredients in the liquids that maintain freshness of cut flowers include sucrose, bactericide, organic acids, inorganic salts, and plant growth regulators.

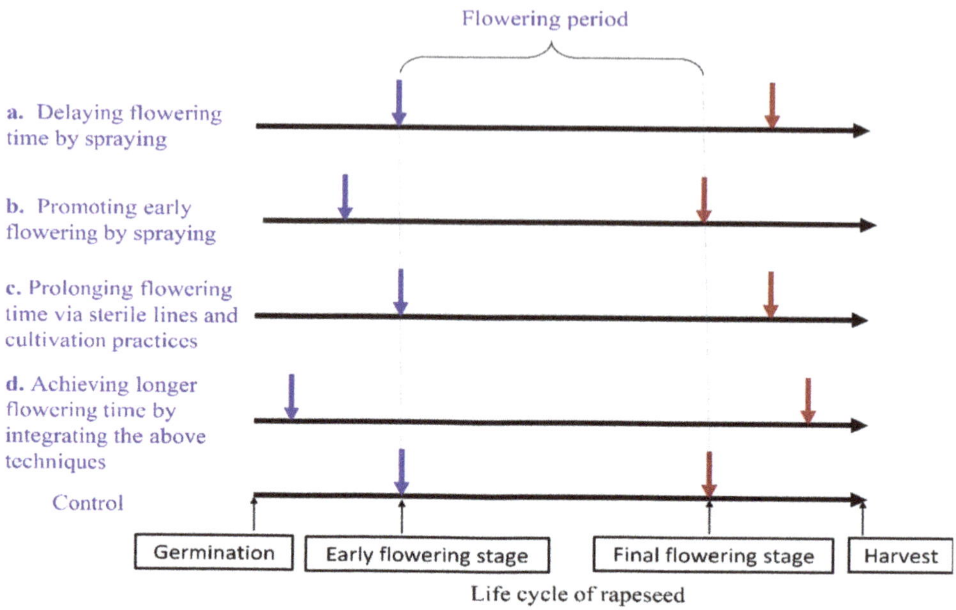

Figure 4. Strategies for prolonging the flowering period. To prolong the flowering period, there are four major methods: (**a**) spraying reagents which can delay flowering; (**b**) spaying reagents which can promote early flowering; (**c**) prolonging the flowering period by using sterile lines and cultivation practices; and (**d**) prolonging the flowering period by integrating techniques (**a**–**c**). The blue and red arrows represent the "point" of the early flowering stage and "point" of final flowering stage, respectively. "Control", at the bottom of the diagram, exhibits the planting cycle of rapeseed under ordinary planting and management conditions.

(1) Ethylene

In a large number of ornamental species, flower longevity can be efficiently enhanced by hindering the plant response to ethylene [40]. Erysimum linifolium (wallflower) senescence can be accelerated by rises in the endogenous ethylene level, and by application of exogenous ethylene released by 2-chloroethylphosphonic acid, but can be delayed by exogenously applied cytokinin, 6-methyl purine (which is an inhibitor of cytokinin oxidase), or treatments with the ethylene signaling inhibitor silver thiosulphate [37]. Silver nano-particles (SNP) and silver thiosulfate (STS) can prolong the vase life of cut carnation flowers by decreasing oxidative stress, enhancing anti-oxidant systems, lowering bacterial populations and delaying flowering [41]. Similarly, silver nanoparticles and chlorophenol may be used to prolong vase life and increase the postharvest quality of cut gerbera flowers [42]. In the same species, either 50 or 100 mg/L carvacrol and either 1 or 2 mg/L silver nanoparticles (SNP) can prolong the vase life of gerbera flowers from 8.3 to 16 days [43]. In addition, either by itself or in combination with silver nano-particles (2–5 nm diameter) and antimicrobial agents, boric acid, as an inhibitor of ethylene production, can significantly extend the vase life and decrease ethylene production, with 9.7 days as the highest cut flower longevity achieved [44]. Moreover, a gaseous compound, 1-methylcyclopropene (1-MCP), hindered a series of plant responses to ethylene, with the result of extending the senescence of cut flowers [40]. In conclusion, although regulating ethylene by treatment with various chemicals is efficient in prolonging vase life, many of these chemicals (in particular heavy metals such as silver nanoparticles) are not suitable for use in prolonging flowering time in the field, as these chemicals are costly to produce and can result in environmental pollution. Therefore, selecting environmental-friendly, low cost, and low toxicity reagents is necessary for prolonging flowering life in the field.

(2) 8-HQC

8-hydroxyquinoline citrate has a bactericidal effect that prevents degradation of cut flowers by bacteria and fungi. Cut carnation flowers treated with 8-HQC, copper coin, and leaf extracts of P. guajava and P. betle exhibit longer vase life and larger flower diameter [45]. Treatment with the extract of Mentha pulegium and 8-hydroxy quinoline sulphate (8-HQS) can increase the vase life of cut rose (*R. hybrida* L.) flowers and lower total chlorophyll (SPAD value), with the greatest vase life longevity reaching 11.2 and 10.3 days using 400 mg/L 8-HQS and 10% of Mentha pulegium extract [46]. Similarly, the application of 8-hydroxyquinoline sulfate (8-HQS) with 2% sucrose can retard the degradation of chlorophyll and carbohydrate, delay flower senescence and extend the vase-life of sweet pea cut flowers up to 17 days [47].

(3) Citric acid

Citric acid is a common ingredient in vase solution formulations, but pre-harvest use of citric acid is a novel method to extend vase life of cut flowers which has previously been reported in tuberose [48]. Treatment with 0.15% citric acid significantly extended the vase life of Lilium flowers from 11.8 days to 14 days [49].

(4) 6-BA and sucrose

In petunia, 6-benzylaminopurine (BA) treatment can prolong flower life for 2.3 to 3.3 days, and promotes increased concentrations of phenols and anthocyanins, but lower total carotenoids [36]. In addition, flower longevity in isolated flowers of Dianthus chinensis was successfully extended by treatment with sucrose for 5 days followed by 3 days in d-glucose [50]. Sucrose treatment also promoted lily flower opening and extended flower life, but did not affect tepal abscission [51].

Since many chemical reagents efficiently prolong the flowering period in various plant species, it is necessary to assess the effects of these reagents on prolonging the flowering period, in addition to the optimal reagent dosage, treatment, time of application, and the possible adverse effects of reagent application. Subsequently, the best reagent can be chosen according to the following standards: simple formulation, low cost, low residuals, high solubility, long-lasting effects, low toxicity and ease of application. Optimal management practices also need to be determined, including dosage, timing, method, location and frequency of spraying. Both the reagents and the spraying management practices can be effectively combined for best effect.

Under the financial support of the key research and development program of Jiangxi Province, our group screened 15 types of chemical reagents for effect in prolonging the flowering period of *B. napus*. Rapeseed flowering was prolonged by 4 days with 2,4-D(2,4-Dichlorophenoxyacetic acid), by 5 days with indole acetic acid, and by 6 days with growth retardants (a mixture of uniconazole, nitrogen, potassium, phosphate, sulphur, zinc, calcium and magnesium fertilizers), and the optimal spraying time was one week before flowering. However, 2,4-D led to a small amount of rapeseed withering. Therefore, a combination of indole acetic acid and growth retardant reagents was found to be optimal for extending rapeseed flowering for 6 days without side effects, an important guideline for the chemical regulation of rapeseed flowering [52–54].

This research is beneficial for prolonging flowering time in an ornamental context short-term, but there are several problems which need to be addressed in the long run. Firstly, the effect and feasibility of spraying with reagents is affected by the rapeseed developmental stage, reagent dosage and frequency of spraying, and environmental conditions such as temperature, weather, terrain, and, in particular, labor. Certainly, unmanned planes are commonly used to spray reagents, but specialized conditions such as terraces still require more labor. Environmental conditions can also significantly reduce the efficacy of spraying, for example, in the spring in the south of China, too much rain often reduces the effectiveness of reagent sprays. Secondly, chemical reagents may cause environmental pollution, including soil and water pollution, despite the selection of reagents with low toxicity. Repeated applications may also cause reagent residue. It is therefore necessary to screen for reagents without environmental residue, and which have permissible environmental impacts under the scope of national law. Therefore, developing long-flowering cultivars is a more efficient goal in the long term to prolong flowering in rapeseed than spraying with chemical reagents.

3. Development of a Range of Flower Colors in Rapeseed

3.1. Problem

Most rapeseed petals are yellow, with a few varieties of white (such as green-white, milk-white and pure white) and orange-red [55]. In contrast to white flowers, yellow flowers are more attractive to pollinating insects, and have a higher utilization ratio of light energy. Yellow flower color in rapeseed is recessive compared to white flower color in rapeseed, and is maintained in evolution. The petals of almost all rapeseed cultivars are yellow, which can result in aesthetic fatigue from an ornamental perspective, and is hence not beneficial to sustained development of the ornamental rapeseed industry. The addition of another bright color system, e.g., pure red, purple or even blue, would greatly improve the prospects for development of rapeseed as an ornamental attraction. The development of additional flower colors in rapeseed is therefore a high priority.

3.2. The Molecular Mechanisms of Flower Color

Pigments can generally be classified into three groups: carotenoids, flavonoids, and alkaloids. The presence of pigments, such as blue to red anthocyanins, yellow to reddish carotenoids, or red betalains, make flowers exhibit different colors [56]. Carotenoids may exhibit brilliant red, orange and yellow [57]. The processes of both anthocyanin synthesis and carotenoid synthesis and degradation are highly conserved across angiosperms [58]. The presence, type, and amount of carotenoid pigment contribute to substantial variation in flower color [59]. Petals of many different species contain carotenoids, but the exact compositions vary between and sometimes within species [60]. Carotenoids and flavonoids are deposited in the cytoplasmic plastids, and in vacuoles, respectively. Carotenoids are one of the most widely distributed pigments, and are found in many organs in higher plants, such as flowers, fruits, leaves and roots. Carotenoids contain two major categories: carotene and lutein. In Osmanthus fragrans, yellow petals have less β-carotene, golden yellow petals contain a large quantity of lutein, but a small quantity of α-carotene and β-carotene, and orange–red petals contain a high level of α-carotene and β-carotene [61].

Flavonoids, which are widely distributed in plants, belong to a large class of secondary metabolites [62]. Water soluble flavonoids, as one of most important pigment groups, show the full spectrum of colors, varying from pale yellow to blue-purple [60]. Anthocyanin, as a major class of flavonoid, exhibits a wide range of colors, from red and pink series to blue-violet [60]. Other flavonoids produce a pure yellow series, from deep yellow, controlled by chalcone and aurone, to light yellow or nearly colorless, regulated by flavones, flavonols and flavanones [60].

The regulation of key rate-limiting enzymes (e.g., chalcone synthase, CHS) may change flower colors in the flavonoid biosynthetic pathway. Transgenic petunia expressing *FhCHS1* from Freesia hybrida changed flower color from white to pink [63]. Similarly, transgenic tobacco plants constitutively expressing *McCHS* from Malus (crabapple) had increased anthocyanin accumulation and a deeper red petal color in contrast to untransformed control lines [64].

Delphinidin-based anthocyanins are the major constituents of violet and blue flowers. Six major classes of anthocyanidins are predominant in nature, including pelargonidin, cyanidin, peonidin, delphinidin, petunidin and malvidin [65]. More than 600 anthocyanins have been found, with core anthocyanidins varying in their side chain decorations [65]. The anthocyanin color is affected by the number of hydroxyl groups on the B-ring of the anthocyanidins: the more hydroxyl groups present, the bluer the color. A *flavonoid 3′,5′-hydroxylase* gene, as a key enzyme controlling delphinidin biosynthesis, has previously been genetically engineered to establish an exclusive accumulation of delphinidin (Dp)-type anthocyanins and make petals bluish in some plants, such as carnations, roses and chrysanthemums, but appearance of a true blue color was developed by modifying anthocyanins with multiple aromatic acyl groups (often referred to as polyacylated anthocyanins) [66]. Two cytochrome P450s, *flavonoid 3′-hydroxylase* (*F3′H*) and *flavonoid 3′,5′-hydroxylase* (*F3′5′H*), determine the hydroxylation pattern which controls flower color. Most *F3′H* and *F3′5′H* are CYP75B and CYP75A, respectively, with the exception of the *F3′5′Hs* arising from gene duplication and variation of CYP75B in Compositae [67]. The lack of delphinidin-related *F3′5′Hs* results in a lack of naturally occurring blue/violet flower colors in roses and carnations. Therefore, expressing *F3′5′H* coding regions can result in the emergence of carnations and roses with novel blue colors [67]. Similarly, expressing violet *F3′5′H* genes in some Rosa hybrida cultivars led to the accumulation of numerous delphinidins and the appearance of a novel bluish flower color [68]. Suppression of an enzyme (*flavonoid 3′, 5′-hydroxylase* (*F3′5′H*)) in the flavonoid pathway of cyclamen flowers via antisense inhibition led to reduction of delphinidin-derived pigment levels, but an increase in cyanidin-derived pigment content, resulting in a shift of the petal color from purple to red/pink [69]. In addition, RNAi-mediated suppression of *anthocyanin 5,3′-aromatic acyltransferase* (5/3′AT) and *flavonoid 3′,5′-hydroxylase* (F3′5′H) expression led to lilac and pale-blue flower colors [70].

The color stability of anthocyanins can also be affected by light, temperature, pH, oxidants and reducing reagents [71,72]. For instance, acidic pH makes anthocyanins red, neutral or nearly neutral pH leads to lack of color and alkaline pH produces blue colors. A vacuolar iron transporter, TgVit1, plays an important role in blue coloration in tulip petals through iron accumulation [73]. Mutation of *PH5* (a P(3A)-ATPase proton pump) decreased vacuolar acidification in *petunia* petals, also resulting in a blue flower color [74].

3.3. *How to Breed Rapeseed of Different Colors*

To date, research related to flower color in rapeseed has mainly focused on the genetics of white and orange flowers, with a few reports related to genetic mechanisms and gene identification. For example, in Brassica species, the insertion of a *CACTA*-like transposable element in a coding region of a *carotenoid cleavage dioxygenase 4* gene was found to lead to disruption of gene expression, switching flower color from white to yellow [75]. In addition, the major volatile apocarotenoid in white rapeseed petals was determined to be alpha-ionone, which is not the case for yellow petals [75]. In the molecular mechanism of yellow to orange flower formation in rapeseed, a zeaxanthin epoxidase encoding gene was discovered, which catalyzes the conversion of zeaxanthin to antherxanthin and violaxanthin, knocked out this gene by CRISPR/Cas9 technique, it was found that the accumulation of lutein in the petals increased, and the petals were changed from yellow to orange [76]. A mutation in a *phytoene desaturase 3* gene in the yellow-white petal mutant interrupted the carotenoid biosynthesis pathway in *ywf*, resulting in a decrease in carotenoid content and a yellow-white petal phenotype [55].

There are two major avenues to produce new flower colors in rapeseed. The first method is to alter the intrinsic pigment composition and content in the petals. For this method, it is important to know how the genes in the carotenoid synthesis pathway mutate or can be mutated to modify flower color without affecting basic plant physiology [59]. Many ornamental plant species show a limited range of flower color because of the presence of limited kinds of flavonoids. Overexpressing heterologous genes and/or down regulating endogenous genes in flavonoid biosynthesis pathways can be used to alter flower color [77,78]. The well-characterized flavonoid biosynthesis pathway is often metabolically engineered to change anthocyanin profiles and then to enrich flower color, as previously discussed. However, little research in this area has been reported in rapeseed.

The second major method for altering flower color in rapeseed is to import new genes or alleles affecting petal pigmentation using wide hybridization or transgenic techniques. The success of quite distant hybridization is critical for the introgression of flower color into rapeseed, as only wilder relatives contain useful variation in flower color. Hence, some additional biotechnological means are required to enhance the success rate of hybridization, such as embryo rescue, multiple pollinations and pollination after grafting. The crucifer *Orychophragmus violaceus* is widely cultivated as an ornamental plant for its beautiful purple flowers, and can be hybridized with rapeseed with some difficulty [22]. A homolog of *AtPAP2* (*Arabidopsis Production of Anthocyanin Pigment 2*) was identified by transcriptomic analysis using a *B. napus—O. violaceus* disomic chromosome addition line exhibiting slightly red petals. The ectopic expression of the *OvPAP2* gene can produce red-flowering oilseed rape, enhancing its ornamental value [22]. This approach successfully produced red-flowering rapeseed, but more research is still required to improve the color intensity, and application of this line in practice is limited by transgenic regulations.

According to the Announcement of Variety Rights on 1 September 2017 (No. 109) from the Office for the Protection of New Plant Varieties, the Ministry of Agriculture of China, the group of Xiaolan Li first developed rapeseed with red and purple flower colors, although with variation in not only the depth and shade of color observed, but also in production of normal pollen grains and in seed setting ability. They crossed one purple-flowering radish (*Raphanus*) landrace as a donor with rapeseed, and used backcrossing and self-pollination to obtain one plant with slightly red flower color. This plant was self-pollinated, crossed with common yellow and white *B. napus* with excellent comprehensive properties, and continuous rounds of selection were undertaken on the progeny to finally obtain more than ten colors, including purple, red, deep red and mottled flower colors (Figure 5). This process has resulted in five new protected plant varieties (Announcement of Variety Rights on 1 September 2017 (No. 109), which constitute the first rapeseed varieties with varying flower colors which can be widely applied in production.

Figure 5. Colorful, non-genetically-modified *B. napus* accessions with normal seed setting and stable heritability of flower color. (**a**) Control (natural color), (**b–u**): diverse colorful accessions bred by Donghui Fu from crosses among common cultivars, with colorful rapeseed varieties provided by Mr. Xiaonan Li and his colleagues from Zaojiao Agricultural Science Park, Shifang, Sichuan, China.

Certainly, there are several other important Cruciferae species that are also used as ornamental plants (Figure 6), such as *Matthiola incana*, which has purple, pink or white flowers [79], *Aubrieta* x *cultorum*, with purple, violet or white flowers [80], *Cheiranthus allionii*, with yellow and orange flower types, *Hesperis matronalis* with white, *lilac*, or purple flowers [80], *Iberis umbellata*, with red and purple-red flowers and spherical inflorescences [81], radish (*Raphanus sativus*) with red, purple, and white flower types [82], and *Heliophila coronopifolia* with blue flowers [81]. These species can be used as donors of elite flowering color germplasm into rapeseed by hybridization or transgenic modification. As red- and purple-flowering rapeseed have now been developed, the major focus should be on creating blue-flowering rapeseed, although this is difficult because of the lack of pure blue flowers in Cruciferae species. One option could be to produce blue colors via manipulating the delphinidin-related *F3′5′H* gene derived from the pure blue flower of *Delphinium ajacis* (Figure 7).

Figure 6. Ornamental Cruciferae species. (**a**) *Orychophragmus violaceus;* (**b**) *Hesperis matronalis;* (**c**) *Iberis amara;* (**d**) *Matthiola incana;* (**e**) *Aubrieta cultorum;* (**f**) *Cheiranthus allionii;* (**g–i**) *Raphanus sativus* (radish). (**a–f**) photos are provided by by Mr. Liming Zhao (Jiuquan, Gansu Province, China).

Figure 7. Diverse *Delphinium ajacis* (photos provided by Mr. Liming Zhao, Jiuquan, Gansu Province, China).

4. Conclusions

Rapeseed has developed rapidly as an ornamental plant, and its economic value has risen sharply, especially in rural areas of China, bringing considerable economic value to rapeseed growers. However, there are two major hindrances to further development: a short flowering period and a monotonous flower color (Figure 8). To make full use of the ornamental value of rapeseed, the guiding principle is to produce rapeseed with optimal vegetative and flowering growth. First and foremost, this review discusses the factors that affect the flowering period, and proposes that each rapeseed variety produced as an ornamental plant should be grown in the optimal environment correspondingly, the flowering period can be adjusted by temperature, light, water and soil conditions, cultivation and planting methods, and cultivation systems (including sowing time, fertilization amount and fertilization time, thinning time and planting density). Secondly, cultivating rapeseed varieties with a long flowering period and rich color variety is the most direct and effective means to enhance the ornamental value of rapeseed. Creating new flower colors can be achieved through gene pyramiding, wide hybridization with other Cruciferae ornamental plants, or transgenic approaches. Red- and purple-flowering rapeseed varieties have already been developed (Xiaolan Li, pers. comm, 2018). The greatest difficulty is expected in creating blue-flowering rapeseed, but this may be possible by manipulating the delphinidin-related *F3′5′H* gene derived from blue-flowering *Delphinium ajacis*. More information is also required about the genetic determination of flower color in Brassica species, such as the number of genes controlling flower color, the dominant or recessive effects of genes and alleles, heritability and the corresponding molecular mechanisms. With the solution of these problems and the release of colorful and long-flowering cultivars, rapeseed can greatly increase its economic value as an ornamental as a novel functional utilization.

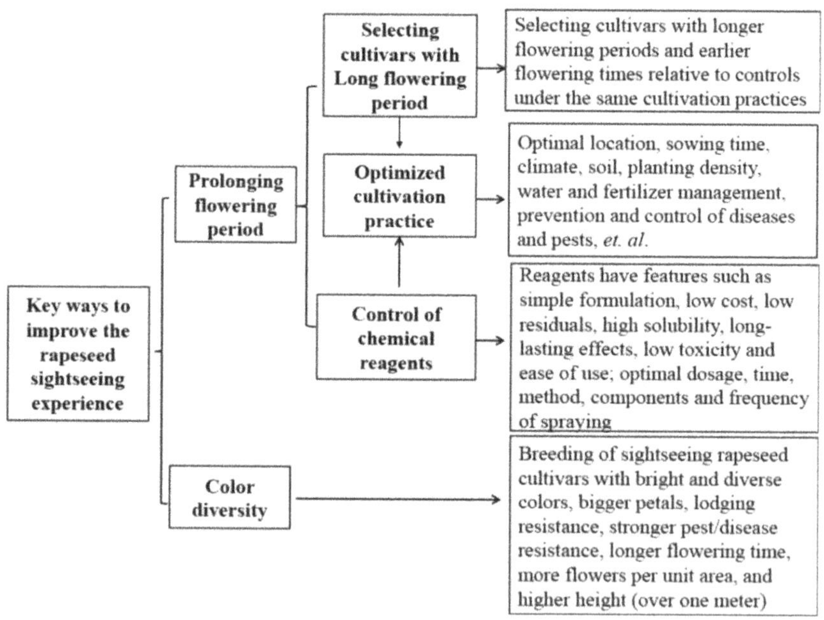

Figure 8. Key points for rapeseed as a tourist attraction.

Author Contributions: Conceptualization, M.X. and D.F.; resources, X.L.; writing—original draft preparation, D.F.; writing—review and editing, H.W.; visualization, A.S.M.; supervision,; funding acquisition, D.F. All authors have read and agreed to the published version of the manuscript.

Funding: This work was supported financially by National Natural Science Foundation of China (Code: 31860417 and 32060494); and Emmy Noether DFG grant MA 6473/1-1.

Institutional Review Board Statement: Not applicable.

Informed Consent Statement: Not applicable.

Data Availability Statement: This study did not report any data.

Conflicts of Interest: The authors declare no conflict of interest.

References

1. Pu, G.; Zheng, M.; Lu, S.; Huang, J. Study on the Use of Cooking Oil in Chinese Dishes. *Int. J. Environ. Res. Public Health* **2019**, *16*, 3367. [CrossRef]
2. Liu, C.; Huang, J.; Leng, B.; Feng, Z.; Li, J. Current situation, development difficulties and suggestions of Chinese rape industry. *J. China Agric. Univ.* **2017**, *22*, 203–210.
3. Wigge, P.A. Ambient temperature signalling in plants. *Curr. Opin. Plant Biol.* **2013**, *16*, 661–666. [CrossRef] [PubMed]
4. Li, B.; Suzuki, J.-I.; Hara, T. Latitudinal variation in plant size and relative growth rate in *Arabidopsis thaliana*. *Oecologia* **1998**, *115*, 293–301. [CrossRef] [PubMed]
5. Moharekar Lokhande, S.; Moharekar, S.; Kobayashi, T.; Ishii, H.; Sumida, A.; Hara, T. Phenotypic plasticity and ecotypic variations in growth and flowering time of *Arabidopsis thaliana* (L.) under different light and temperature conditions. *Indian J. Exp. Biol.* **2014**, *52*, 344–351.
6. Stavang, J.A.; Gallego-Bartolome, J.; Gomez, M.D.; Yoshida, S.; Asami, T.; Olsen, J.E.; Garcia-Martinez, J.L.; Alabadi, D.; Blazquez, M.A. Hormonal regulation of temperature-induced growth in Arabidopsis. *Plant J.* **2009**, *60*, 589–601. [CrossRef] [PubMed]
7. Zhou, Q.; Han, D.; Mason, A.S.; Zhou, C.; Zheng, W.; Li, Y.; Wu, C.; Fu, D.; Huang, Y. Earliness traits in rapeseed (*Brassica napus*): SNP loci and candidate genes identified by genome-wide association analysis. *DNA Res.* **2017**, *25*, 229–244. [CrossRef]
8. Xu, L.; Hu, K.; Zhang, Z.; Guan, C.; Chen, S.; Hua, W.; Li, J.; Wen, J.; Yi, B.; Shen, J.; et al. Genome-wide association study reveals the genetic architecture of flowering time in rapeseed (*Brassica napus* L.). *DNA Res.* **2016**, *23*, 43–52. [CrossRef] [PubMed]
9. Lobell, D.B.; Cahill, K.N.; Field, C.B. Historical effects of temperature and precipitation on California crop yields. *Clim. Chang.* **2007**, *81*, 187–203. [CrossRef]
10. Ausin, I.; Alonso-Blanco, C.; Martinez-Zapater, J.M. Environmental regulation of flowering. *Int. J. Develop. Biol.* **2005**, *49*, 689–705. [CrossRef]
11. Pigliucci, M.; Kolodynska, A. Phenotypic plasticity and integration in response to flooded conditions in natural accessions of *Arabidopsis thaliana* (L.) Heynh (Brassicaceae). *Ann. Botany* **2002**, *90*, 199–207. [CrossRef] [PubMed]
12. Pigliucci, M.; Whitton, J.; Schlichting, C.D. Reaction norms of Arabidopsis. I. Plasticity of characters and correlations across water, nutrient and light gradients. *J. Evolut. Biol.* **1995**, *8*, 421–438. [CrossRef]
13. Kang, D.-J.; Futakuchi, K. Effect of Moderate Drought-Stress on Flowering Time of Interspecific Hybrid Progenies (*Oryza sativa* L. × *Oryza glaberrima* Steud.). *J. Crop Sci. Biotechnol.* **2019**, *22*, 75–81. [CrossRef]
14. Lee, J.H.; Lee, J.S.; Ahn, J.H. Ambient temperature signaling in plants: An emerging field in the regulation of flowering time. *J. Plant Biol.* **2008**, *51*, 321–326. [CrossRef]
15. Liang, M.; Xiao, S.; Cai, J.; Ow, D.W. OXIDATIVE STRESS 3 regulates drought-induced flowering through APETALA 1. *Biochem. Biophys. Res. Commun.* **2019**, *519*, 585–590. [CrossRef]
16. Zeng, Y. Effects of different fertilizer and density on the growth and yield of rapeseed. *Huazhong Agric. Univ.* **2011**, *6*, 26–30.
17. Yuan, S.; Zhang, Z.W.; Zheng, C.; Zhao, Z.Y.; Wang, Y.; Feng, L.Y.; Niu, G.; Wang, C.Q.; Wang, J.H.; Feng, H.; et al. Arabidopsis cryptochrome 1 functions in nitrogen regulation of flowering. *Proc. Nat. Acad. Sci. USA* **2016**, *113*, 7661–7666. [CrossRef]
18. Shao, L. Effects of planting density of the development and yield of rapessed (*Brassica napus* L.) under different sowing dates. *Huazhong Agric. Univ.* **2009**, *4*, 12–14.
19. Yantai, G.; Harker, K.N.; Kutcher, H.R.; Gulden, R.H.; Irvine, B.; May, W.E.; O'Donovan, J.T. Canola seed yield and phenological responses to plant density. *Can. J. Plant Sci.* **2016**, *96*, 151–159. [CrossRef]
20. Helland, S.J.J.D.; Gradworks, T. *Effects of Environment and Planting Density on Plant Stature, Flowering Time, and Ear Set in IBM Populations of Maize*; Theses and Dissertation; Iowa State University,: Ames, IA, USA, 2012.
21. Gao, W. A study of stalk picking to postpone the flowering period of *Brassica napus*. *Guizhou Agric. Sci.* **1990**, *7*, 37–40.
22. Fu, W.; Chen, D.; Pan, Q.; Li, F.; Zhao, Z.; Ge, X.; Li, Z. Production of red-flowered oilseed rape via the ectopic expression of *Orychophragmus violaceus* OvPAP2. *Plant Biotechnol. J.* **2017**, *16*, 367–380. [CrossRef]
23. Rubinstein, B. Regulation of cell death in flower petals. *Plant Mol. Biol.* **2000**, *44*, 303–318. [CrossRef]
24. Taylor, J.E.; Whitelaw, C.A. Signals in abscission. *New Phytol.* **2001**, *151*, 323–340. [CrossRef]
25. Tripathi, S.K.; Tuteja, N. Integrated signaling in flower senescence: An overview. *Plant Signal. Behav.* **2007**, *2*, 437–445. [CrossRef]
26. van Doorn, W.G. Effect of ethylene on flower abscission: A survey. *Ann. Botany* **2002**, *89*, 689–693. [CrossRef]
27. Shibuya, K.; Shimizu, K.; Niki, T.; Ichimura, K. Identification of a NAC transcription factor, EPHEMERAL1, that controls petal senescence in Japanese morning glory. *Plant J.* **2014**, *79*, 1044–1051. [CrossRef]

28. Yin, J.; Chang, X.; Kasuga, T.; Bui, M.; Reid, M.S.; Jiang, C.Z. A basic helix-loop-helix transcription factor, PhFBH4, regulates flower senescence by modulating ethylene biosynthesis pathway in petunia. *Hortic. Res.* **2015**, *2*, 15059. [CrossRef] [PubMed]
29. Jensen, L.; Hegelund, J.N.; Olsen, A.; Lutken, H.; Muller, R. A natural frameshift mutation in Campanula EIL2 correlates with ethylene insensitivity in flowers. *BMC Plant Biol.* **2016**, *16*, 117. [CrossRef] [PubMed]
30. Aida, R.; Yoshida, T.; Ichimura, K.; Goto, R.; Shibata, M. Extension of flower longevity in transgenic torenia plants incorporating ACC oxidase transgene. *Plant Sci.* **1998**, *138*, 91–101. [CrossRef]
31. Shaw, J.-F.; Chen, H.-H.; Tsai, M.-F.; Kuo, C.-I.; Huang, L.-C. Extended flower longevity of *Petunia hybrida* plants transformed with boers, a mutated ERS gene of *Brassica oleracea*. *Mol. Breed.* **2002**, *9*, 211–216. [CrossRef]
32. Bradley, D.; Ratcliffe, O.; Vincent, C.; Carpenter, R.; Coen, E. Inflorescence commitment and architecture in Arabidopsis. *Science* **1997**, *275*, 80–83. [CrossRef]
33. Ratcliffe, O.J.; Amaya, I.; Vincent, C.A.; Rothstein, S.; Carpenter, R.; Coen, E.S.; Bradley, D.J. A common mechanism controls the life cycle and architecture of plants. *Development* **1998**, *125*, 1609–1615. [CrossRef] [PubMed]
34. Prinsi, B.; Negri, A.S.; Quattrocchio, F.M.; Koes, R.E.; Espen, L. Proteomics of red and white corolla limbs in petunia reveals a novel function of the anthocyanin regulator ANTHOCYANIN1 in determining flower longevity. *J. Proteom.* **2016**, *131*, 38–47. [CrossRef] [PubMed]
35. Trivellini, A.; Ferrante, A.; Vernieri, P.; Serra, G. Effects of abscisic acid on ethylene biosynthesis and perception in Hibiscus rosa-sinensis L. flower development. *J. Exp. Botany* **2011**, *62*, 5437–5452. [CrossRef]
36. Trivellini, A.; Cocetta, G.; Vernieri, P.; Mensuali-Sodi, A.; Ferrante, A. Effect of cytokinins on delaying petunia flower senescence: A transcriptome study approach. *Plant Mol. Biol.* **2015**, *87*, 169–180. [CrossRef]
37. Salleh, F.M.; Mariotti, L.; Spadafora, N.D.; Price, A.M.; Picciarelli, P.; Wagstaff, C.; Lombardi, L.; Rogers, H. Interaction of plant growth regulators and reactive oxygen species to regulate petal senescence in wallflowers (*Erysimum linifolium*). *BMC Plant Biol.* **2016**, *16*, 77. [CrossRef]
38. Stead, A.D. Pollination-induced flower senescence: A review. *Plant Growth Regul.* **1992**, *11*, 13–20. [CrossRef]
39. Fu, D.; Zhu, L.; Xiao, M.; Zhou, Q. Methods, Medicament and Preparation Methods for Prolonging the Flowering Period of Ornamental Rapeseed. Patent CN201610209467.6, 6 April 2016.
40. Serek, M.; Sisler, E.C.; Reid, M.S. 1-Methylcyclopropene, a novel gaseous inhibitor of ethylene action, improves the life of fruits, cut flowers and potted plants. *Plant Bioregul. Hortic.* **1995**, *394*, 337–346. [CrossRef]
41. Hashemabadi, D.; Liavali, M.H.; Kaviani, B.; Mousavi, M.; Keyghobadi, S.; Zahiri, S. Effect of nano-silver and boric acid on extending the vase life of cut rose (*Rosa hybrida* L.). *J. Environ. Biol.* **2014**, *35*, 833–838. [PubMed]
42. Safa, Z.; Hashemabadi, D.; Kaviani, B.; Nikchi, N.; Zarchini, M. Studies on quality and vase life of cut Gerbera jamesonii cv. 'Balance' flowers by silver nanoparticles and chlorophenol. *J. Environ. Biol.* **2015**, *36*, 425–431.
43. Solgi, M.; Kafi, M.; Taghavi, T.S.; Naderi, R. Essential oils and silver nanoparticles (SNP) as novel agents to extend vase-life of gerbera (Gerbera jamesonii cv. 'Dune') flowers. *Postharvest Biol. Technol.* **2009**, *53*, 155–158. [CrossRef]
44. Hashemabadi, D. The role of silver nano-particles and silver thiosulfate on the longevity of cut carnation (Dianthus caryophyllus) flowers. *J. Environ. Biol.* **2014**, *35*, 661–666. [PubMed]
45. Rahman, M.M.; Ahmad, S.H.; Lgu, K.S. Psidium guajava and Piper betle leaf extracts prolong vase life of cut carnation (Dianthus caryophyllus) flowers. *Sci. World J.* **2012**, *2012*, 102805. [CrossRef] [PubMed]
46. Hashemabadi, D.; Torkashvand, A.M.; Kaviani, B.; Bagherzadeh, M.; Rezaalipour, M.; Zarchini, M. Effect of Mentha pulegium extract and 8-hydroxy quinoline sulphate to extend the quality and vase life of rose (Rosa hybrid) cut flower. *J. Environ. Biol.* **2015**, *36*, 215–220.
47. Elhindi, K.M. Evaluation of several holding solutions for prolonging vase-life and keeping quality of cut sweet pea flowers (*Lathyrus odoratus* L.). *Saudi J. Biolog. Sci.* **2012**, *19*, 195–202. [CrossRef]
48. Darandeh, N.; Shoor, E.H.M. Extention of post-harvest vase life of Lilium cv. Brunello by Citric Acid sprays during plant growth [Conference Abstract]. In Proceedings of the Symplitaly, Pesica, Italy, 30 July 2010.
49. Darandeh, N.; Hadavi, E. Effect of Pre-Harvest Foliar Application of Citric Acid and Malic Acid on Chlorophyll Content and Post-Harvest Vase Life of Lilium cv. Brunello. *Front. Plant Sci.* **2012**, *2*, 106. [CrossRef] [PubMed]
50. Dar, R.A.; Tahir, I.; Ahmad, S.S. Sugars and sugar alcohols have their say in the regulation of flower senescence in *Dianthus chinensis* L. *Sci. Hortic.* **2014**, *174*, 24–28. [CrossRef]
51. Arrom, L.; Munne-Bosch, S. Sucrose accelerates flower opening and delays senescence through a hormonal effect in cut lily flowers. *Plant Sci.* **2012**, *188–189*, 41–47. [CrossRef]
52. Fu, D.; Zhu, L.; Xiao, M.; Zhou, Q. A Method of Prolonging the Flowering Period of Rapeseed and the Seed Production Method of the Sterile Line Rapeseed. Patent CN201610250303.8, 20 April 2016.
53. Fu, D.; Zhu, L.; Xiao, M.; Zhou, Q. A Reagent of Prolonging the Flowering Period and the Corresponding Usage Method. Patent CN201610249992.0, 27 March 2016.
54. Zhu, L.; Liu, B.; Xiao, M.; Zhou, Q.; Fu, D. A Study on Chemical Regulation of Flowering in Brassica napus. *Acta Agric. Univ. Jiangxiensis* **2017**, *39*, 1057–1066.
55. Zhao, C.; Safdar, L.B.; Xie, M.; Shi, M.; Liu, S.J.T.C.J. Mutation of the *PHYTOENE DESATURASE 3* gene causes yellowish-white petals in *Brassica napus*. *Crop J.* **2021**, *9*, 1124–1134. [CrossRef]

56. Wan, H.; Yu, C.; Luo, L.; Han, Y.; Pan, H.; Zhang, Q.J.M.P.B. Extraction and Determination of Flavonoids and Carotenoids in Petals of Roses (*Rosa spp.*). *Mol. Plant Breeding* **2019**, *17*, 20–24.
57. Britton, G.; Liaaen-Jensen, S.; Pfander, H. *Carotenoids Handbook*; Birkhäuser: Basel, Switzerland, 2004. [CrossRef]
58. Grotewold, E. The genetics and biochemistry of floral pigments. *Ann. Review Plant Biol.* **2006**, *57*, 761–780. [CrossRef]
59. Wessinger, C.A. A genetic route to yellow flowers. *New Phytol.* **2015**, *206*, 1193–1195. [CrossRef]
60. Zhao, D.; Tao, J. Recent advances on the development and regulation of flower color in ornamental plants. *Front. Plant Sci.* **2015**, *6*, 261. [CrossRef] [PubMed]
61. Han, Y.; Wang, X.; Chen, W.; Dong, M.; Yuan, W.; Liu, X.; Shang, F. Differential expression of carotenoid-related genes determines diversified carotenoid coloration in flower petal of Osmanthus fragrans. *Tree Genet. Genomes* **2014**, *10*, 329–338. [CrossRef]
62. Lysanne, A.; Dietmar, K.; Florian, S.; Otmar, S.J.I.J.o.M.S. Comparative Metabolite Profiling of Triterpenoid Saponins and Flavonoids in Flower Color Mutations of *Primula veris* L. *Int. J. Mol. Sci.* **2017**, *18*, 153.
63. Sun, W.; Meng, X.; Liang, L.; Jiang, W.; Huang, Y.; He, J.; Hu, H.; Almqvist, J.; Gao, X.; Wang, L. Molecular and Biochemical Analysis of Chalcone Synthase from Freesia hybrid in flavonoid biosynthetic pathway. *PLoS ONE* **2015**, *10*, e0119054. [CrossRef]
64. Tai, D.; Tian, J.; Zhang, J.; Song, T.; Yao, Y. A Malus crabapple chalcone synthase gene, *McCHS*, regulates red petal color and flavonoid biosynthesis. *PLoS ONE* **2014**, *9*, e110570. [CrossRef] [PubMed]
65. Zhang, Y.; Butelli, E.; Martin, C. Engineering anthocyanin biosynthesis in plants. *Curr. Opin. Plant Biol.* **2014**, *19*, 81–90. [CrossRef] [PubMed]
66. Sasaki, N.; Nakayama, T. Achievements and perspectives in biochemistry concerning anthocyanin modification for blue flower coloration. *Plant Cell Physiol.* **2015**, *56*, 28–40. [CrossRef]
67. Tanaka, Y.; Brugliera, F. Flower colour and cytochromes P450. *Philos. Trans. R. Soc. B Biolog. Sci.* **2013**, *368*, 20120432. [CrossRef] [PubMed]
68. Katsumoto, Y.; Fukuchi-Mizutani, M.; Fukui, Y.; Brugliera, F.; Holton, T.A.; Karan, M.; Nakamura, N.; Yonekura-Sakakibara, K.; Togami, J.; Pigeaire, A.; et al. Engineering of the rose flavonoid biosynthetic pathway successfully generated blue-hued flowers accumulating delphinidin. *Plant Cell Physiol.* **2007**, *48*, 1589–1600. [CrossRef]
69. Boase, M.R.; Lewis, D.H.; Davies, K.M.; Marshall, G.B.; Patel, D.; Schwinn, K.E.; Deroles, S.C. Isolation and antisense suppression of flavonoid 3′, 5′-hydroxylase modifies flower pigments and colour in cyclamen. *BMC Plant Biol.* **2010**, *10*, 107. [CrossRef] [PubMed]
70. Nakatsuka, T.; Mishiba, K.; Kubota, A.; Abe, Y.; Yamamura, S.; Nakamura, N.; Tanaka, Y.; Nishihara, M. Genetic engineering of novel flower colour by suppression of anthocyanin modification genes in gentian. *J. Plant Physiol.* **2010**, *167*, 231–237. [CrossRef] [PubMed]
71. Bordignon-Luiz, M.T.; Gauche, C.; Gris, E.F.; Falcao, L.D. Colour stability of anthocyanins from Isabel grapes (*Vitis labrusca* L.) in model systems. *LWT-Food Sci. Technol.* **2007**, *40*, 594–599. [CrossRef]
72. Zhao, X.; Sheng, F.; Zheng, J.; Liu, R. Composition and stability of anthocyanins from purple Solanum tuberosum and their protective influence on Cr (VI) targeted to bovine serum albumin. *J. Agric. Food Chem.* **2011**, *59*, 7902–7909. [CrossRef] [PubMed]
73. Momonoi, K.; Yoshida, K.; Mano, S.; Takahashi, H.; Nakamori, C.; Shoji, K.; Nitta, A.; Nishimura, M. A vacuolar iron transporter in tulip, TgVit1, is responsible for blue coloration in petal cells through iron accumulation. *Plant J.* **2009**, *59*, 437–447. [CrossRef]
74. Verweij, W.; Spelt, C.; Di Sansebastiano, G.P.; Vermeer, J.; Reale, L.; Ferranti, F.; Koes, R.; Quattrocchio, F. An H+ P-ATPase on the tonoplast determines vacuolar pH and flower colour. *Nat. Cell Biol.* **2008**, *10*, 1456–1462. [CrossRef]
75. Zhang, B.; Liu, C.; Wang, Y.; Yao, X.; Wang, F.; Wu, J.; King, G.J.; Liu, K. Disruption of a *CAROTENOID CLEAVAGE DIOXYGENASE 4* gene converts flower colour from white to yellow in Brassica species. *New Phytol.* **2015**, *206*, 1513–1526. [CrossRef]
76. Liu, Y.; Ye, S.; Yuan, G.; Ma, X.; Heng, S.; Yi, B.; Ma, C.; Shen, J.; Tu, J.; Fu, T.; et al. Gene silencing of *BnaA09.ZEP* and *BnaC09.ZEP* confers orange color in Brassica napus flowers. *Plant J.* **2020**, *104*, 932–949. [CrossRef]
77. Luo, P.; Ning, G.; Wang, Z.; Shen, Y.; Jin, H.; Li, P.; Huang, S.; Jian, Z.; Bao, M.J.F.i.P.S. Disequilibrium of Flavonol Synthase and Dihydroflavonol-4-Reductase Expression Associated Tightly to White vs. Red Color Flower Formation in Plants. *Front. Plant Sci.* **2015**, *6*, 1257–1269. [CrossRef] [PubMed]
78. Tanaka, Y.; Brugliera, F.; Kalc, G.; Senior, M.; Chandler, S.J.B.B.B. Flower Color Modification by Engineering of the Flavonoid Biosynthetic Pathway: Practical Perspectives. *Biosci. Biotechnol. Biochem.* **2010**, *74*, 1760–1769. [CrossRef] [PubMed]
79. Nuraini, L.; Ando, Y.; Kawai, K.; Tatsuzawa, F.; Tanaka, K.; Ochiai, M.; Suzuki, K.; Aragones, V.; Daros, J.A.; Nakatsuka, T. Anthocyanin regulatory and structural genes associated with violet flower color of *Matthiola incana*. *Planta* **2020**, *251*, 61. [CrossRef] [PubMed]
80. Tatsuzawa, F.; Aiba, Y.; Morino, T.; Saito, N.; Shinoda, K.; Kato, K.; Toki, K.; Honda, T.J.J.-J.S.f.H.S. Copigmentation with Acylated Anthocyanin and Kaempferol Glycosides in Violet and Purple Flower Cultivars of Aubrieta × cultorum (Brassicaceae). *J. Jpn. Soc. Hortic. Sci.* **2012**, *81*, 275–284. [CrossRef]
81. Saito, N.; Tatsuzawa, F.; Toki, K.; Shinoda, K.; Shigihara, A.; Honda, T. The blue anthocyanin pigments from the blue flowers of *Heliophila coronopifolia* L. (Brassicaceae). *Phytochemistry* **2011**, *72*, 2219–2229. [CrossRef] [PubMed]
82. Naing, A.H.; Kang, H.H.; Hui, Y.J.; Soe, M.T.; Chang, K.K.J.M.B. Overexpression of the Raphanus sativus *RsMYB1* Using the Flower-specific Promoter (*InMYB1*) Enhances Anthocyanin Accumulation in Flowers of Transgenic Petunia and Their Hybrids. *Mol. Breed.* **2020**, *40*, 97–107. [CrossRef]

horticulturae

MDPI

Article

Assessment of Brassicaceae Seeds Quality by X-ray Analysis

Farhad Musaev [1], Nikolay Priyatkin [2,*], Nikolay Potrakhov [3], Sergey Beletskiy [4] and Yuri Chesnokov [2]

1 Laboratory Analysis Center, Federal Scientific Vegetable Center, 14 Selektsionnaya Street, 143080 Odintsovo, Russia; musayev@bk.ru
2 Plant Lightphysiology and Agroecosystem Bioproductivity Department, Agrophysical Research Institute, 14 Grazhdanskiy pr., 195220 Saint Petersburg, Russia; yuv_chesnokov@agrophys.ru
3 Electronic Instruments and Devices Department, Saint Petersburg Electrotechnical University "LETI", Ul. Professora Popova, 5, 197022 Saint Petersburg, Russia; kzhamova@gmail.com
4 All-Russian Scientific Research Institute of the Confectionery Industry—Branch of the V. M. Gorbatov Federal Research Center for Food Systems, 20 Elektrozavodskaya St., 107076 Moscow, Russia; conditerprom@mail.ru
* Correspondence: prini@mail.ru

Abstract: A serious problem of vegetable production is the quality of sown seeds. In this regard, assessment of seed quality before sowing and storage is of great practical interest. The modern level of scientific research requires the use of instrumental automated methods of seed quality evaluation, allowing to obtain more information and in a shorter time. The material for the study was a variety of samples from the collection of *Brassica oleracea* L., var. *capitata*, *Raphanus sativus* L., var. *radicula*, and *Lepidium sativum* L. seeds from the Federal Scientific Center of Vegetable Breeding and the Timofeev Selection Station. Digital X-ray images of seeds were obtained using a mobile X-ray diagnostic device PRDU-02. Automatic analysis of digital X-ray images was performed in the software "VideoTesT-Morphology 5.2." The following latent defects of cabbage seeds of economic importance were revealed and identified: irregular darkening, significant "patterning" with deep separation of embryo parts, "angularity of seeds" leading to the loss of their viability. Automatic analysis of digital X-ray images of seeds confirmed the informativeness of brightness indices of digital X-ray images, as well as shape indices. Their connection with sowing qualities of the studied seeds was established.

Keywords: *Brassica oleracea* L.; var. *capitata*; *Raphanus sativus* L.; var. *radicula*; *Lepidium sativum* L.; microfocus X-ray; seed quality; seed image analysis

Citation: Musaev, F.; Priyatkin, N.; Potrakhov, N.; Beletskiy, S.; Chesnokov, Y. Assessment of Brassicaceae Seeds Quality by X-ray Analysis. *Horticulturae* **2022**, *8*, 29. https://doi.org/10.3390/horticulturae8010029

Academic Editors: Xiaowu Wang, Jian Wu and Xu Cai

Received: 30 November 2021
Accepted: 24 December 2021
Published: 28 December 2021

Publisher's Note: MDPI stays neutral with regard to jurisdictional claims in published maps and institutional affiliations.

Copyright: © 2021 by the authors. Licensee MDPI, Basel, Switzerland. This article is an open access article distributed under the terms and conditions of the Creative Commons Attribution (CC BY) license (https://creativecommons.org/licenses/by/4.0/).

1. Introduction

Poor seed quality sometimes delays the development of domestic crop production. Modern technologies that provide direct and precise sowing of seeds in the ground, eliminating the thinning of crops, as well as methods for adjusting plant stands, require uniform seeds with high field germination.

During the period of generative development, plants are particularly demanding to the conditions of light, heat and moisture supply [1]. However, climatic conditions in most regions of our country are unfavorable for seed production, poor technical equipment of seed production and post-harvest treatment has a negative impact.

The situation in seed production of vegetable crops is particularly acute. Large morphobiological diversity, increased heat demand of the majority of species requires a careful approach to the organization of seed production.

In practice, seed growers constantly have to deal with the heterogeneity of seeds. On the one hand, seed variety, or heterogeneity is a biologically useful phenomenon, developed in the process of evolution, which ensures the stability and reliability of the population, necessary for the survival of species [2,3]. At the same time, seed variability is often undesirable for the practice of agricultural production. Unevenness of seedlings, different productivity of plants and heterogeneity of products in terms of quality is largely determined by the poor quality of seeds.

Cabbage vegetable crops, white cabbage in particular, play a leading role in vegetable growing in Russia, both traditionally and in modern conditions. More than 20% of the sown areas under vegetables are occupied by cabbage in Russia [4,5]. Cabbage accounts for one fourth of the vegetable ration of the Russian population. The demand for white cabbage seeds in the country is more than 50 tons annually. Both domestic and foreign companies compete for the market. Seed production is carried out all over the world, where there are optimal natural and economic conditions.

The seeds of cruciferous crops (*Brassicaceae* family) are high in fat oil (33–40%) and protein (25–36%). They are the main nutrients of the seed [1]. At the same time, seeds have a thin, hydrophilic shell, which is easily broken in a humid atmosphere, as well as in case of indelicate drying, which does not contribute to their "durability". Fats become accessible to oxidation, the seed becomes open to fungal infection, its seeding and productive properties are reduced. In this regard, it is of great practical interest to assess the quality of seeds before sowing and putting in storage.

Traditional methods of seed quality control are mostly labor-intensive, time-consuming and lead to destruction of the analyzed sample. Modern scientific research enables the use of automated methods of seed quality evaluation, allowing to obtain more information and in a shorter time.

The method of radiography of seeds favorably differs from others simplicity of use, expeditiousness and non-destructive nature, which allows applying other, morphometric methods for the analysis of seeds quality. The method is standardized: ISO 6639/4-87; GOST 28666.4-90; GOST R 59603-2021 [6–8].

In recent years, in a joint work of the staff of the Federal Scientific Vegetable Center (FSVC), Agrophysical Research Institute (ARI) and St. Petersburg Electrotechnical University (SPbETU), "Methodological guidelines for radiographic analysis of vegetable crops seed quality" were developed based on a large-scale experiment on the seeds of 25 species of vegetable crops belonging to 10 botanical families [9–11].

Large-scale studies on the radiographic analysis of seeds, including vegetable crops, are also conducted abroad. Various radiographic characteristics of seeds of cucumber [12], watermelon [13], tomato [14,15], *Capsicum annuum* L. [16] and broccoli cabbage [17], using automated digital X-ray image analysis techniques [15,16,18].

Research objective: study of latent deficiency of *Brassicaceae* seeds and its relation with sowing qualities.

2. Materials and Methods

For the radiographic analysis, the collection of seeds from Federal Scientific Vegetable Center and Timofeev Breeding Station of N.N. Timofeev of Timiryazev Russian State Agrarian Academy (RSAA) was used (Table 1).

Analysis of the internal structure of seeds was performed according to the "Methodology of radiography in agriculture and crop production" [19] and "Methodological guidelines for radiographic analysis of vegetable seeds quality" [9] at the Department of Electronic Instruments and Devices, SPbETU. Radiographic imaging of seeds was performed on a mobile X-ray diagnostic device PRDU-02 and an X-ray microscope RM-1, produced by JSC "ELTECH-Med", (Russia). Seed sampling: 50 seeds (5 rows of 10 seeds) of each analyzed sample (Figure 1).

Table 1. Seed collection for radiographic analysis.

Sample	Variety	Maintainer	Production Area
Brassica oleracea L., var. *capitata*	Amager 611	FSVC	Derbent, Russia
Brassica oleracea L., var. *capitata*	Belorusskaya 455	FSVC	Moscow Region, Russia
Brassica oleracea L., var. *capitata*	Moskovskaya pozdnyaya 15	FSVC	Moscow Region, Russia
Brassica oleracea L., var. *capitata*	Podarok 2500	FSVC	Derbent, Russia
Brassica oleracea L., var. *capitata*	F1 Malakhit	RSAA	Italy
Brassica oleracea L., var. *capitata*	F1 Transfer	RSAA	Italy
Brassica oleracea L., var. *capitata*	F1 Valentina	RSAA	Italy
Brassica oleracea L., var. *capitata*	F1 Valentina	RSAA	Australia
Raphanus sativus L., var. *radicula*	Ariya	FSVC	Moscow Region, Russia
Raphanus sativus L., var. *radicula*	Pink-red with a white tip	FSVC	Moscow Region, Russia
Raphanus sativus L., var. *radicula*	Niger, population I_3	FSVC	Moscow Region, Russia
Raphanus sativus L., var. *radicula*	Inbred line I_4	FSVC	Moscow Region, Russia
Lepidium sativum L.,	Prestige	FSVC	Moscow Region, Russia
Lepidium sativum L.,	Flagman	FSVC	Moscow Region, Russia
Lepidium sativum L.,	Mechta Derbenta	RSAA	Derbent, Russia

(**a**) (**b**)

Figure 1. Preparation of seeds for radiographic study: (**a**) white cabbage, (**b**) radish.

Shooting mode is set depending on the size of the seeds: voltage 16–20 kV, current strength 98–105 μA, exposure 3–5 s. Microfocus imaging, unlike contact imaging, allows obtaining sharp, contrast X-ray images with high magnification, without loss of quality. The latent image on the plate is digitized in a special scanner "DIGORA PCT" (SOREDEX, Finland), from where the image is transmitted to the computer screen for editing, analyzing and archiving (Figure 2).

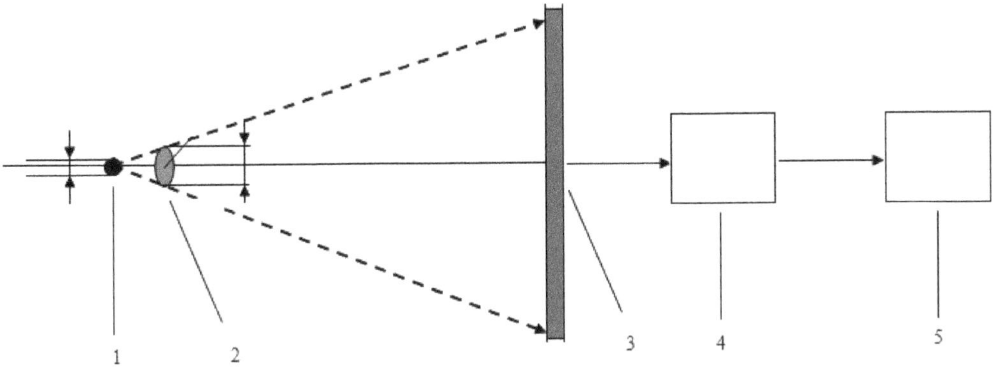

Figure 2. Functional scheme of receiving and processing of X-ray images: 1—radiation source (X-ray tube), 2—object of irradiation (seed), 3—receiver (CCD), 4—signal processing unit (special scanner), 5—personal computer.

Identification and classification of hidden seed defects was performed by visual transcription by an operator radiographer. The following parameters were used in the visual transcription of seed X-rays: normal, empty, germ partitioning of germ parts with soft partitioning, patterned, angulated, with irregular shading.

Additionally, software processing of digital X-ray images of seeds was carried out using morphometry software "VideoTesT-Morphology 5.2. ("Argus-BIO"), produced by "ArgusSoft", Ltd., Saint-Petersburg, Russia. The following parameters were analyzed: roundness, (dimensionless value) (1), circle factor (dimensionless value) (2), average brightness (brightness units) and brightness deviation (brightness units).

$$F_r = \frac{4A}{f^2\pi}, \quad (1)$$

where A—Area, f—maximal Feret diameter

$$F_c = \frac{4\pi A}{P^2}, \quad (2)$$

where A—Area, P—perimeter

Then, individual germination of seeds was assessed using filter paper and Petri dishes under controlled temperature according to GOST 12038-84 [4]. Seeds were germinated on moistened filter paper at 25 °C in the light. The emergence rate of white cabbage seeds was determined after three days, seed germination—8 days; the emergence rate of garden cress seeds was determined after 3 days, seed germination—5 days. We took pictures of seeds, seedlings and plants with a professional camera CANON-5D with a macro lens CANON-100 with a resolution of 12–24 megapixels in the FSVC photo laboratory, combined with digital morphometry of seedlings (root and sprout length), carried out with the use of morphometry software "VideoTesT-Morphology 5.2. ("Argus-BIO").

Statistical analysis (calculation of Spearman correlation coefficients) was performed in Statistica 10, TIBCO, Palo Alto, CA, USA.

3. Results

3.1. Inspection of *Brassica oleracea* L., var. Capitata Seeds

The results of radiographic analysis of seeds of eight samples of white cabbage of different origin and production sites, revealed a great heterogeneity of their internal structure, (Figure 3). Even a general glance at the radiograph is able to determine the heterogeneity of seeds. The details of the internal structure of the seeds are differently arranged and give out different drawings in the image.

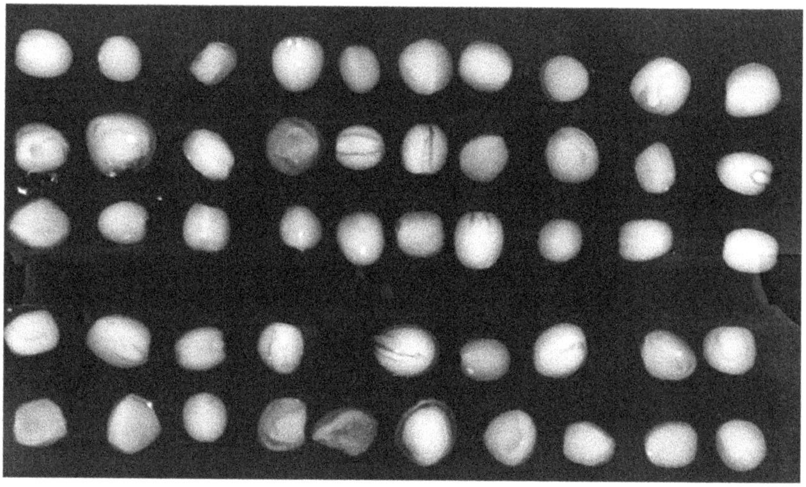

Figure 3. Radiographic image of *Brassica oleracea* L., var. *capitata* seeds.

Analysis of Table 2 allows us to indicate a general tendency of connection between X-ray features and seed viability. It consists in the following: as a rule, seeds, which look uniformly light or with insignificant detailing on the images, sprout well. Basically, the number of normal seeds from the point of view of radiographic analysis coincides with the number of germinated seeds. Seeds from the group "germ partitioning of germ parts with soft partitioning" germinated (Table 2).

Table 2. Comparison of the results of radiographic analysis and laboratory germination of *Brassica oleracea* L., var. *capitata* seeds.

Parameters	Varieties			
	Amager	**Belorusskaya**	**Moskovskaya Pozdnyaya**	**Podarok**
	Trait according to the results of radiographic analysis, %			
Normal	45	41	42	56
Empty	0	0	14	0
Germ partitioning of germ parts with soft partitioning	11	8	8	9
Patterned	17	16	5	22
Angulated	13	23	22	10
With irregular shading	14	12	9	3
	Germination, %			
Germinated	54	52	48	62
Sprouting	16	14	11	18
Not germinated	30	34	39	20

X-ray seed quality indices were established visual analysis of numerous of different varieties of cabbage seeds and their comparison with the results of laboratory germination.

We present a selection of radiographic images of seeds and photos of sprouts, where the X-ray image of a particular seed and its behavior under conditions favorable for germination are compared.

The obvious separation of the embryo parts and irregular shadows on the projection of seed #1 were reflected in its delayed germination and in the abnormality of the seedling (Figure 4).

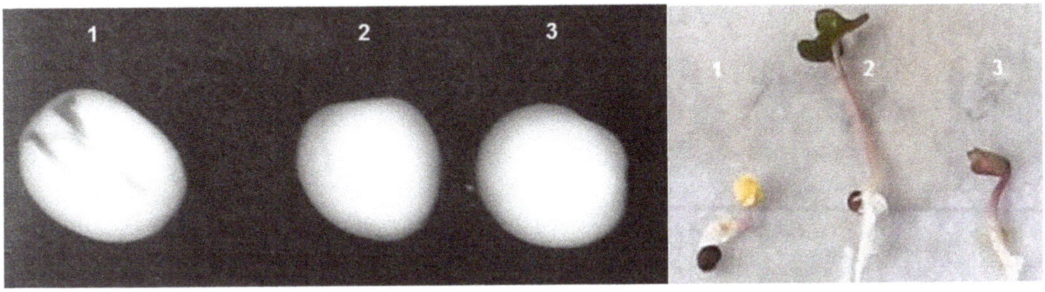

Figure 4. Radiography of seeds and photograph of white cabbage sprouts with "separation of embryo parts" clearly visible in seed number 1.

A softer separation of the embryo parts of seed #8 without additional shadows does not lead to a deterioration in the quality of the seedling (Figure 5).

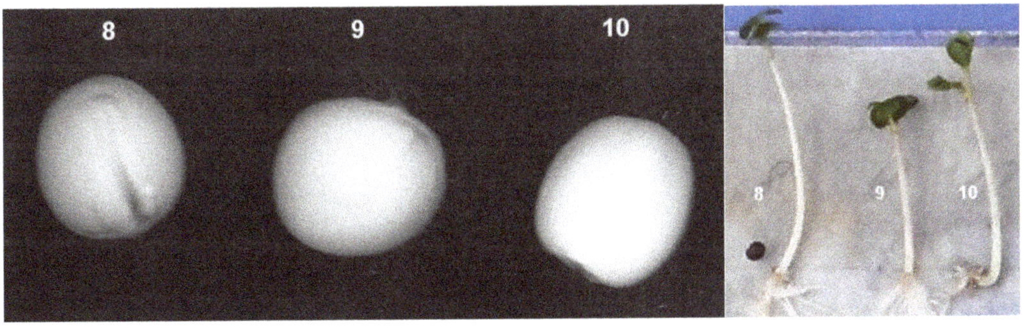

Figure 5. Radiography of seeds and photograph of white cabbage seedlings "soft separation of embryo elements".

The pronounced "patterning" with deep separation of seed embryo parts and reduced values of the index "average brightness" # 38, 40 indicate their non-viability (Figure 6, Table 3).

Table 3. Analysis of brightness parameters of digital X-ray images of *Brassica oleracea* L., var. *capitata* seeds with the trait "patterning of internal structure".

Parameter	Seed ID		
	38	39	40
Average Brightness, brightness units	162	184	162
Brightness Deviation, brightness units	39	42	45

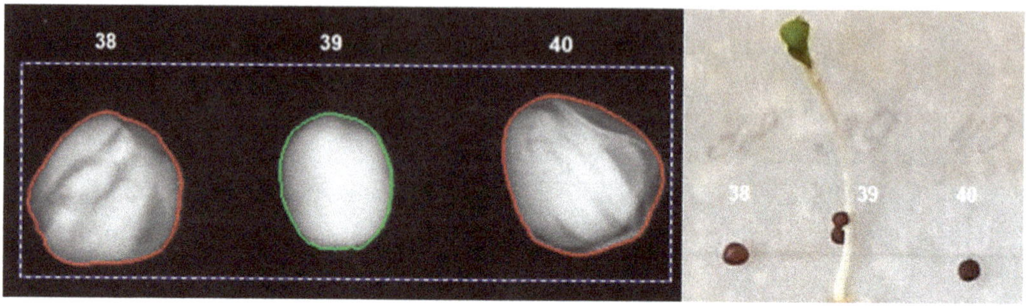

Figure 6. Radiography of seeds and photograph of white cabbage sprouts with "patterning of internal structure" (seed number 38 and 40 only).

Irregular shading on the projection of seed # 46, confirmed by low values of the indices "average brightness" and "standard deviation of brightness" are also evidence of its non-viability (Figure 7, Table 4).

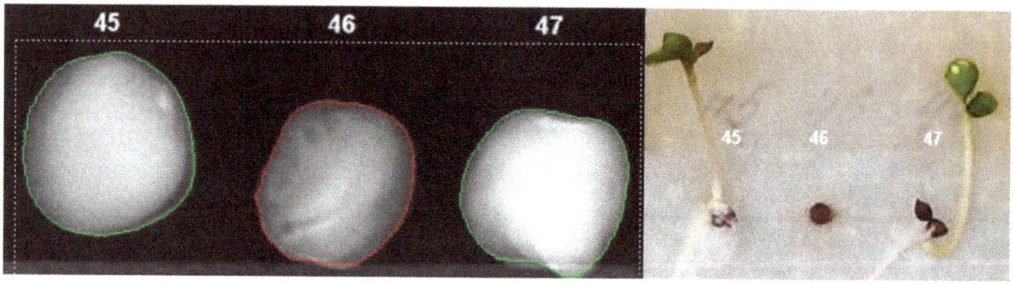

Figure 7. Radiography of seeds and photograph of *Brassica oleracea* L., var. *capitata* seedlings with "irregular shading" trait.

Table 4. Analysis of brightness parameters of digital X-ray images of *Brassica oleracea* L., var. *capitata* seeds with the trait "irregular shading".

Parameter	Seed ID		
	45	46	47
Average Brightness, brightness units	165	142	193
Brightness Deviation, brightness units	42	35	47

Angular shape of a seed in most cases indicates its non-viability (Figure 8). As a result of automatic analysis of digital X-ray images of white cabbage seeds, it was found that the trait "angularity" of the seed can be described by such indicators as roundness and circle factor.

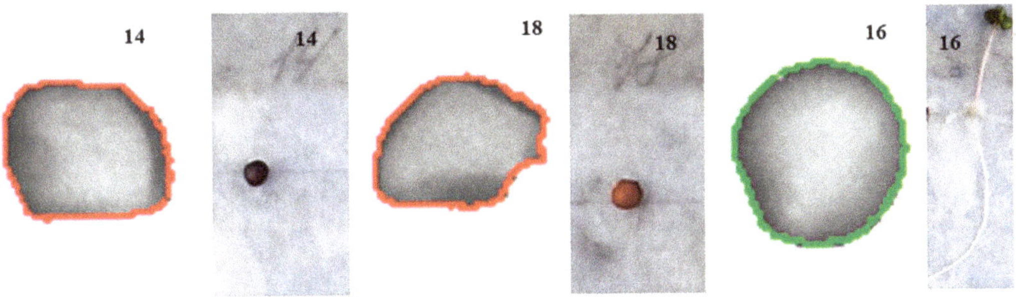

Figure 8. Radiography of seeds and photograph of *Brassica oleracea* L., var. *capitata* seedlings with angular seed shape.

In seeds with the trait "angularity" the values of these indicators are significantly lower than in normally formed seeds (Table 5). Seeding qualities of angular seeds in most cases are impaired (Figure 8).

Table 5. Analysis of the shape of digital X-ray images of angular *Brassica oleracea* L., var. *capitata* seeds.

Parameter	Seed ID		
	14	18	16
Roundness, nondimensional value	0.734	0.647	0.890
Circle factor, nondimensional value	0.927	0.907	0.991

Surface mycota of seeds are clearly distinguished on radiographs due to their dense consistency (Figure 9). Although they do not particularly affect seed germination, their negative effect may appear in the further development of the plant.

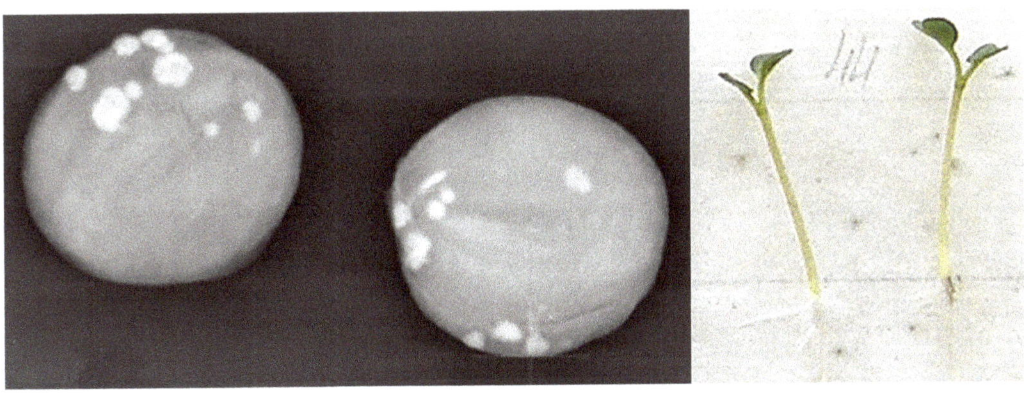

Figure 9. Radiography of infected seeds and photograph of *Brassica oleracea* L., var. *capitata* seedlings.

Thus, the main X-ray traits of cabbage seeds, which can be used to judge about their seeding qualities, were revealed. They are as follows: clear separation of embryo details indicates reduced seed viability (Figure 4). Irregular darkening (Figure 7) in the images also indicates poor seed quality. Expressed "patterning" with deep separation of seed embryo parts (Figure 6) as an indicator of loss or reduction of nutritive tissue, respectively, leads to seed death. The images often show angular seeds (Figure 8). They look as such only on the X-ray images, mainly due to some desiccation of the cotyledons, while outwards they have almost regular round shape. In our experiment, about 75% of such seeds do not germinate. The patterns revealed are probabilistic. Further research is required to find out the reasons for this relationship.

The high information value of the method of seed radiography provides an opportunity for multidimensional analysis. Radiography of seeds as a "non-destructive" method is especially convenient in breeding practice, when working with small lots of breeding or collection material.

3.2. Inspection of *Raphanus sativus* L., var. *Radicula* Seeds

Populations and homogeneous radish seeds were analyzed and fully preserved for further work. As a rule, seeds of inbred lines are obtained piecemeal and it is not allowed to waste them for analysis. In such cases, the method of radiography is indispensable, as it provides complete preservation of the analyzed seed sample. Seeds of variety Aria (I_0) in the X-ray image (Figure 10a) have low, but sufficiently uniform optical density of projection and during germination gave complete root crops. On the contrary, the internal structure of seeds of inbred generation I_3 looks more "patterned," with pronounced darkening revealing voids and loss of embryo body density (Figure 10b). When germinated, such seeds were unable to form complete root crops, an inbred depression expressed by reproductive dysfunction.

(a) (b)

Figure 10. Reproductive dysfunction of *Raphanus sativus* L., var. *radicula* plants due to defects in internal seed structure: (**a**) population seeds (I_0) and their complete crop; (**b**) seeds of inbred line (I_3) and unformed product swelling.

3.3. Inspection of *Lepidium sativum* L., Seeds

The manifestation of defects, anomalies, and flaws on the X-ray projection of seeds can be very diverse. Some of them are even species-specific. Here is an example of one of them. Seeds of cress varieties Prestige, Flagman of FSVC selection and "Mechta Derbenta" obtained at the Breeding Station named after N.N. Timofeev were taken for radiographic analysis. Seeds of Prestige and Flagman varieties were multiplied in the FSVC experimental field (Moscow Region), and "Mechta Derbenta" in Derbent district of the Republic of Dagestan.

Visual analysis of X-ray radiographs of cress seeds showed the following. Seeds of Prestige and Flagman varieties propagated in conditions of temperate zone (Moscow region) look mostly light on X-ray projections, indicating their completeness and fullness (Figure 11). Regular longitudinal shadows are associated with the anatomical structure of the seed. In laboratory germination, they showed a high degree of germination.

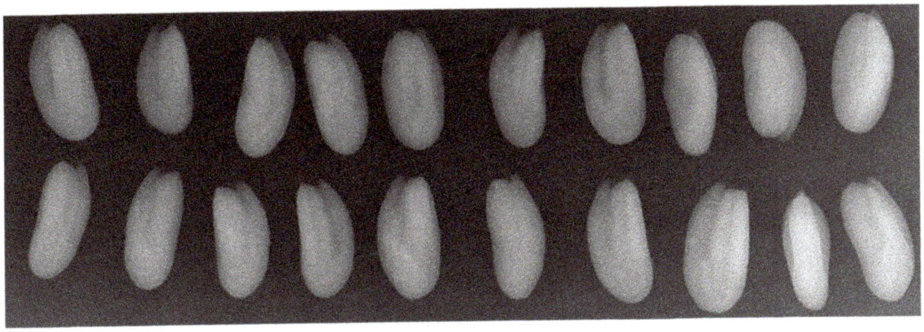

Figure 11. Fragment of X-ray image of *Lepidium sativum* L. seeds of Prestige variety.

A scanned sample of cress seeds is shown in Figure 12. Analysis of X-ray images of seeds of variety "Mechta Derbenta" revealed unusual "patterning" of images not revealed on seeds of other crops. In the X-ray image presented, 10 of 50 analyzed seeds were found (Figure 13). As a result of laboratory germination, 12 seeds did not germinate (Figure 14). All seeds with "patterned projections" (# 9, 14, 16, 18, 19, 22, 27, 42, 43, 45) failed to germinate. Two seeds (# 11 and 34) with normal dense projection but relatively small size did not germinate yet. It is noteworthy that such significant internal injuries are not reflected in the external appearance of the seeds (Figure 12), hence, have a hidden nature.

Figure 12. Photo of *Lepidium sativum* L. seeds of Mechta Derbenta variety.

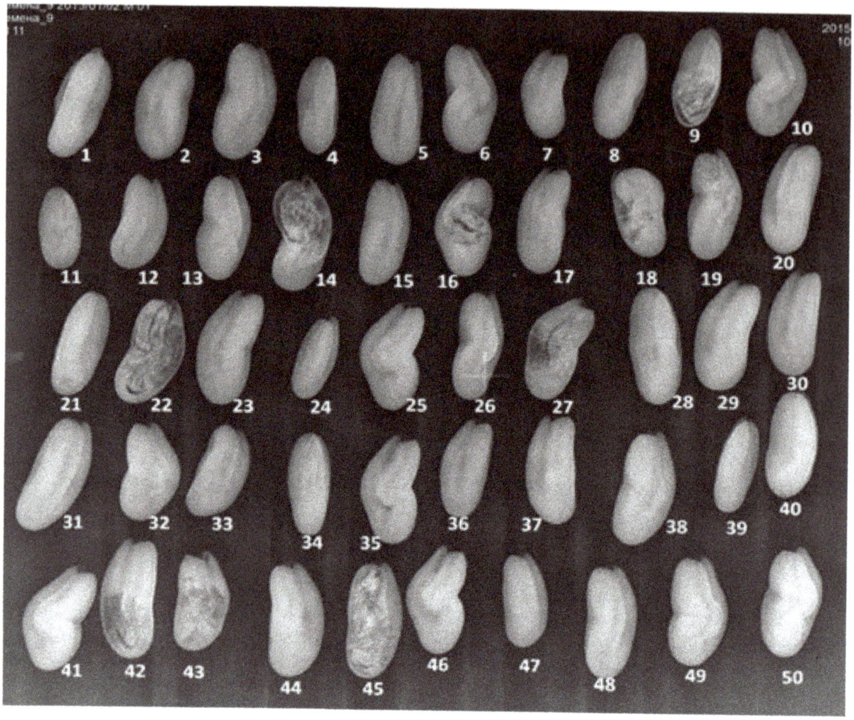

Figure 13. X-ray image of *Lepidium sativum* L. seeds of Mechta Derbenta variety.

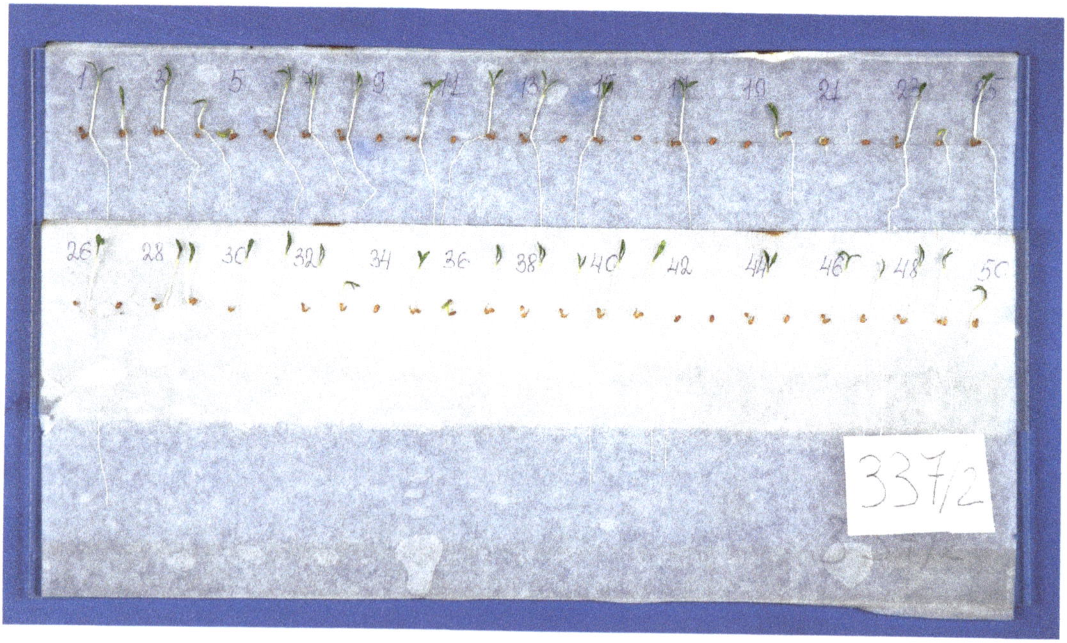

Figure 14. Photo of *Lepidium sativum* L. seedlings of Mechta Derbenta variety.

The results of automatic analysis of digital X-ray images of garden cress seeds revealed that the indicator "average brightness" is quite informative with respect to the trait "patterning of internal structure" of seeds. With the increase of "patterning" the average brightness of digital X-ray images decreases (Tables 6 and S1), and the sowing quality of seeds (Figure 15) deteriorates. The green border shows the automatic classification of the seeds as the best in brightness of the digital X-ray images, the yellow is the intermediate class.

Table 6. Results of automatic analysis of digital X-ray images of *Lepidium sativum* L. seeds Mechta Derbenta variety.

Parameters of X-ray Pattern of Seed	Seed ID									
	41	**42**	**43**	**44**	**45**	**46**	**47**	**48**	**49**	**50**
	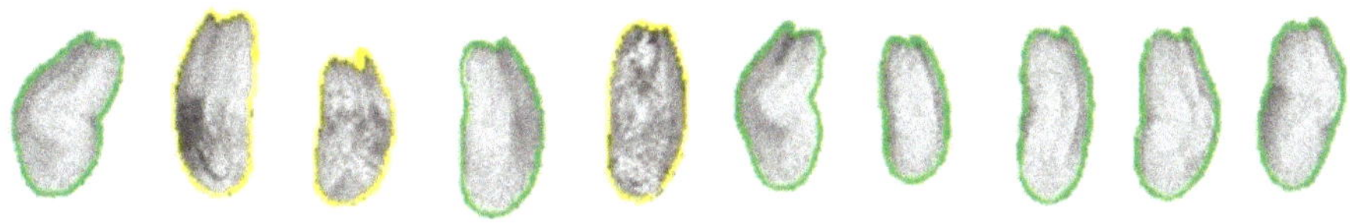									
Average Brightness, Brightness Units	184	182	175	185	168	184	184	188	187	184
Brightness Standard Deviation, Brightness Units	25	33	23	26	27	28	28	26	26	30

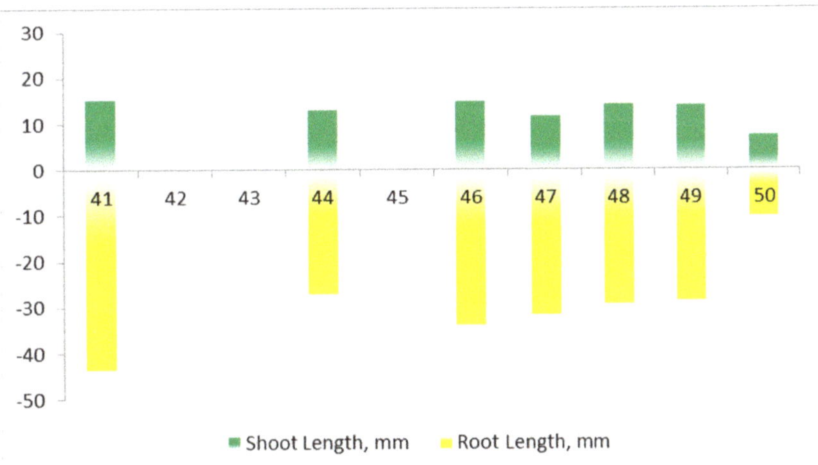

Figure 15. Results of digital morphometry of *Lepidium sativum* L. seedlings Mechta Derbenta variety, preliminary analyzed by X-ray technique.

Spearman correlation coefficients ($p < 0.05$) were, respectively: between average brightness of digital X-ray images and shoot length r = 0.34; standard deviation of brightness and shoot length r = −0.38; average brightness of digital X-ray images and root length r = 0.38; standard deviation of brightness and root length r = −0.40.

So, according to the results of analysis of a large number of X-ray images, X-ray trait directly related to the viability of seeds was established. Regular "patterning" of the internal structure of garden cress seeds on the X-ray projection is evidence of their non-viability. By means of X-ray quality analysis, it is possible to quickly and without loss of the seeds themselves to give a conclusion about the viability of a batch of damaged seeds. According to Table 7, the results of radiographic analysis of seeds almost coincide with the results of their laboratory germination: 74 and 77% of germination, respectively. In this case, the informative and fast performance of the radiographic method is obvious.

Table 7. Comparison of methods for analyzing the quality of *Lepidium sativum* L. seeds.

Replications	X-ray Analysis, %			Laboratory Germination, %
	Normal	Insects Damaged	Irregular Shadings	
1	76	14	10	80
2	72	20	8	74
3	64	36	0	72
4	85	12	3	82
Mean	74.0	20.5	5.0	77.0

It is noteworthy that this X-ray trait has so far only been detected on cress seeds and only on the lot of southern reproduction. This is the result of the work of insect pests. The rapeseed moth (*Meligethes aeneus* (Fabricius, 1775) pests on the seeds of cultivated cruciferous plants [20,21]. During bud formation, they invade plants and feed on the inner parts of buds and flowers, eating stamens, pistils, petals and pollen. As a result, incomplete seeds develop and seed production decreases.

4. Discussion

The possibilities of radiographic analysis of seeds are not limited to the above examples. With the help of the method, it is possible to determine the degree of seed fullness, injuries, the presence of internal (hidden) germination and other defects and abnormalities of the internal structure.

The experimental data obtained demonstrate the effectiveness of the method of microfocus X-ray radiography for the analysis of the germination of cabbage seeds related to the peculiarities of their internal structure. The radiographic analysis of the internal structure of wheat and barley grain [22], seeds of ornamental apple tree forms [23], tomato [15], pepper [16], eggplant [24] tomato and melon [18] showed high information value of the method and speed of its execution.

Analysis of images of seed material, in particular, evaluation of various shape indices for classification of seed species diversity, is widely enough applied in seed science [25–27]. Regarding the application of the X-ray method, it should be noted that the works of other researchers are mainly devoted to the visualization and identification of individual, but quite common types of hidden defects, for example—incomplete seed [28], hidden insect infestation [29] and embryo defects [14].

At the same time, cabbage seeds have specific X-ray traits, such as, "angularity of seeds", indicating the incompleteness of hoarding tissue. Just as specific is the trait of "regular patterning" of the internal structure of cress seeds, a consequence of damage by insect pests, directly related to their viability. It should be noted that earlier the study of the features of the internal structure of cabbage seeds was of a trial nature [30]. Thus, in the work of Gusakova (1997) [31], Derunov (2004) [30] the sign of "unfulfilled" cabbage seeds, determined by the radiographic method, indicating the non-viability of seeds, was shown. From the few modern works we can cite the example of the study of Abud et al. (2018) [17], which shows the influence of features and defects of the internal structure of broccoli seeds on their vigor. We tried to give the work a more systematic character and analyzed the defectiveness of cabbage seeds not only visually, but also using the means of computer morphometry. Defects and deficiencies in the internal structure of cabbage seeds were identified and classified by us, the degree of their impact on seed viability was shown [10]. This approach allows us to objectify the express-analysis of cabbage seeds quality when they are fully preserved, to extract information about defects and disadvantages of internal structure, which significantly complements the traditional methods used in seed control.

5. Conclusions

The method of seed radiography makes it possible to identify and record defects and abnormalities of the internal structure of the seeds of vegetable cruciferous crops according to a number of economically significant features.

The detected defects and abnormalities of seed development are well correlated with their viability, which increases the practicality of the method. The high informative value of the method makes it possible not only to establish the degree of viability of the analyzed seeds, but also to identify the causes of its reduction.

Automatic analysis of digital X-ray images of seeds is an effective tool in the objectification of such traits of their hidden defects as irregular darkening, angularity and patterning of the internal structure.

The advantage of the method is its rapid application, integrity and safety of the studied material, which is especially important when working with small lots of seeds of collection and breeding material. By recording and archiving the results of analysis, it is possible to trace the change in the quality of seeds over the entire period of their storage.

Supplementary Materials: The following supporting information can be downloaded at: https://www.mdpi.com/article/10.3390/horticulturae8010029/s1, Table S1. Results of automatic analysis of digital X-ray images of *Lepidium sativum* L. seeds Mechta Derbenta variety.

Author Contributions: Conceptualization and methodology F.M. and N.P. (Nikolay Priyatkin); software, N.P. (Nikolay Priyatkin); validation F.M. and N.P. (Nikolay Potrakhov); formal analysis, F.M and N.P. (Nikolay Priyatkin); investigation, F.M.; resources, N.P. (Nikolay Potrakhov); data curation, F.M.; writing—original draft preparation, F.M. and S.B.; writing—review and editing, S.B.; visualization, F.M., N.P. (Nikolay Priyatkin) and N.P. (Nikolay Potrakhov); supervision, project administration and funding acquisition, Y.C. All authors have read and agreed to the published version of the manuscript.

Funding: The study was financially supported by the Ministry of Education and Science of the Russian Federation (agreement no. 075-15-2020-805 from 2 October 2020).

Informed Consent Statement: Not applicable.

Data Availability Statement: Not applicable.

Acknowledgments: We express our gratitude to V.O. Kosov for help in X-ray imaging of seed samples and to L.P. Velikanov for help in visual analysis of latent seed defects.

Conflicts of Interest: The authors declare no conflict of interest.

References

1. Ludilov, V.A.; Alekseev, Y.B.; Timina, O.O. *Practical Seed Production of Vegetable Crops with the Basics of Seed Science*; Ludilov, V.A., Alekseev, Y.B., Eds.; Scientific Publishing House KMMC: Moscow, Russia, 2011; p. 220. (In Russian)
2. Bukharov, A.F. Heterogeneity of seeds: Theory and practice (review). *Veg. Russ.* **2020**, 2, 23–31. (In Russian)
3. Probert, R.J.; Hay, F. *Keeping Seeds Alive/Seed Technology and Its Biological Basis*; Black, M., Bewley, D., Eds.; Sheffield Academic Press: Sheffield, UK, 2000; pp. 184–225.
4. Litvinov, S.S.; Razin, A.F.; Mescheryakova, R.A.; Lebedeva, N.N.; Razin, A.A. Safety of vegetable products. *Potatoes Veg.* **2017**, *4*, 22–23. (In Russian)
5. Monakhos, S.G.; Monakhos, G.F. Selection of white cabbage (*Brassica oleracea* L.) for resistance to keel. *Rep. Timiryazev Agric. Acad.* **2016**, *288*, 473–477. (In Russian)
6. ISO6639/4 -87. Cereals and legumes. In *Determination of Hidden Insect Infestations. Part. 4. Accelerated Methods*; Standards Publishers: Moskow, Russia, 1987. (In Russian)
7. GOST 28666.4-90. Cereals and legumes. In *Determination of Latent Insect Infestation. Part. 4. Accelerated Methods*; Standards Publishers: Moskow, Russia, 1990. (In Russian)
8. GOST R 59603-2021. *Agricultural Seeds. Methods of Digital X-ray*; Standartinform: Moscow, Russia, 2021; p. 16. (In Russian)

9. Musaev, F.B.; Antoshkina, M.S.; Arkhipov, M.V.; Velikanov, L.P.; Gusakova, L.P.; Bessonov, V.B.; Gryaznov, A.Y.; Zhamova, K.K.; Kosov, V.O.; Potrakhov, E.N.; et al. *X-ray Quality Analysis of Seeds of Vegetable Crops/Methodological Guidelines*; Publishing house of St. Petersburg State Electrotechnical University "LETI": Moscow, Russia, 2015; p. 42. (In Russian)
10. Musaev, F.B.; Potrakhov, N.N.; Beletskiy, S.L. *BriefAtlas of Radiographic Traits of Seeds of Vegetable Crops*; Federal Scientific Center for Vegetable Growing: Moscow, Russia, 2017; p. 40. (In Russian)
11. Musaev, F.B.; Soldatenko, A.V.; Krivenkov, L.V.; Bukharov, A.F.; Beletskiy, S.L.; Kezimana, P. Assessment of vegetable seeds quality by micro-focus X-ray analysis. *Res. Crops* **2020**, *21*, 604–610. [CrossRef]
12. Gomes-Junior, F.G.; Chiquito, A.A.; Marcos-Filho, J. Semi-automated assessment of the embryonic area of cucumber seeds and its relationship to germination and seedling length. *J. Seed Sci.* **2013**, *35*, 183–189. [CrossRef]
13. Ahmed, M.R.; Yasmin, J.; Park, E.; Kim, G.; Kim, M.S.; Wakholi, C.; Mo, C.; Cho, B.K. Classification of watermelon seeds using morphological patterns of X-ray imaging: A comparison of conventional machine learning and deep learning. *Sensors* **2020**, *20*, 6753. [CrossRef] [PubMed]
14. Van der Burg, W.J.; Aartse, J.W.; van Zwol, R.A.; Bino, R.J. Predicting tomato seedling morphology by X-ray analysis of seeds. *J. Amer. Soc. Hort. Sci.* **1994**, *119*, 258–263. [CrossRef]
15. Marcos-Filho, J.; Gomes-Junior, F.G.; Bernett, M.A.; Wells, A.A.; Stieve, S. Using tomato analyzer software to determine embryo size in X-rayed seeds. *Rev. Bras. Sementes* **2010**, *32*, 146–153. [CrossRef]
16. Dell'Aquila, A. Pepper seed germination assessed by combined X-radiography and computer-aided imaging analysis. *J. Biologia. Plantarum.* **2007**, *51*, 777–781. [CrossRef]
17. Abud, H.F.; Cicero, S.M.; Gomes Junior, F.G. Radiographic images and relationship of the internal morphology and physiological potential of broccoli seeds. *Acta Scientiarum. Agron.* **2018**, *40*, e34950. [CrossRef]
18. Bruggink, H.; van Duijn, B. X-Ray based seed image analysis. *Seed Test. Int.* **2017**, *153*, 45–50.
19. Arkhipov, M.V.; Alekseeva, D.I.; Batygin, N.F.; Velikanov, L.P.; Gusakova, L.P.; Derunov, I.V.; Zheludkov, A.G.; Nikolenko, V.F.; Nikitina, L.I.; Savin, V.N.; et al. *Method of Radiography in Agriculture and Crop Production*; Manual; Publishing House of Russian Academy of Agricultural Sciences: Moskov, Russia, 2001; p. 93. (In Russian)
20. Lychkovskaya, I.Y.; Slukin, A.S. Distribution of phytophage insects on oil-bearing cabbage crops under conditions of dry vegetation period. *Rep. Russ. Acad. Agric. Sci.* **2013**, *1*, 28–30. (In Russian)
21. Hokkanen, H.M.T. Biological and agrotechnical control of the rape blossom beetle Meligethes aeneus (Coleoptera, Nitidulidae). *Actaentomol. Fenn.* **1989**, *53*, 25–29.
22. Arkhipov, M.V.; Priyatkin, N.S.; Gusakova, L.P.; Potrakhov, N.N.; Gryaznov, A.Y.; Bessonov, V.B.; Obodovsky, A.V.; Staroverov, N.E. X-Ray Computer Methods for Studying the Structural Integrity of Seeds and Their Importance in Modern Seed Science. *Tech. Phys.* **2019**, *64*, 582–592. [CrossRef]
23. Tkachenko, K.G.; Staroverov, N.E.; Gryaznov, A.Y. Radiographic study of fruit and seed quality. *Hortus Bot.* **2018**, *13*, 52–66. (In Russian) [CrossRef]
24. Silva, V.N.; Cicero, S.M.; Bennett, M. Relationship between eggplant seed morphology and germination. *Rev. Bras. Sementes* **2012**, *34*, 597–604. [CrossRef]
25. Orru, M.; Grillo, O.; Venora, G.; Bacchetta, G. Computer vision as a method complementary to molecular analysis: Grapevine cultivar seeds case study. *Comptes. Rendus. Biol.* **2012**, *335*, 602–615. [CrossRef]
26. Lo Bianco, M.; Grillo, O.; Canadas, E.; Venora, G.; Bacchetta, G. Inter- and intraspecic diversity in Cistus L. (Cistaceae) seeds, analysed with computer vision techniques. *Plant Biol.* **2017**, *19*, 183–190. [CrossRef] [PubMed]
27. Loddo, A.; Loddo, M.; Di Ruberto, C. A Novel deep learning based approach for seed image classification and retrieval. *Comput. Electron. Agric.* **2021**, *187*, 106269. [CrossRef]
28. Severiano, R.L.; Pinheiro, P.R.; Gomes Junior, F.G.; de Medeiros, A.D.; Pereira, M.D. X-ray test on passion fruit seeds submitted to different aryl removal methods. *Comun. Sci.* **2018**, *9*, 356–362. [CrossRef]
29. De Medeiros, A.D.; Zavala-León, M.J.; de Oliveira Araújo, J.; Pereira, M.D.; dos Santos, D.C.F.; Silva, D.L.J. Relationship between internal morphology and physiological quality of Leucaena leucocephala seeds using image analysis. *Rev. Árvore* **2019**, *43*, 1–9. [CrossRef]
30. Derunov, I.V. *Radiographic Study of Seeds of Various Agricultural Crops and Products of Their Processing: Abstract of Ph*; Candidate of Biological Sciences ARI: Saint Petersburg, Russia, 2004; p. 22. (In Russian)
31. Gusakova, L.P. *X-Ray and Cytophotometric Analysis of the Viability of Agricultural Crop Seeds: Extended Abstract of Cand*; Science Dissertation: Saint-Petersburg, Russia, 1997; p. 20. (In Russian)

MDPI
St. Alban-Anlage 66
4052 Basel
Switzerland
www.mdpi.com

Horticulturae Editorial Office
E-mail: horticulturae@mdpi.com
www.mdpi.com/journal/horticulturae

Disclaimer/Publisher's Note: The statements, opinions and data contained in all publications are solely those of the individual author(s) and contributor(s) and not of MDPI and/or the editor(s). MDPI and/or the editor(s) disclaim responsibility for any injury to people or property resulting from any ideas, methods, instructions or products referred to in the content.

www.ingramcontent.com/pod-product-compliance
Lightning Source LLC
Chambersburg PA
CBHW040828250926
51850CB00007BA/1285
* 9 7 8 3 7 2 5 8 0 0 7 9 7 *